# Lecture Notes in Mathematics

Olivier Catoni

# Statistical Learning Theory and Stochastic Optimization

Ecole d'Eté de Probabilités de Saint-Flour XXXI - 2001

Editor: Jean Picard

Springer

Author

Olivier Catoni

Laboratoire de Probabilités
et Modèles Aléatoires
UMR CNRS 7599, Case 188
Université Paris 6
4, place Jussieu
75252 Paris Cedex 05
France

*e-mail: catoni@ccr.jussieu.fr*

Editor

Jean Picard

Laboratoire de Mathématiques Appliquées
UMR CNRS 6620
Université Blaise Pascal Clermont-Ferrand
63177 Aubière Cedex, France

*e-mail: Jean.Picard@math.univ-bpclermont.fr*

The lectures of this volume are the second part of the St. Flour XXXI-2001 volume that has appeared as LNM 1837.

Cover picture: Blaise Pascal (1623-1662)

Library of Congress Control Number: 2004109143

Mathematics Subject Classification (2000):
62B10, 68T05, 62C05, 62E17, 62G05, 62G07, 62G08, 62H30, 62J02, 94A15, 94A17, 94A24, 68Q32, 60F10, 60J10, 60J20, 65C05, 68W20.

ISSN 0075-8434
ISBN 3-540-22572-2 Springer Berlin Heidelberg New York
DOI: 10.1007/b99352



Springer is a part of Springer Science + Business Media
www.springeronline.com

Printed in Germany



Typesetting: Camera-ready TEX output by the authors

41/3142/du - 543210 - Printed on acid-free paper

# Preface

Three series of lectures were given at the 31st Probability Summer School in Saint-Flour (July 8–25, 2001), by the Professors Catoni, Tavaré and Zeitouni. In order to keep the size of the volume not too large, we have decided to split the publication of these courses into two parts. This volume contains the course of Professor Catoni. The courses of Professors Tavaré and Zeitouni have been published in the *Lecture Notes in Mathematics*. We thank all the authors warmly for their important contribution.

55 participants have attended this school. 22 of them have given a short lecture. The lists of participants and of short lectures are enclosed at the end of the volume.

Finally, we give the numbers of volumes of Springer *Lecture Notes* where previous schools were published.

*Lecture Notes in Mathematics*

| | | | |
|---|---|---|---|
| 1971: vol 307 | 1973: vol 390 | 1974: vol 480 | 1975: vol 539 |
| 1976: vol 598 | 1977: vol 678 | 1978: vol 774 | 1979: vol 876 |
| 1980: vol 929 | 1981: vol 976 | 1982: vol 1097 | 1983: vol 1117 |
| 1984: vol 1180 | 1985/86/87: vol 1362 | 1988: vol 1427 | 1989: vol 1464 |
| 1990: vol 1527 | 1991: vol 1541 | 1992: vol 1581 | 1993: vol 1608 |
| 1994: vol 1648 | 1995: vol 1690 | 1996: vol 1665 | 1997: vol 1717 |
| 1998: vol 1738 | 1999: vol 1781 | 2000: vol 1816 | 2001: vol 1837 |
| 2002: vol 1840 | | | |

*Lecture Notes in Statistics*

1986: vol 50

Jean Picard, Université Blaise Pascal
Chairman of the summer school

# Contents

# Introduction[1]

The main purpose of these lectures will be to estimate a probability distribution $P \in \mathcal{M}_+^1(\mathcal{Z})$ from an observed sample $(Z_1, \dots, Z_N)$ distributed according to $P^{\otimes N}$. (The notation $\mathcal{M}_+^1(\mathcal{Z}, \mathcal{F})$ will stand throughout these notes for the set of probability distributions on the measurable space $(\mathcal{Z}, \mathcal{F})$ — the sigma-algebra $\mathcal{F}$ will be omitted when there is no ambiguity about its choice). In a regression estimation problem, $Z_i = (X_i, Y_i) \in \mathcal{X} \times \mathcal{Y}$ will be a set of two random variables, and the distribution to be estimated will rather be the conditional probability distribution $P(dY \mid X)$, or even only its mode (when $\mathcal{Y}$ is a finite set) or its mean (when $\mathcal{Y} = \mathbb{R}$ is the real line). A large number of pattern recognition problems could be formalized within this framework. In this case, the random variable $Y_i$ takes a finite number of values, representing the different "labels" into which the "patterns" $X_i$ are to be classified. The patterns may for instance be digital signals or images.

A major role will be played in our study by the risk function

$$R(Q) = \mathcal{K}(P, Q) \overset{\text{def}}{=} \begin{cases} \mathbb{E}_P\left(\log \frac{P}{Q}\right) & \text{if } P \ll Q \\ +\infty & \text{otherwise} \end{cases}, \qquad Q \in \mathcal{M}_+^1(\mathcal{Z}).$$

Let us remind that the function $\mathcal{K}$ is known as the Kullback divergence function, or relative entropy, that it is non negative and cancels only on the set $P = Q$. To see this, it is enough to remember that, whenever it is finite, the Kullback divergence can also be expressed as

$$\mathcal{K}(P, Q) = \mathbb{E}_Q\left[1 - \frac{P}{Q} + \frac{P}{Q} \log\left(\frac{P}{Q}\right)\right]$$

and that the map $r \mapsto 1 - r + r \log(r)$ is non negative, strictly convex on $\mathbb{R}_+$ and cancels only at point $r = 1$.

[1] I would like to thank the organizers of the Saint-Flour summer school for making possible this so welcoming and rewarding event year after year. I am also grateful to the participants for their kind interest and their useful comments.

In the case of regression estimation and pattern recognition, we will also use risk functions of the type

$$R(f) = \mathbb{E}\big[d\big(f(X), Y\big)\big], \qquad f : \mathcal{X} \to \mathcal{Y},$$

where $d$ is a non negative function measuring the discrepancy between $Y$ and its estimate $f(X)$ by a function of $X$. We will more specifically focuss on two loss functions : the quadratic risk $d\big(f(X),Y\big) = (f(X)-Y)^2$ in the case when $\mathcal{Y} = \mathbb{R}$, and the error indicator function $d\big(f(X),Y\big) = \mathbb{1}\big(f(X) \neq Y\big)$ in the case of pattern recognition.

Our aim will be to prove, for well chosen estimators $\hat{P}(Z_1,\dots,Z_N) \in \mathcal{M}_+^1(\mathcal{Z})$ [resp. $\hat{f}(Z_1,\dots,Z_N) \in \mathcal{L}(\mathcal{X},\mathcal{Y})$], non asymptotic oracle inequalities. Oracle inequalities is a point of view on statistical inference introduced by David Donoho and Iain Johnstone. It consists in making no (or few) restrictive assumptions on the nature of the distribution $P$ of the observed sample, and to restrict instead the choice of an estimator $\hat{P}$ to a subset $\{P_\theta : \theta \in \Theta\}$ of the set $\mathcal{M}_+^1(\mathcal{Z})$ of all probability distributions defined on $\mathcal{Z}$ [resp. to restrict the choice of a regression function $\hat{f}$ to a subset $\{f_\theta : \theta \in \Theta\}$ of all the possible measurable functions from $\mathcal{X}$ to $\mathcal{Y}$]. The estimator $\hat{P}$ is then required to approximate $P$ almost as well as the best distribution in the estimator set $\{P_\theta : \theta \in \Theta\}$ [resp. The regression function $\hat{f}$ is required to minimize as much as possible the risk $R(f_\theta)$, within the regression model $\{f_\theta : \theta \in \Theta\}$]. This point of view is well suited to "complex" data analysis (such as speach recognition, DNA sequence modeling, digital image processing, ...) where it is crucial to get quantitative estimates of the performance of approximate and simplified models of the observations.

Another key idea of this set of studies is to adopt a "pseudo-Bayesian" point of view, in which $\hat{P}$ is not required to belong to the reference model $\{P_\theta : \theta \in \Theta\}$ [resp. $\hat{f}$ is not required to belong to $\{f_\theta : \theta \in \Theta\}$]. Instead $\hat{P}$ is allowed to be of the form $\hat{P}(Z_1,\dots,Z_N) = \mathbb{E}_{\hat{\rho}(Z_1,\dots,Z_N)(d\theta)}(P_\theta)$, [resp. $\hat{f}$ is allowed to be of the form $\hat{f} = \mathbb{E}_{\hat{\rho}(d\theta)}(f_\theta)$], where $\hat{\rho}_{(Z_1,\dots,Z_N)}(d\theta) \in \mathcal{M}_+^1(\mathcal{Z})$ is a *posterior* parameter distribution, that is a probability distribution on the parameter set depending on the observed sample.

We will investigate three kinds of oracle inequalities, under different sets of hypotheses. To simplify notations, let us put

$$\hat{R}(Z_1,\dots,Z_N) = R\big(\hat{P}(Z_1,\dots,Z_N)\big) \qquad \text{[resp. } R\big(\hat{f}(Z_1,\dots,Z_N)\big)\text{]},$$

and $R_\theta = R(P_\theta)$ [resp. $R(f_\theta)$].

- **Upper bounds on the cumulated risk of individual sequences of observations.** In the pattern recognition case, these bounds are of the type :

$$\frac{1}{N+1}\sum_{k=0}^{N} \mathbb{1}\big[Y_{k+1} \neq \hat{f}(Z_1,\dots,Z_k)(X_{k+1})\big]$$

$$\leq \inf_{\theta \in \Theta} \Bigl\{ C \frac{1}{N+1} \sum_{k=0}^{N} \mathbb{1}\left[ Y_{k+1} \neq f_\theta(X_{k+1}) \right] + \gamma(\theta, N) \Bigr\}.$$

Similar bounds can also be obtained in the case of least square regression and of density estimation. Integrating with respect to a product probability measure $P^{\otimes(N+1)}$ leads to

$$\frac{1}{N+1} \sum_{k=0}^{N} \mathbb{E}_{P^{\otimes k}} \left[ \hat{R}(Z_1, \dots, Z_k) \right] \leq \inf_{\theta \in \Theta} \left\{ C R_\theta + \gamma(\theta, N) \right\}.$$

Here, $\gamma(\theta, N)$ is an upper bound for the estimation error, due to the fact that the best approximation of $P$ within $\{P_\theta : \theta \in \Theta\}$ is not known to the statistician. From a technical point of view, the size of $\gamma(\theta, N)$ depends on the *complexity* of the model $\{P_\theta : \theta \in \Theta\}$ in which an estimator is sought. In the extreme case when $\Theta$ is a one point set, it is of course possible to take $\gamma(\theta, N) = 0$. The constant $C$ will be equal to one or greater, depending on the type of risk function to be used and on the type of the estimation bound $\gamma(\theta, N)$. These inequalities for the cumulated risk will be deduced from lossless data compression theory, which will occupy the first chapter of these notes.

- **Upper bounds for the mean non cumulated risk**, of the type

$$\mathbb{E}\bigl[\hat{R}(Z_1, \dots, Z_N)\bigr] \leq \inf_{\theta \in \Theta} \bigl\{ C R_\theta + \gamma(\theta, N) \bigr\}.$$

Obtaining such inequalities will not come directly from compression theory and will require to build specific estimators. Proofs will use tools akin to statistical mechanics and bearing some resemblance to deviation (or concentration) inequalities for product measures.

- **Deviation inequalities**, of the type

$$P^{\otimes N} \Biggl\{ \hat{R}(Z_1, \dots, Z_N) \geq \inf_{\theta \in \Theta} \Bigl[ C R_\theta + \gamma(\theta, N, \epsilon) \Bigr] \Biggr\} \leq \epsilon.$$

These inequalities, obtained for a large class of *randomized* estimators, provide an *empirical* measure $\gamma(\theta, N, \epsilon)$ of the *local* complexity of the model around some value $\theta$ of the parameter. Through them, it is possible to make a link between randomized estimators and the method of penalized likelihood maximization, or more generaly penalized empirical risk minimization.

In chapter 7, we will study the behaviour of Markov chains with "rare" transitions. This is a clue to estimate the convergence rate of stochastic simulation and optimization methods, such as the Metropolis algorithm and simulated annealing. These methods are part of the statistical learning program sketched above, since the posterior distributions on the parameter space

$\hat{\rho}_{(Z_1,\dots,Z_N)}$ we talked about have to be estimated in practice and cannot, except in some special important cases, be computed exactly. Therefore we have to resort to approximate simulation techniques, which as a rule consist in simulating some Markov chain whose invariant probability distribution is the one to be simulated. Those posterior distributions used in statistical inference are hopefully sharply concentrated around the optimal values of the parameter when the observed sample size is large enough. Consequently, the Markov chains under which they are invariant have uneven transition rates, some of them being a function of the sample size converging to zero at exponential speed. This is why they fall into the category of (suitably generalized) Metropolis algorithms. Simulated annealing is a variant of the Metropolis algorithm where the rare transitions are progressively decreased to zero as time flows, resulting in a nonhomogeneous Markov chain which may serve as a stochastic (approximate) maximization algorithm and is useful to compute in some cases the mode of the posterior distributions we already alluded to.

# 1

# Universal lossless data compression

## 1.1 A link between coding and estimation

### 1.1.1 Coding and Shannon entropy

We consider in this chapter a finite set $E$, called in this context the *alphabet*, and a $E$ valued random process $(X_n)_{n\in\mathbb{N}}$.

The problem of lossless compression is to find, for each input length $N$, a "code" $c$ that is a one to one map

$$c : E^N \longrightarrow \{0,1\}^*$$

where $\{0,1\}^* = \bigcup_{n=1}^{+\infty}\{0,1\}^n$ stands for the set of finite sequences of zeros and ones with arbitrary length. Given any $s \in \{0,1\}^*$, its length will be noted $\ell(s)$. It is defined by the relation $s \in \{0,1\}^{\ell(s)}$.

We will look for codes with the lowest possible *mean length*

$$\mathbb{E}\Bigl(\ell\bigl(c(X_1, \ldots, X_N)\bigr)\Bigr). \tag{1.1.1}$$

If no other requirements are imposed on $c$, the optimal solution to this problem is obviously to sort the blocks of length $N$, $(x_1, \ldots, x_N)$ in decreasing order according to their probability to be equal to $(X_1, \ldots, X_N)$. Let $(b_i)_{i=1}^{|E|^N}$ be such an ordering of $E^N$, which satisfies

$$\mathbb{P}\Bigl((X_1, \ldots, X_N) = b_i\Bigr) \leq \mathbb{P}\Bigl((X_1, \ldots, X_N) = b_{i-1}\Bigr), \qquad i = 2, \ldots, |E|^N.$$

Let us introduce

$$\mathcal{B}(i) \overset{\text{def}}{=} \left(\left\lfloor \frac{i+1}{2^j} \right\rfloor \bmod 2\right)_{j=0}^{\lfloor \log_2(i+1)\rfloor - 1}, \qquad i = 1, \ldots, |E|^N,$$

the binary representation of $i+1$ from which the leftmost bit (always equal to 1) has been removed. The code

$$c(b_i) \stackrel{\text{def}}{=} \mathcal{B}(i) \tag{1.1.2}$$

obviously minimizes (1.1.1). Indeed,

$$\begin{aligned} \mathbb{N}^* &\longrightarrow \{0,1\}^* \\ i &\longmapsto \mathcal{B}(i) \end{aligned}$$

is a bijection and $i \mapsto \ell(\mathcal{B}(i))$ is non decreasing. Starting with any given code, we can modify it to take the same values as $c$ by exchanging binary words with shorter ones taken from the values of (1.1.2). We can then exchange pairs of binary words without increasing the code mean length to make it eventually equal to (1.1.2), which is thus proved to be optimal.

The *mean length* of the optimal code, defined by equation (1.1.1), is linked with Shannon's entropy, defined below.

**Definition 1.1.1.** The Shannon entropy $H(p)$ of a probability distribution $p$, defined on a finite set $\mathcal{X}$ is the quantity

$$H(p) = -\sum_{x \in \mathcal{X}} p(x) \log_2\bigl(p(x)\bigr).$$

The notation $\log_2$ stands for the logarithm function with base 2. Entropy is thus measured in *bits.* It is a concave function of $p$, linked with the Kullback Leibler divergence function with respect to the uniform distribution $\mu$ on $\mathcal{X}$ by the identity

$$H(p) = \log_2(|\mathcal{X}|) - \frac{1}{\log(2)} \mathcal{K}(p, \mu).$$

It cancels on Dirac masses and is equal to $\log_2(|\mathcal{X}|)$ for $\mu$.

Let us recall a basic fact of ergodic theory :

**Proposition 1.1.1.** *For any stationary source $(X_n)_{n \in \mathbb{N}}$, the map $N \mapsto H\bigl(\mathbb{P}(dX_1^N)\bigr)$ is* sub-additive, *proving the existence of the limit*

$$\lim_{N \to +\infty} \frac{H\bigl(\mathbb{P}(dX_1^N)\bigr)}{N} = \inf_{N \in \mathbb{N}^*} \frac{H\bigl(\mathbb{P}(dX_1^N)\bigr)}{N} \stackrel{\text{def}}{=} \bar{H}\bigl(\mathbb{P}(dX_1^{+\infty})\bigr).$$

*which is called the (Shannon)* entropy *of the source $(X_n)_{n \in \mathbb{N}}$.*

Next proposition shows that Shannon's entropy measures in first approximation the optimal compression rate.

**Proposition 1.1.2.** *For any finite source $X_1^N$ distributed according to $\mathbb{P}$, the mean length of the optimal code is such that*

$$\begin{aligned} &H\bigl(\mathbb{P}(dX_1^N)\bigr)(1 - 1/N) - 1 - \log_2(N) \\ &\leq \sup_{\alpha > 1} \frac{1}{\alpha} H\bigl(\mathbb{P}(dX_1^N)\bigr) - 1 + \frac{\log_2(\alpha - 1)}{\alpha} \leq \mathbb{E}\Bigl(\ell\bigl(c(X_1^N)\bigr)\Bigr) \leq H\bigl(\mathbb{P}(dX_1^N)\bigr) + 1. \end{aligned}$$

*Thus, for any infinite stationary source* $(X_n)_{n\in\mathbb{N}}$ *with distribution* $\mathbb{P}$,

$$\lim_{N\to+\infty}\frac{1}{N}\mathbb{E}\Big(\ell(c(X_1^N))\Big)=\bar{H}(\mathbb{P}).$$

*Proof.* Let us adopt the short notation

$$\mathbb{P}\Big((X_1,\,\ldots,\,X_N)=b_i\Big)\overset{\text{def}}{=}p(b_i).$$

The chain of inequalities

$$p(b_i)\leq i^{-1}$$

$$\begin{aligned}\mathbb{E}\Big(\ell(c(X_1^N))\Big)&=\sum_{i=1}^{|E|^N}p(b_i)\left(\lfloor\log_2(i+1)\rfloor\right)\\&\leq\sum_{i=1}^{|E|^N}-p(b_i)\log_2(p(b_i))+1\\&=H(p)+1\end{aligned}$$

shows that the mean length of the optimal code is upper-bounded by the Shannon entropy of the distribution of blocks of length $N$ (up to one bit at most).

On the other hand, for any $\alpha>1$,

$$\begin{aligned}\mathbb{E}\Big\{\ell\left[c(X_1^N)\right]\Big\}&\geq\sum_{i=1}^{|E|^N}p(b_i)\Big(\log_2(i+1)-1\Big)\\&\geq-\frac{1}{\alpha}\sum_{i=1}^{|E|^N}p(b_i)\log_2\left(\frac{\alpha-1}{(i+1)^\alpha}\right)-1+\frac{\log_2(\alpha-1)}{\alpha}.\end{aligned}$$

We can then notice that $b_i\mapsto\frac{\alpha-1}{(i+1)^\alpha}$ is a sub-probability distribution. This shows, along with the fact that the Kullback divergence is non-negative, that

$$-\sum_{i=1}^{|E|^N}p(b_i)\log_2\left(\frac{\alpha-1}{(i+1)^\alpha}\right)\geq-\sum_{i=1}^{|E|^N}p(b_i)\log_2(p(b_i)),$$

and consequently that

$$\mathbb{E}\Big(\ell(c(X_1^N))\Big)\geq-\frac{1}{\alpha}\sum_{i=1}^{|E|^N}p(b_i)\log_2(p(b_i))-1+\frac{\log_2(\alpha-1)}{\alpha}.$$

Then we can e.g. choose $\alpha=(1-1/N)^{-1}$, to obtain

$$\mathbb{E}\Big\{\ell\left[c(X_1^N)\right]\Big\}\geq H(\mathbb{P}(dX_1^N))(1-1/N)-1-\log_2(N).$$

□

It is to be remembered from this first discussion about lossless data compression that it requires, to be done efficiently, a fairly precise knowledge of the distribution of the source, and that for any stationary source of infinite length, the optimal compression rate per symbol tends towards the entropy $\bar{H}$.

### 1.1.2 Instantaneous codes

In the previous sections, we have considered arbitrary binary codes, our purpose being to show that the compression rate could not be significantly lower than Shannon's entropy, whatever choice of code is made. We are now going to focus on the restricted family of *prefix* codes, which share the property to be *instantaneously* separable into words, when the codes for many blocks are concatenated together before being sent through a transmission channel.

The optimal code (1.1.2), described previously, indeed suffers from a major drawback : it cannot be used to code more than a single block. Indeed, if the codes for two successive blocks of length $N$ (or more) are concatenated to be sent through a transmission channel, the receiver of this message will have no mean to find out how it should be decomposed into block codes. Such a code is called non separable, or non uniquely decodable or decipherable. Moreover, even if only one block is to be transmitted, the receiver has no mean to know whether the received message is completed, or whether more data should be waited for (such a code is said to be non *instantaneous*).

Instantaneous codes have a very simple characterization : a code is instantaneous if and only if no codeword is the beginning of another codeword. For this reason, such a code is also said to be a *prefix* code.

**Definition 1.1.2.** We will say that a finite subset $\mathcal{D} \subset \{0,\ 1\}^*$ of finite binary words is a *prefix dictionary* if any two distinct words $a_1^r$ and $b_1^s$ of $\mathcal{D}$ of respective lengths $r \leq s$ are always such that $a_1^r \neq b_1^r$. In other words we require that no word of the dictionary should be the prefix of another one. We will say that the map

$$c : E^N \longrightarrow \{0,\ 1\}^*$$

is a prefix code if it is one to one with values in a prefix dictionary.

**Proposition 1.1.3 (Kraft inequality).** *For any prefix dictionary* $\mathcal{D}$

$$\sum_{m \in \mathcal{D}} 2^{-\ell(m)} \leq 1.$$

*Proof.* This inequality can be proved using a construction which will lead also to *arithmetic coding*, to be described further below.

The set of finite binary sequences $\{0,1\}^* \cup \{\varnothing\}$ (where $\varnothing$ is the "void sequence" of null length) can be put into one to one correspondence with the set $\mathfrak{D}$ of *dyadic intervals*, defined by

$$\mathfrak{D} = \{[k\,2^{-n},\, (k+1)\,2^{-n}[ : \, n \in \mathbb{N},\; k \in \mathbb{N} \cap [0, 2^n[\},$$

putting for any sequence

$$s = (s_i)_{i=1}^{\ell(s)} \in \{0,1\}^*,$$

$$\mathfrak{I}(s) \overset{\text{def}}{=} [k\,2^{-n},\; (k+1)\,2^{-n}[,$$

where $n = \ell(s)$, $k = \sum_{i=1}^{\ell(s)} s_i 2^{n-i}$ and putting moreover $\mathfrak{I}(\varnothing) = [0,1[$.

It is then easy to see that no one of two codewords $s$ and $s'$ is the prefix of the other if and only if the corresponding dyadic intervals $\mathfrak{I}(s)$ and $\mathfrak{I}(s')$ are disjoint. It immediately follows that the sum of the lengths of the intervals attached to the words of a prefix dictionary cannot exceed one (because all these dyadic intervals are subsets of the unit interval): this is precisely the Kraft inequality. This proof also shows that the Kraft inequality remains true for infinite dictionaries. □

**Proposition 1.1.4 (Inverse Kraft inequality).** *For any sequence* $(r_i)_{i=1}^T \in (\mathbb{N}^*)^T$ *of positive integers such that*

$$\sum_{i=1}^{T} 2^{-r_i} \leq 1,$$

*there exists a prefix code*

$$c : \{1, \ldots, T\} \longrightarrow \{0,1\}^*$$

*such that*

$$\ell\big(c(i)\big) = r_i, \qquad i = 1, \ldots, T.$$

*Proof.* Without loss of generality, the sequence $(r_i)_1^T$ may be assumed to be non decreasing. It is then easy to check that the intervals $[\alpha_i,\, \beta_i[$ defined by

$$\alpha_i = \sum_{j=1}^{i-1} 2^{-r_j},$$

$$\beta_i = \alpha_i + 2^{-r_i},$$

are dyadic in the sense defined in the proof of the Kraft inequality. Moreover, they are obviously non overlaping. The code $c(i) \overset{\text{def}}{=} \mathfrak{I}^{-1}([\alpha_i,\, \beta_i[)$ is therefore a prefix code. Let us remark here again that this proof holds also for infinite sequences. □

**Definition 1.1.3.** A prefix dictionary $\mathcal{D}$ is said to be complete (as well as any code using such a dictionary) if it is maximal for the inclusion relation, that is if any prefix dictionary $\mathcal{D}'$ such that $\mathcal{D} \subset \mathcal{D}'$ is in fact equal to it.

To a prefix dictionary $\mathcal{D}$ corresponds a binary tree, whose vertex set and edge set are defined as

$$\mathcal{N} = \bigcup_{s \in \mathcal{D}} \{I \in \mathfrak{D} \,:\, \mathcal{I}(s) \subset I\},$$
$$\mathcal{A} = \left\{(I, I') \in \mathcal{N}^2 \,:\, I' \subset I\right\}.$$

It is the same thing to say that the dictionary $\mathcal{D}$ is complete or that the corresponding binary tree $(\mathcal{N}, \mathcal{A})$ is complete in the usual sense that its vertices have either zero or two sons (in other words, any interior vertex has exactly two sons).

**Proposition 1.1.5 (Kraft equality).** *A prefix dictionary $\mathcal{D}$ is complete if and only if*

$$\sum_{m \in \mathcal{D}} 2^{-\ell(m)} = 1.$$

*Proof.* The tree $(\mathcal{N}, \mathcal{A})$ is complete if and only if the set $\mathcal{F} = \left\{\mathcal{I}(s) \,:\, s \in \mathcal{D}\right\}$ of its leaves is a partition of the unit interval $[0, 1[$.

Indeed, the two sons of $I \in \mathcal{N}$ are the two half length dyadic intervals into which it decomposes.

An interval $I$ belongs to $\mathcal{N}$ if and only if it contains a leaf $I' \in \mathcal{F}$. In the case when $\mathcal{F}$ is a partition, it is either a leaf itself, or its two sons belong to the tree.

On the other hand, let us assume that some point $x \in [0, 1[$ does not belong to any leaf. Let $I$ be the largest dyadic interval containing $x$ and not overlaping with $\mathcal{F}$. The "father" of this interval thus meets $\mathcal{F}$, therefore its other son belongs to $\mathcal{N}$, proving that the tree $(\mathcal{N}, \mathcal{A})$ is not complete. □

**Proposition 1.1.6.** *Any prefix code satisfies*

$$\mathbb{E}\Big\{\ell\big[c(X_1^N)\big]\Big\} \geq H\big[\mathbb{P}(dX_1^N)\big].$$

*Proof.* It is enough to notice that, from the Kraft inequality,

$$\begin{aligned} E^N &\longrightarrow [0, 1] \\ x_1^N &\longmapsto 2^{-\ell\left(c(x_1^N)\right)} \end{aligned}$$

defines a subprobability, and to follow the same reasoning as in the proof of proposition 1.1.2. □

**Theorem 1.1.1.** *There exists a complete prefix code $c$ such that*

$$H\Big[\mathbb{P}\big(dX_1^N\big)\Big] \leq \mathbb{E}\Big\{\ell\big[c(X_1^N)\big]\Big\} < H\Big[\mathbb{P}\big(dX_1^N\big)\Big] + 1.$$

*Proof.* The sequence of positive integers

$$\left\{r(b) = \left\lceil -\log_2\bigl(p(b)\bigr)\right\rceil \ : b \in E^N, p(b) > 0\right\}$$

satisfies the Kraft inequality. Thus, there exists a prefix code $c'$ defined on the support of $p$ such that

$$\ell\bigl[c'(b)\bigr] = \left\lceil -\log_2\bigl[p(b)\bigr]\right\rceil, \qquad b \in E^N,\ p(b) > 0.$$

From any prefix code $c'$, a complete prefix code $c$ can be built in such a way that $\ell\bigl[c(b)\bigr] \leq \ell\bigl[c'(b)\bigr]$, for any block $b \in E^N$ such that $p(b) > 0$, by "erasing the non coding bits" (the $i$th bit of $(s_1^r) \in \mathcal{D}'$ is "non coding" if $\mathfrak{I}\bigl[(s_1^{i-1}, 1 - s_i)\bigr] \notin \mathcal{N}'$). This code $c$ is therefore such that

$$\mathbb{E}\Bigl\{\ell\bigl[c(X_1^N)\bigr]\Bigr\} < \sum_{b \in E^N} p(b)\Bigl[-\log_2\bigl(p(b)\bigr) + 1\Bigr] = H\Bigl[\mathbb{P}\bigl(dX_1^N\bigr)\Bigr] + 1.$$

□

*Remark* 1.1.1. One sees that looking for an optimal prefix code is, up to some rounding errors, strictly equivalent to estimating the distribution of the blocks of length $N$.

**Proposition 1.1.7 (Huffman code).** *Let $p$ be a probability distribution on the set $\{1, \ldots, M\}$ giving a positive weight to each point. Let $i$ and $j$ be any two indices satisfying*

$$\max\{p(i), p(j)\} \leq \min\{p(k) : k \notin \{i, j\}\}.$$

*Let $c$ be an optimal prefix code for the probability vector $\bigl[p(k)_{k \notin \{i,j\}}, p(i) + p(j)\bigr]$. The prefix code $c' : \{1, \ldots, M\} \to \{0, 1\}^*$ defined by*

$$\begin{cases} c'(k) &= c(k), \qquad k \notin \{i, j\}, \\ c'(i) &= \Bigl[c\bigl((i, j)\bigr), 0\Bigr], \\ c'(j) &= \Bigl[c\bigl((i, j)\bigr), 1\Bigr] \end{cases} \tag{1.1.3}$$

*is optimal.*

*Proof.* Let $c$ be a code on $\{1, \ldots, M\} \setminus \{i, j\} \cup \{(i, j)\}$. Let $c'$ be the code on $\{1, \ldots, M\}$ defined by (1.1.3). Obviously

$$\mathbb{E}\bigl[\ell(c')\bigr] = \mathbb{E}\bigl[\ell(c)\bigr] + p(i) + p(j).$$

The code $c$ is thus optimal if and only if the code $c'$ is optimal within the set $\mathcal{C}'$ of prefix codes such that the codewords for $i$ and $j$ differ only in their last bit. In the case when $p(i) + p(j)$ is minimal, $\mathcal{C}'$ contains a subset of the set of optimal prefix codes. Indeed, let $c''$ be an optimal prefix code for $(p_k)_{k=1,\ldots,M}$

and $i$ and $j$ two indices such that $p(i)+p(j)$ is minimal. Exchanging codewords if necessary, we can build from $c''$ an optimal prefix code $c'$ such that $c'(i)$ and $c'(j)$ are of maximal length and differ only in their last bit ($c''$ being necessarily complete, there is to be a pair of codewords of maximal length satisfying this property). □

Huffman codes are not easily built in practice for long codewords, since it requires to sort appropriately the block probabilities. We are going to describe next arithmetic coding, which is almost as efficient and does not share this weakness.

**Proposition 1.1.8.** *A vector of positive probabilities $p(i)_{i=1}^{T}$ being given (and ordered in some arbitrary way), let us build the corresponding partition of $[0,1[$ in the following way*

$$\xi(i) = \sum_{j=1}^{i-1} p(j), \qquad i = 1, \ldots, T+1.$$

*An arithmetic (or Shannon-Fano-Elias) code is then defined by putting*

$$I(i) = \max\Big\{ I \in \mathfrak{D} : I \subset \big[\xi(i), \xi(i+1)\big[ \Big\},$$
$$c(i) = \mathfrak{I}^{-1}\big(I(i)\big).$$

*Arithmetic codes are prefix codes satisfying*

$$\mathbb{E}_p\Big(\ell\big(c(i)\big)\Big) < H(p) + 2.$$

*Proof.* The dyadic intervals defined along with arithmetic coding are non overlaping, therefore it is a prefix code. Moreover any interval of length $p(i)$ contains at least one dyadic interval of length $2^{-n}$, as long as $2^{-n} \leq p(i)/2$. The smallest valid integer $n$ thus satisfies $2^{-n+1} > p(i)/2$, or $2^{-n} > p(i)/4$. □

*Remark* 1.1.2. Arithmetic coding of a block $x_1^N \in \{1, \ldots, M\}^N$ of length $N$ is fast when the blocks are ordered lexicographically. Indeed one gets

$$\xi(x_1^N) = \sum_{k=1}^{N} \sum_{y < x_k} p\big[(x_1^{k-1}, y)\big].$$

This quantity can be computed by the following induction :

$$\pi_0 = 1,$$
$$\xi_0 = 0;$$
$$\vdots$$
$$\pi_k = \pi_{k-1}\, p\big(x_k \,|\, x_1^{k-1}\big),$$

$$
\begin{aligned}
\xi_k &= \xi_{k-1} + \sum_{y_k < x_k} \pi_{k-1}\, p\big(y_k \,|\, x_1^{k-1}\big);\\
&\vdots\\
\xi(x_1^N) &= \xi_N;\\
p(x_1^N) &= \pi_N;\\
I(x_1^N) &= 2^{-L(x_1^N)}\Big(\Big\lceil \xi(x_1^N) 2^{L(x_1^N)} \Big\rceil + [0,1[\Big), \text{ where}\\
L(x_1^N) &= \lceil -\log_2\big(p(x_1^N)\big)\rceil + 1.
\end{aligned}
$$

In the worst case this computation requires $N\,|E|$ computations of conditional probabilities, $N\,|E|$ multiplications and $N|E|$ sums.

## 1.2 Universal coding and mixture codes

We have seen that building a prefix code for blocks of length $N$ was up to some details equivalent to choosing a probability distribution $E^N$. Such a distribution is called a *coding* distribution, or an "ideal code".

If a source with a known distribution is to be coded, the highest mean compression rate will be obtained using this distribution as the coding distribution. There are anyhow many cases where the distribution of the source is unknown, and the only sample of the source to be observed is the one to be coded and presumably transmitted.

"Universal coding theory" is devoted to the compression of a source with an unknown — or at least partially unknown — distribution.

**Definition 1.2.1.** The performance of the coding distribution $\mathbb{Q}$ applied to a source distributed according to $\mathbb{P}$ is measured by the *redundancy*

$$
\begin{aligned}
\mathcal{R}(\mathbb{P},\mathbb{Q}) &= \frac{1}{\log(2)} \mathcal{K}(\mathbb{P},\mathbb{Q})\\
&= \begin{cases} \mathbb{E}_{\mathbb{P}}\left(\log_2\left(\dfrac{\mathbb{P}}{\mathbb{Q}}\right)\right) & \text{when } \mathbb{P} \ll \mathbb{Q}\\ +\infty & \text{otherwise.}\end{cases}
\end{aligned}
$$

*Remark* 1.2.1. Up to rounding errors, redundancy measures the difference between the mean length of a prefix code built from $\mathbb{Q}$ and an optimal prefix code.

In these matters of universal coding, *mixtures* of distributions play a prominent role. Let us begin with a definition.

**Definition 1.2.2.** Let $\{\mathbb{P}_\theta \in \mathcal{M}_1^+(\mathcal{X}) : \theta \in \Theta\}$ be a family of probability distributions on the finite set $\mathcal{X}$ (which in the applications to compression will be the set $E^N$ of blocs of length $N$ produced by some random source). Let us assume that the parameter set is a measurable space $(\Theta, \mathfrak{T})$. Let us assume also that the map

$$\theta \longmapsto \mathbb{P}_\theta(x)$$

is measurable for any $x \in \mathcal{X}$. Let $\rho \in \mathcal{M}_+^1(, \mathfrak{T})$ be a probability distribution on the parameter space. Let $\mathbb{P}_\rho$ be the mixture of $\mathbb{P}_\theta$ with respect to the prior distribution $\rho$ on the parameter space :

$$\mathbb{P}_\rho(x) = \int_\Theta \mathbb{P}_\theta(x)\rho(d\theta).$$

We will describe three possible points of view on universal coding: the Bayesian approach, the minimax approach and "oracle" inequalities.

### 1.2.1 The Bayesian approach

Mixtures of coding distributions appear as optimal coding distributions when, like in the Bayesian approach to statistical inference, one minimizes the *mean redundancy* with respect to some prior distribution $\rho$ defined on the parameter set $\Theta$.

**Definition 1.2.3.** Let $\rho$ be a probability distribution defined on the parameter space $(\Theta, \mathfrak{T})$ (where $\mathfrak{T}$ is a sigma-algebra) of a parametric model $\{\mathbb{P}_\theta \in \mathcal{M}_+^1(\mathcal{X}) : \theta \in \Theta\}$ consisting in a family of distributions on the finite set $\mathcal{X}$. Let us assume that for any $x \in \mathcal{X}$, $\theta \longmapsto \mathbb{P}_\theta(x)$ is measurable. For any coding distribution $\mathbb{Q}$, the *mean redundancy with respect to* $\rho$ will be defined as

$$\mathcal{R}_\rho(\mathbb{Q}) = \mathbb{E}_{\rho(d\theta)}\Big(\mathcal{R}(\mathbb{P}_\theta, \mathbb{Q})\Big).$$

The following proposition is straightforward :

**Proposition 1.2.1.** *With the notations and under the hypotheses of definitions 1.2.2 and 1.2.3, the mixture $\mathbb{P}_\rho$ is the unique coding distribution which minimizes the mean redundancy with respect to $\rho$ :*

$$\mathcal{R}_\rho(\mathbb{P}_\rho) = \inf\{\mathcal{R}_\rho(\mathbb{Q}) : \mathbb{Q} \in \mathcal{M}_+^1(\mathcal{X})\}.$$

*Proof.* It is easy to check that

$$\begin{aligned}\mathcal{R}_\rho(\mathbb{Q}) &= \mathbb{E}_{\rho(d\theta)}\big[\mathcal{R}(\mathbb{P}_\theta, \mathbb{P}_\rho)\big] + \mathcal{R}(\mathbb{P}_\rho, \mathbb{Q}) \\ &= \mathcal{R}_\rho(\mathbb{P}_\rho) + \mathcal{R}(\mathbb{P}_\rho, \mathbb{Q}).\end{aligned}$$

We conclude using the fact that the redundancy function is positive and cancels only on the diagonal. □

The optimal mean redundancy $\mathcal{R}_\rho(\mathbb{P}_\rho)$ is equal to the *mutual information* between $\theta$ and $X$ under the joint distribution

$$\mathbb{P}_\rho\big(d(\theta, x)\big) = \rho(d\theta)\mathbb{P}_\theta(dx).$$

**Definition 1.2.4.** *The mutual information* between two random variables $X$ and $Y$ defined on the same probability space is defined as

$$\begin{aligned}\mathcal{I}(X,Y) &= \frac{1}{\log(2)}\mathcal{K}(\mathbb{P}(dX,dY),\mathbb{P}(dX)\otimes\mathbb{P}(dY))\\ &= \frac{1}{\log(2)}\mathbb{E}_{\mathbb{P}(dY)}\mathcal{K}(\mathbb{P}(dX\,|\,Y),\mathbb{P}(dX)).\end{aligned}$$

**Proposition 1.2.2.** *As for the Kullback Leibler divergence function, there is a decomposition formula for mutual information : let $(X_1,\dots,X_n,Y)$ be a vector of random variables defined on the product $\mathcal{X}_1\times\cdots\times\mathcal{X}_n\times\mathcal{Y}$ of two Polish spaces equiped with their Borel sigma-algebras. (This implies the existence of regular conditional probabilities.) Then*

$$\mathcal{I}((X_1,\dots,X_n),Y) = \sum_{k=1}^{n}\mathbb{E}_{\mathbb{P}(dX_1^{k-1})}\Bigl(\mathcal{I}(X_k,Y\,|\,X_1^{k-1})\Bigr),$$

*where $\mathcal{I}(X_k,Y\,|\,X_1^{k-1})$ is the mutual information of the conditional joint distribution of $(X_k,Y)$ knowing $(X_1,\dots,X_{k-1})$ :*

$$\begin{aligned}&\mathcal{I}(X_k,Y\,|\,X_1^{k-1})\\ &\quad= \frac{1}{\log(2)}\mathcal{K}\bigl[\mathbb{P}(dX_k,dY\,|\,X_1^{k-1}),\mathbb{P}(dX_k\,|\,X_1^{k-1})\otimes\mathbb{P}(dY\,|\,X_1^{k-1})\bigr]\\ &\quad= \frac{1}{\log(2)}\mathbb{E}_{\mathbb{P}(dY\,|\,X_1^{k-1})}\Bigl\{\mathcal{K}\bigl[\mathbb{P}(X_k\,|\,X_1^{k-1},Y),\,\mathbb{P}(X_k\,|\,X_1^{k-1})\bigr]\Bigr\}.\end{aligned}$$

This is an easy consequence of the decomposition formula for the Kullback divergence function, which is recalled and proved in proposition 1.7 of appendix 1.5.4. See also [31, Appendice D].

**Proposition 1.2.3.** *Let $(X,Y)$ be a couple of random variables taking their values in the product $\mathcal{X}\times\mathcal{Y}$ of two Polish spaces equiped with their Borel sigma algebras. For any probability distribution $\mathbb{Q}$ on $\mathcal{Y}$,*

$$\mathcal{I}(X,Y)+\mathcal{R}(\mathbb{P}(dY),\mathbb{Q}) = \mathbb{E}_{\mathbb{P}(dX)}\Bigl[\mathcal{R}(\mathbb{P}(dY\,|\,X),\mathbb{Q})\Bigr].$$

*Consequently, in the case when $\mathcal{X}\times\mathcal{Y}$ is a finite set,*

$$\mathcal{I}(X,Y) = H(\mathbb{P}(dY)) - \mathbb{E}_{\mathbb{P}(dX)}\Bigl(H(\mathbb{P}(dY\,|\,X))\Bigr).$$

### 1.2.2 The minimax approach

Another point of view on universal coding is to assume that the source distribution belongs to a parametric family $\{\mathbb{P}_\theta : \theta\in\Theta\}$ of possible distributions, without assuming anyhow that the parameter $\theta$ itself is a random variable.

The question to be solved is to find a coding distribution $\mathbb{Q}$, independent of $\theta$, which minimizes the *worst case redundancy*

$$\mathcal{R}(\mathbb{Q}) = \sup_{\theta \in \Theta} \mathcal{R}(\mathbb{P}_\theta, \mathbb{Q}).$$

This is similar to the minimax approach to parametric statistical inference : the source (which takes the place of the observed sample) follows a distribution belonging to some parametric family, and the aim is to find the code which "performs the best in the worst case".

The following property introduces the notion of "least favourable prior". It shows how to get lower and upper bounds for the minimax redundancy with the help of mixtures of coding distributions.

**Proposition 1.2.4.** *Let $\{\mathbb{P}_\theta \in \mathcal{M}_+^1(\mathcal{X}) : \theta \in \Theta\}$ be a family of probability distributions on $\mathcal{X}$. Let us assume that the parameter space $\Theta$ is equiped with a sigma algebra $\mathfrak{T}$ and that $\theta \mapsto \mathbb{P}_\theta(x)$ is measurable for any $x \in \mathcal{X}$. Then for any probability distribution $\rho \in \mathcal{M}_+^1(\Theta, \mathfrak{T})$ on the parameter space,*

$$\begin{aligned}\operatorname*{ess\,inf}_{\rho(d\theta)} \mathcal{R}(\mathbb{P}_\theta, \mathbb{P}_\rho) &\leq \mathbb{E}_{\rho(d\theta)}\bigl(\mathcal{R}(\mathbb{P}_\theta, \mathbb{P}_\rho)\bigr) \\ &\leq \inf_{\mathbb{Q} \in \mathcal{M}_+^1(\mathcal{X})} \sup_{\theta \in \Theta} \mathcal{R}(\mathbb{P}_\theta, \mathbb{Q}) \leq \sup_{\theta \in \Theta} \mathcal{R}(\mathbb{P}_\theta, \mathbb{P}_\rho).\end{aligned}$$

*Consequently, if $\widehat{\rho}\bigl[\mathcal{R}(\mathbb{P}_\theta, \mathbb{P}_{\widehat{\rho}}) = \sup_{\theta \in \Theta} \mathcal{R}(\mathbb{P}_\theta, \mathbb{P}_{\widehat{\rho}})\bigr] = 1$, that is if $\theta \mapsto \mathcal{R}(\mathbb{P}_\theta, \mathbb{P}_{\widehat{\rho}})$ is constant $\widehat{\rho}$ almost surely, and everywhere not greater than this value, $\mathbb{P}_{\widehat{\rho}}$ is the (unique) minimax coding distribution. The distribution $\widehat{\rho}$ is then called a "least favourable prior". It indeed solves the following maximization problem:*

$$\mathcal{R}_{\widehat{\rho}}(\mathbb{P}_{\widehat{\rho}}) = \sup_{\rho \in \mathcal{M}_+^1(\Theta)} \mathcal{R}_\rho(\mathbb{P}_\rho).$$

*Proof.* For any probability distribution $\rho \in \mathcal{M}_+^1(\Theta)$ on the parameter space,

$$\begin{aligned}\operatorname*{ess\,inf}_{\rho(d\theta)} \mathcal{R}(\mathbb{P}_\theta, \mathbb{P}_\rho) &\leq \mathbb{E}_{\rho(d\theta)}\bigl(\mathcal{R}(\mathbb{P}_\theta, \mathbb{P}_\rho)\bigr) \\ &= \inf_{\mathbb{Q} \in \mathcal{M}_+^1(\mathcal{X})} \mathbb{E}_{\rho(d\theta)}\bigl(\mathcal{R}(\mathbb{P}_\theta, \mathbb{Q})\bigr) \leq \inf_{\mathbb{Q} \in \mathcal{M}_+^1(\mathcal{X})} \sup_{\theta \in \Theta} \mathcal{R}(\mathbb{P}_\theta, \mathbb{Q}) \\ &\qquad \leq \sup_{\theta \in \Theta} \mathcal{R}(\mathbb{P}_\theta, \mathbb{P}_\rho).\end{aligned}$$

The coding distribution $\mathbb{P}_{\widehat{\rho}}$ is indeed unique. Assume that $\widehat{Q}$ is another minimax coding distribution. Then

$$\sup_{\theta \in \Theta} \mathcal{R}(\mathbb{P}_\theta, \widehat{Q}) = \mathbb{E}_{\widehat{\rho}(d\theta)}\bigl[\mathcal{R}(\mathbb{P}_\theta, \mathbb{P}_{\widehat{\rho}})\bigr],$$

and therefore

$$\mathbb{E}_{\widehat{\rho}(d\theta)}\bigl[\mathcal{R}(\mathbb{P}_\theta, \widehat{Q})\bigr] \leq \mathbb{E}_{\widehat{\rho}(d\theta)}\bigl[\mathcal{R}(\mathbb{P}_\theta, \mathbb{P}_{\widehat{\rho}})\bigr].$$

As moreover

$$\mathbb{E}_{\widehat{\rho}(d\theta)}\big[\mathcal{R}(\mathbb{P}_\theta, \widehat{Q})\big] = \mathbb{E}_{\widehat{\rho}(d\theta)}\big[\mathcal{R}(\mathbb{P}_\theta, \mathbb{P}_{\widehat{\rho}})\big] + \mathcal{R}(\mathbb{P}_{\widehat{\rho}}, \widehat{Q}),$$

this shows that

$$\mathcal{R}(\mathbb{P}_{\widehat{\rho}}, \widehat{Q}) = 0,$$

and consequently that $\widehat{Q} = \mathbb{P}_{\widehat{\rho}}$. □

In the minimax approach, mixtures of distributions from the model used for the source outperform any other choice of coding distribution. This is shown in the following theorem, which further explores the properties of the "least favourable prior".

**Theorem 1.2.1 (Csiszar, Körner, Gallager).** *Let $\{\mathbb{P}_\theta \in \mathcal{M}_+^1(\mathcal{X}) : \theta \in \Theta\}$ be a parametric family of probability distributions. Let, as above, $\mathcal{X}$ be a finite set, which will be equal to $E^N$, the set of values of blocks of length $N$, in the data compression problem. Let us assume that $\Theta$ is a measurable set for some sigma algebra $\mathfrak{T}$ and that for any $x \in \mathcal{X}$*

$$\begin{aligned} \Theta &\longrightarrow [0,1] \\ \theta &\longmapsto \mathbb{P}_\theta(x) \end{aligned}$$

*is measurable. Let us also assume that the set $\{\mathbb{P}_\theta : \theta \in \Theta\}$ is closed (and therefore compact, because the simplex of all probability distributions on the finite set $\mathcal{X}$ is compact). The set of all mixtures, that is of all the distributions of the form $\mathbb{P}_\rho$ where $\rho$ is a probability distribution on $(\Theta, \mathfrak{T})$ is then also a compact set (generated by the distributions $\rho$ which can be written as the convex combinations of at most $|\mathcal{X}|$ Dirac masses, from Carathéodory's theorem).*

*Let $\mathbb{Q}$ be a coding distribution which dominates at least one of the distributions $\mathbb{P}_\theta$, $\theta \in \Theta$, and $\mathbb{P}_{\rho^*}$ its projection on the mixtures $\{\mathbb{P}_\rho : \rho \in \mathcal{M}_1^+(\Theta, \mathfrak{T})\}$ with respect to Kullback divergence, that is the unique probability distribution such that*

$$\mathcal{K}(\mathbb{P}_{\rho^*}, \mathbb{Q}) = \inf_{\rho \in \mathcal{M}_1^+(\Theta)} \mathcal{K}(\mathbb{P}_\rho, \mathbb{Q}).$$

*(Existence comes from compactness, uniqueness from the fact that the Kullback divergence function is strictly convex in its first argument.)*

*The coding distribution $\mathbb{P}_{\rho^*}$ then performs strictly better than $\mathbb{Q}$ as soon as it differs from it, in the sense that, for any $\theta \in \Theta$*

$$\mathcal{R}(\mathbb{P}_\theta, \mathbb{Q}) \geq \mathcal{R}(\mathbb{P}_\theta, \mathbb{P}_{\rho^*}) + \mathcal{R}(\mathbb{P}_{\rho^*}, \mathbb{Q}) > \mathcal{R}(\mathbb{P}_\theta, \mathbb{P}_{\rho^*}).$$

*Let $\hat{\rho} \in \mathcal{M}_+^1(\Theta)$ be some distribution maximizing*

$$\begin{aligned} \mathcal{M}_+^1(\Theta) &\longrightarrow \mathbb{R}_+ \\ \rho &\longmapsto \mathbb{E}_{\rho(d\theta)}\mathcal{R}(\mathbb{P}_\theta, \mathbb{P}_\rho). \end{aligned}$$

*The existence of* $\hat{\rho}$ *comes from the weak compactness of* $\mathcal{M}_+^1\Big(\{\mathbb{P}_\theta : \theta \in \Theta\}\Big)$. *The map*

$$\begin{aligned} \Theta &\longrightarrow \mathbb{R}_+ \\ \theta &\longmapsto \mathcal{R}\big[\mathbb{P}_\theta, \mathbb{P}_{\hat{\rho}}\big] \end{aligned}$$

*is constant* $\hat{\rho}$ *almost surely. More precisely for any* $\theta \in \Theta$

$$\mathcal{R}(\mathbb{P}_\theta, \mathbb{P}_{\hat{\rho}}) \leq \mathbb{E}_{\hat{\rho}(d\theta)}\big[\mathcal{R}(\mathbb{P}_\theta, \mathbb{P}_{\hat{\rho}})\big].$$

*This implies that* $\mathbb{P}_{\hat{\rho}}$ *is the (unique) minimax coding distribution :*

$$\sup_{\theta\in\Theta} \mathcal{R}(\mathbb{P}_\theta, \mathbb{P}_{\hat{\rho}}) = \inf_{\mathbb{Q}\in\mathcal{M}_+^1(\mathcal{X})} \sup_{\theta\in\Theta} \mathcal{R}(\mathbb{P}_\theta, \mathbb{Q}).$$

*Proof.* We will more generally prove that for any closed convex subset $C$ of the set $\mathcal{M}_1^+(\mathcal{X})$ of probability distributions, for any distribution $\mathbb{Q} \in \mathcal{M}_1^+(\mathcal{X})$ which dominates at least one distribution of $C$, the projection $\mathbb{P}^*$ of $\mathbb{Q}$ on $C$ according to the Kullback divergence function, namely the only probability distribution $\mathbb{P}^* \in C$ such that

$$\mathcal{K}(\mathbb{P}^*, \mathbb{Q}) = \inf_{\mathbb{P}\in C} \mathcal{K}(\mathbb{P}, \mathbb{Q}),$$

satisfies the following triangular inequality : for any probability distribution $\mathbb{P} \in C$

$$\mathcal{K}(\mathbb{P}, \mathbb{Q}) \geq \mathcal{K}(\mathbb{P}, \mathbb{P}^*) + \mathcal{K}(\mathbb{P}^*, \mathbb{Q}).$$

To see this, let us first notice that the above inequality is trivial in the case when $\mathcal{K}(\mathbb{P}, \mathbb{Q}) = +\infty$. Otherwise $\mathcal{K}(\mathbb{P}^*, \mathbb{Q}) \leq \mathcal{K}(\mathbb{P}, \mathbb{Q}) < +\infty$, therefore both $\mathbb{P}$ and $\mathbb{P}^*$, and consequently $\lambda\mathbb{P} + (1-\lambda)\mathbb{P}^*$ for any $\lambda \in [0, 1]$, are absolutely continuous with respect to $\mathbb{Q}$. The base space $\mathcal{X}$ being finite, the convex function

$$\lambda \mapsto \mathcal{K}(\lambda\mathbb{P} + (1-\lambda)\mathbb{P}^*, \mathbb{Q}) : [0, 1] \to \mathbb{R}_+$$

is a finite sum of elementary functions, and the computation of its right-hand derivative at point $\lambda = 0$ is straightforward. This derivative may be finite or equal to $-\infty$, due to the convexity of the function, and is indeed non-negative, because $\mathcal{K}(\mathbb{P}^*, \mathbb{Q})$ is minimal :

$$0 \leq \frac{\partial}{\partial\lambda}_{\lambda=0} \mathcal{K}\big[\lambda\mathbb{P} + (1-\lambda)\mathbb{P}^*,\ \mathbb{Q}\big] = \mathcal{K}(\mathbb{P}, \mathbb{Q}) - \mathcal{K}(\mathbb{P}, \mathbb{P}^*) - \mathcal{K}(\mathbb{P}^*, \mathbb{Q}).$$

Let us come back to the existence of $\mathbb{P}_{\hat{\rho}}$. As we assume that the parametrization is measurable, one can without loss of generality assume that $\mathbb{P}_\theta$ is indexed by itself, namely that $\Theta = \{\mathbb{P}_\theta : \theta \in \Theta\}$ and that $\theta \longmapsto \mathbb{P}_\theta$ is the identity map. It is then possible to write the optimal Bayesian redundancy as

$$\mathbb{E}_{\rho(d\theta)}\big(\mathcal{R}(\mathbb{P}_\theta, \mathbb{P}_\rho)\big) = H(\mathbb{P}_\rho) - \mathbb{E}_{\rho(d\theta)} H(\mathbb{P}_\theta).$$

This expression shows that it is a concave function of $\rho$. It is moreover continuous for the weak topology of $\mathcal{M}_+^1(\Theta)$ (the topology associated with the continuity of the integrals of continuous functions defined on the compact set $\Theta$). Indeed $\theta \mapsto \mathbb{P}_\theta$ and $\theta \mapsto H(\mathbb{P}_\theta)$ are continuous, thus by definition of weak topology, $\rho \mapsto \mathbb{P}_\rho$ and $\rho \mapsto \mathbb{E}_{\rho(d\theta)} H(\mathbb{P}_\theta)$ are weakly continuous, therefore $\rho \mapsto \mathcal{R}_\rho(\mathbb{P}_\rho) = H(\mathbb{P}_\rho) - \mathbb{E}_\rho H(\mathbb{P}_\theta)$ is also weakly continuous. Moreover the set $\mathcal{M}_+^1(\Theta)$ is weakly compact, thus $\rho \mapsto \mathcal{R}_\rho(\mathbb{P}_\rho)$ reaches its maximum on this set at some point $\widehat{\rho}$ (at least).

It is then possible to write

$$\mathcal{R}_\nu(\mathbb{P}_\nu) - \mathcal{R}_\rho(\mathbb{P}_\rho) = \bigl(\mathbb{E}_{\nu(d\theta)} - \mathbb{E}_{\rho(d\theta)}\bigr)\Bigl(\mathcal{R}(\mathbb{P}_\theta, \mathbb{P}_\rho)\Bigr) - \mathcal{R}(\mathbb{P}_\nu, \mathbb{P}_\rho). \quad (1.2.1)$$

Applying this inequality to $\rho = \hat{\rho}$ and to $\nu' = \lambda\nu + (1-\lambda)\hat{\rho}$, and differentiating at point $\lambda = 0$, one gets

$$\bigl(\mathbb{E}_{\nu(d\theta)} - \mathbb{E}_{\hat{\rho}(d\theta)}\bigr)\Bigl(\mathcal{R}(\mathbb{P}_\theta, \mathbb{P}_{\hat{\rho}})\Bigr) \leq 0.$$

Choosing $\nu = \delta_\theta$ shows that

$$\mathcal{R}(\mathbb{P}_\theta, \mathbb{P}_{\hat{\rho}}) \leq \mathbb{E}_{\hat{\rho}(d\theta)}\bigl(\mathcal{R}(\mathbb{P}_\theta, \mathbb{P}_{\hat{\rho}})\bigr), \qquad \theta \in \Theta.$$

Thus, from proposition 1.2.4, $\mathbb{P}_{\widehat{\rho}}$ is the unique minimax coding probability distribution. □

This theorem shows that when universal data compression is addressed in a parametric framework, it is enough to restrict to coding distributions obtained as mixtures of model distributions.

### 1.2.3 Oracle inequalities

Another point of view on universal compression is to start with a parametric family of coding distributions, and not with a parametric family of source distributions. Given such a family $\{\mathbb{Q}_\theta : \theta \in \Theta\}$, a coding distribution $\mathbb{Q}_u$ may be sought which leads in any case to a mean code length of the same order as the mean length of the best coding distribution in the considered family. More precisely, we will look for a code $\mathbb{Q}_u$ such that for any sequence $(x_1, \dots, x_N) \in E^N$,

$$-\log_2\Bigl(\mathbb{Q}_u(x_1^N)\Bigr) \leq \inf_{\theta \in \Theta}\Bigl\{-\log_2\Bigl(\mathbb{Q}_\theta(x_1^N)\Bigr) + \gamma(\theta)\Bigr\},$$

where the loss function $\gamma(\theta)$ is as small as possible. This third point of view is analogous to the framework of oracle inequalities in statistics.

The mixtures of a countable family of coding distributions satisfy a straightforward oracle inequality :

**Proposition 1.2.5.** *Let*

$$\{\mathbb{Q}_\theta \in \mathcal{M}_1^+(\mathcal{X}) : \theta \in \Theta\}$$

*be a countable family of coding distributions on the finite set* $\mathcal{X}$*, let* $\rho \in \mathcal{M}_1^+(\Theta)$ *be a probability distribution on the parameter space and* $\mathbb{Q}_\rho$ *corresponding mixture. The following oracle inequality is satisfied : for any* $x \in \mathcal{X}$,

$$-\log_2\big(\mathbb{Q}_\rho(x)\big) \leq \inf_{\theta \in \Theta}\Big\{-\log_2\big(\mathbb{Q}_\theta(x)\big) - \log_2\big(\rho(\{\theta\})\big)\Big\}.$$

*Remark* 1.2.2. Although the proof is straightforward, it is a far reaching result, both from the practical and the theoretical point of view. We will see on some examples how to generalize this type of inequalities to the case when the parameter space $\Theta$ is continuous.

When a source of length $N$ distributed according to some *arbitrary* probability measure $\mathbb{P} \in \mathcal{M}_1^+(E^N)$ is coded, we get

$$\frac{1}{N}\mathcal{R}(\mathbb{P}, \mathbb{Q}_\rho) \leq \inf_{\theta \in \Theta}\left\{\frac{1}{N}\mathcal{R}(\mathbb{P}, \mathbb{Q}_\theta) - \frac{1}{N}\log_2\big(\rho(\{\theta\})\big)\right\}.$$

The strength of this result comes from the fact that it does not require *any* hypothesis on the distribution $\mathbb{P}$ of the source.

## 1.3 Lower bounds for the minimax compression rate

We will first consider the case when the parameter set is finite.

**Theorem 1.3.1 (Merhav and Feder).** *Let* $\{\mathbb{P}_\theta \in \mathcal{M}_+^1(\mathcal{X}) : \theta \in \Theta\}$ *be a finite set of probability distributions on a finite set* $\mathcal{X}$*. Let* $\hat{\theta}(x)$ *be an estimator of* $\theta$*. Let* $p_e$ *be the mean error rate of this estimator:*

$$p_e = \frac{1}{|\Theta|}\sum_{\theta \in \Theta} \mathbb{P}_\theta\big(\hat{\theta}(X) \neq \theta\big).$$

*For any coding distribution* $\mathbb{Q} \in \mathcal{M}_+^1(\mathcal{X})$*, any* $\epsilon > 0$,

$$\frac{1}{|\Theta|}\sum_{\theta \in \Theta} \mathbb{1}\Big(\mathcal{R}(\mathbb{P}_\theta, \mathbb{Q}) \leq (1-\epsilon)\log_2\big(|\Theta|\big)\Big) \leq \frac{p_e \log_2\big(|\Theta|\big) + 2}{\epsilon \log_2\big(|\Theta|\big)}.$$

*Proof.* Let $\mu$ be the uniform probability distribution on $\Theta$. Let us consider on $\Theta \times \mathcal{X}$ the distribution defined by

$$\mathbb{P}_\mu(\{\theta, x\}) = \mu(\theta)\mathbb{P}_\theta(x).$$

Let $B$ be the set

$$B = \{\theta \in \Theta : \mathcal{R}(\mathbb{P}_\theta, \mathbb{Q}) \leq (1-\epsilon) H(\mu)\}.$$

Let us introduce on the product probability space $\Theta \times \mathcal{X}$ the random variables

$$\begin{aligned} Z &= \mathbb{1}(\theta \in B), \\ T &= \mathbb{1}\big(\hat{\theta}(X) = \theta\big). \end{aligned}$$

Let us notice first that $\mathcal{I}(Z, X \,|\, \theta) = 0$, because $\mathbb{P}(dZ \,|\, \theta)$ always is a Dirac mass. Thus

$$\begin{aligned} \mathcal{I}((\theta, Z), X) &= \mathcal{I}(\theta, X) + \mathbb{E}_{\mu(d\theta)}\Big(\mathcal{I}(Z, X \,|\, \theta)\Big), \\ &= \mathcal{I}(\theta, X). \end{aligned}$$

On the other hand

$$\begin{aligned} \mathcal{I}((\theta, Z), X) &= \mathcal{I}(Z, X) + \mathbb{E}_{\mathbb{P}(dZ)}\Big(\mathcal{I}(\theta, X \,|\, Z)\Big) \\ &\leq H\big(\mathbb{P}(dZ)\big) + \mu(B)\mathcal{I}(\theta, X \,|\, Z=1) + \big(1-\mu(B)\big)\mathcal{I}(\theta, X \,|\, Z=0) \\ &\leq H\big(\mathbb{P}(dZ)\big) + \mu(B)\mathbb{E}_{\mu(d\theta \,|\, Z=1)}\big(\mathcal{R}(\mathbb{P}_\theta, \mathbb{P}_{\mu(d\theta \,|\, Z=1)})\big) \\ &\qquad + \big(1-\mu(B)\big)H(\mu) \\ &\leq H\big(\mathbb{P}(dZ)\big) + \mu(B)\mathbb{E}_{\mu(d\theta \,|\, Z=1)}\big(\mathcal{R}(\mathbb{P}_\theta, \mathbb{Q})\big) + \big(1-\mu(B)\big)H(\mu) \\ &\leq H\big(\mathbb{P}(dZ)\big) + \mu(B)(1-\epsilon)H(\mu) + \big(1-\mu(B)\big)H(\mu) \\ &= H\big(\mathbb{P}(dZ)\big) + \big(1-\epsilon\mu(B)\big)H(\mu). \end{aligned}$$

Let us compute now a lower bound for $\mathcal{I}(\theta, X)$, making use of the random variable $T$:

$$\mathcal{I}(\theta, X) = H\big(\mu(d\theta)\big) - \mathbb{E}_{\mathbb{P}(dX)} H\big(\mathbb{P}(d\theta \,|\, X)\big).$$

$$\begin{aligned} H\big(\mathbb{P}(d\theta, dT \,|\, X)\big) &= H\big(\mathbb{P}(d\theta \,|\, X)\big) + \mathbb{E}_{\mathbb{P}(d\theta \,|\, X)} H\big(\mathbb{P}(dT \,|\, X, \theta)\big) \\ &= H\big(\mathbb{P}(d\theta \,|\, X)\big). \end{aligned}$$

$$\begin{aligned} \mathbb{E}_{\mathbb{P}(dX)}\Big(H\big(\mathbb{P}(d\theta, dT \,|\, X)\big)\Big) &= \mathbb{E}_{\mathbb{P}(dX)}\Big(H\big(\mathbb{P}(dT \,|\, X)\big)\Big) \\ &\qquad + \mathbb{E}_{\mathbb{P}(dX, dT)}\Big(H\big(\mathbb{P}(d\theta \,|\, X, T)\big)\Big) \\ &\leq H\big(\mathbb{P}(dT)\big) + \mathbb{P}(T=0) H\big(\mu(d\theta)\big) \\ &\qquad + \mathbb{P}(T=1)\mathbb{E}_{\mathbb{P}(dX \,|\, T=1)}\Big(\underbrace{H\big(\mathbb{P}(d\theta \,|\, X, T=1)\big)}_{=0}\Big). \end{aligned}$$

Let us notice that this last identity is satisfied only when $\mu$ is the uniform measure, whereas the previous ones are true for any choice of $\mu$. Putting the previous inequalities together, one sees that

$$\big(1-\mathbb{P}(T=0)\big)H(\mu)-H\big(\mathbb{P}(dT)\big)\leq \mathcal{I}(\theta,X)\leq H\big(\mathbb{P}(dZ)\big)+\big(1-\epsilon\mu(B)\big)H(\mu).$$

Upper bounding $H\big(\mathbb{P}(dT)\big)$ and $H\big(\mathbb{P}(dZ)\big)$ by 1, we conclude that

$$\mu(B)\leq \frac{\mathbb{P}(T=0)H(\mu)+2}{\epsilon H(\mu)}.$$

□

**Corollary 1.3.1.** *Let $\mathcal{X}$ be a finite set and $\{\mathbb{P}_\theta:\theta\in\Theta\}$ a family of probability distributions indexed by a measurable set $(\Theta,\mathfrak{T})$. Let us assume that $\theta\longmapsto \mathbb{P}_\theta(x)$ is measurable for any $x\in\mathcal{X}$. Let $\rho\in\mathcal{M}_+^1(\Theta,\mathfrak{T})$ be a probability measure on the parameter space. Let $\mu$ be a distribution on $\big(\Theta^M,\mathfrak{T}^{\otimes M}\big)$, such that*

$$\frac{1}{M}\sum_{i=1}^{M}\mu(\theta_i\in A)=\rho(A),\qquad A\in\mathfrak{T},$$

*(where $(\theta_1,\dots,\theta_M)$ is the canonical process on $\Theta^M$). Let us consider for any $M$-tuple $\theta_1^M$ an estimator*

$$\hat{\theta}_{\theta_1^M}:\mathcal{X}\longrightarrow\{\theta_i:i=1,\dots,M\}.$$

*For any coding distribution $\mathbb{Q}$, the redundancy is lower bounded by*

$$\rho\Big(\mathcal{R}(\mathbb{P}_\theta,\mathbb{Q})\leq(1-\epsilon)\log_2(M)\Big)$$
$$\leq\frac{1}{\epsilon\log_2 M}\left(\mathbb{E}_{\mu(d\theta_1,\dots,d\theta_M)}\left(\frac{1}{M}\sum_{i=1}^{M}\mathbb{P}_{\theta_i}\Big(\mathbb{1}\big(\hat{\theta}_{\theta_1^M}(X)\neq\theta_i\big)\Big)\right)\log_2(M)+2\right).$$

**Theorem 1.3.2 (Rissanen).** *Let us consider a parameter set consisting in a compact subset of $\mathbb{R}^d$ of the type $\Theta=[0,D]^d$. Let $E$ be a finite set. For any integer $N$, let us consider a family of distributions $\{\mathbb{P}_{\theta,N}\in\mathcal{M}_+^1(E^N):\theta\in\Theta\}$ such that $\theta\longmapsto\mathbb{P}_{\theta,N}(x_1^N)$ is measurable for any $x_1^N\in E^N$. Let us assume that there exists an estimator $\hat{\theta}_N:E^N\longrightarrow\Theta$ such that for any $\theta\in\Theta$*

$$\mathbb{P}_{\theta,N}\left(\|\hat{\theta}_N(X_1^N)-\theta\|>\frac{c}{\sqrt{N}}\right)\leq\alpha(c),\tag{1.3.1}$$

*with $\lim\limits_{c\to+\infty}\alpha(c)=0$. Then for any family of coding distributions $\mathbb{Q}_N\in\mathcal{M}_+^1(E^N)$,*

$$\limsup_{N\to+\infty}\frac{1}{\log_2(N)}\mathcal{R}\big(\mathbb{P}_{\theta,N}(dX_1^N),\mathbb{Q}_N\big)\geq\frac{d}{2},\tag{1.3.2}$$

*except maybe on a subset of $\Theta$ of null Lebesgue measure. Moreover*

$$\liminf_{N\to+\infty}\frac{1}{\log_2(N)}\sup_{\theta\in\Theta}\mathcal{R}\big(\mathbb{P}_{\theta,N}(dX_1^N),\mathbb{Q}_N\big)\geq\frac{d}{2}.$$

*Remark* 1.3.1. It is interesting to notice that equation (1.3.2) can be used to derive a weak asymptotic lower bound for the non-cumulated risk. Let us assume indeed that the distributions $\mathbb{P}_{\theta,N}$ are compatible, in the sense that $\mathbb{P}_{\theta,N}(dX_1^n) = \mathbb{P}_{\theta,n}(dX_1^n)$, for $n < N$, so that there exists $\mathbb{P}_\theta \in \mathcal{M}_+^1(E^{\mathbb{N}})$ such that $\mathbb{P}_{\theta,N} = \mathbb{P}_\theta(dX_1^N)$. For each positive integer $n$, let $\mathbb{Q}(dX_n \,|\, X_1^{n-1})$ be some estimator of $\mathbb{P}_\theta(dX_n \,|\, X_1^{n-1})$ based on the observed sample $(X_1, \dots, X_{n-1})$. Let us form $\mathbb{Q}_N(dX_1^N) = \prod_{n=1}^N \mathbb{Q}(dX_n \,|\, X_1^{n-1})$. It is easily seen from inequality (1.3.2) that

$$\limsup_{N\to+\infty} N\mathbb{E}_{\mathbb{P}_\theta(dX_1^{N-1})}\Bigl\{\mathcal{K}\bigl[P_\theta(dX_N|X_1^{N-1}), \mathbb{Q}(dX_N|X_1^{N-1})\bigr]\Bigr\} \geq \frac{d}{2}. \quad (1.3.3)$$

To prove this lower bound, let us put

$$u_N(\theta) = \mathbb{E}_{\mathbb{P}_\theta(dX_1^{N-1})}\Bigl\{\mathcal{K}\bigl[\mathbb{P}_\theta(dX_N \,|\, X_1^{N-1}), \mathbb{Q}(dX_N \,|\, X_1^{N-1})\bigr]\Bigr\}.$$

For any $\theta$ such that (1.3.2) holds, equation (1.3.2) can be written as

$$\limsup_{N\to+\infty} \frac{1}{\log(N)} \sum_{n=1}^N u_n(\theta) \geq \frac{d}{2}. \quad (1.3.4)$$

As $\log(N) \underset{N\to+\infty}{\sim} \sum_{n=1}^N \frac{1}{n}$, this implies that $\limsup_{n\to+\infty} n u_n(\theta) \geq \frac{d}{2}$. Indeed, if it were not the case, there would exist $N_1$ and $\epsilon > 0$ such that

$$u_n(\theta) \leq (1-\epsilon)\frac{d}{2n}, \qquad n \geq N_1,$$

implying the following contradiction with (1.3.4):

$$\begin{aligned} &\limsup_{N\to+\infty} \frac{1}{\log(N)} \sum_{n=1}^N u_n(\theta) \\ &\leq \limsup_{N\to+\infty} \frac{1}{\log(N)} \sum_{n=N_1}^N (1-\epsilon)\frac{d}{2n} + \limsup_{N\to+\infty} \frac{1}{\log(N)} \sum_{n=1}^{N_1-1} u_n(\theta) = (1-\epsilon)\frac{d}{2}. \end{aligned}$$

In particular, in the i.i.d. case when $\mathbb{P}_{\theta,N} = P_\theta^{\otimes N}$, any estimator $\widehat{\theta}(X_1^N)$ satisfying the hypothesis (1.3.1) of the theorem is such that

$$\limsup_{N\to+\infty} N\mathbb{E}_{P_\theta^{\otimes N}(dX_1^N)}\Bigl[\mathcal{K}\bigl(P_\theta, P_{\widehat{\theta}(X_1^N)}\bigr)\Bigr] \geq \frac{d}{2} \quad (1.3.5)$$

for almost any $\theta \in \Theta$. Thus from the quite weak asymptotic upper bound (1.3.1) on the risk of $\widehat{\theta}$ measured by the Euclidean distance in the parameter space (where no precise constant is involved), it is possible to infer an asymptotic lower bound on the risk of $\widehat{\theta}$ measured by the Kullback divergence in the sample distribution space, with a precise constant $\frac{d}{2}$.

*Proof.* Let $\lambda$ be the Lebesgue measure on $\mathbb{R}^d$. Up to some rescaling of the parametrization of $\{\mathbb{P}_{\theta,N}\}$, we may assume without loss of generality that $D = 1$. Let us apply the corollary of the previous theorem, choosing for $\mu$ the probability distribution of a random lattice with step size $1/K$, where $K$ is an integer to be chosen in the following as a function of $N$. Let us consider the value $M = K^d$ and some indexation

$$\{\mathbf{n}_i, i = 1, \dots, M\} = \{0, \dots, K-1\}^d$$

of the finite $d$-dimensional lattice $\{0, \dots, K-1\}^d$. Let us define the probability distribution $\mu$ by the formula

$$\mu(\theta_1^M \in A) = \int_{\theta \in [0,1/K[^d} \mathbb{1}\left[\left(\theta + K^{-1}\mathbf{n}_i\right)_{i=1}^M \in A\right] K^d \lambda(d\theta),$$
$$A \subset \left([0,1]^d\right)^M, A \text{ measurable.}$$

In other words, $\mu$ is the distribution of a regular lattice whose origin is chosen randomly according to the Lebesgue measure. The assumption required by corollary 1.3.1 is therefore satisfied :

$$K^{-d} \sum_{i=1}^{K^d} \mu\Big(\theta_i \in A\Big) = \lambda(A), \qquad A \subset [0,1]^d, A \text{ measurable.}$$

The estimator $\hat{\theta}_{\theta_1^M}(X_1^N)$ will be defined as the closest point $\theta_i$ to $\hat{\theta}(X_1^N)$ :

$$\|\hat{\theta}_{\theta_1^M}(X_1^N) - \hat{\theta}(X_1^N)\| = \min_{i=1,\dots,M} \|\theta_i - \hat{\theta}(X_1^N)\|.$$

Obviously, $\mu(d\theta_1^M)$ almost surely,

$$\mathbb{P}_{\theta_i,N}\big(\hat{\theta}_{\theta_1^M}(X_1^N) \neq \theta_i\big) \leq \mathbb{P}_{\theta_i,N}\left(\|\theta_i - \hat{\theta}(X_1^N)\| > \frac{1}{2K}\right) \leq \alpha\left(\frac{\sqrt{N}}{2K}\right).$$

According to the corollary 1.3.1 of the previous theorem, it follows that

$$\lambda\Big(\mathcal{R}(\mathbb{P}_{\theta,N}, \mathbb{Q}_N) \leq (1-\epsilon) d \log_2(K)\Big) \leq \frac{\alpha\left(\frac{\sqrt{N}}{2K}\right) d \log_2(K) + 2}{\epsilon\, d \log_2(K)}.$$

Let us choose $K = \sqrt{N}/\log(N)$, to get

$$\lambda\Big(\mathcal{R}(\mathbb{P}_{\theta,N}, \mathbb{Q}_N) \leq (1-\epsilon)\Big(\frac{d}{2}\log_2(N) - d\log_2\big(\log(N)\big)\Big)$$
$$\leq \frac{1}{\epsilon}\left(\alpha(\log(N)/2) + \frac{2}{d\log_2\big(\sqrt{N}/\log(N)\big)}\right).$$

Thus

$$\lim_{N \to +\infty} \lambda\Big(\frac{2}{d\log_2(N)} \mathcal{R}(\mathbb{P}_{\theta,N}, \mathbb{Q}_N) \leq (1-2\epsilon)\Big) = 0, \tag{1.3.6}$$

and therefore

$$\lambda\Big(\Big(\limsup_{N \to +\infty} \frac{2}{d\log_2(N)} \mathcal{R}(\mathbb{P}_{\theta,N}, \mathbb{Q}_N)\Big) < (1-2\epsilon)\Big) = 0.$$

The first assertion of the theorem then follows by letting $\epsilon$ tend to zero. The second assertion is also a consequence of equation (1.3.6). Indeed it shows that for $N$ large enough

$$\lambda\Big(\mathcal{R}(\mathbb{P}_{\theta,N}, \mathbb{Q}_N) \leq (1-2\epsilon)\frac{d\log_2(N)}{2}\Big) < 1,$$

and therefore that

$$\sup_{\theta \in \Theta} \frac{2}{d\log_2(N)} \mathcal{R}(\mathbb{P}_{\theta,N}, \mathbb{Q}_N) \geq (1-2\epsilon),$$

whence

$$\liminf_{N \to +\infty} \sup_{\theta \in \Theta} \frac{2}{d\log_2(N)} \mathcal{R}(\mathbb{P}_{\theta,N}, \mathbb{Q}_N) \geq (1-2\epsilon),$$

and the second part of the theorem is obtained by letting $\epsilon$ tend to 0. □

## 1.4 Mixtures of i.i.d. coding distributions

This section is a preliminary to the following one on double mixture codes. Here again, we model a source using a finite alphabet $E$ by the canonical process $(X_n)_{n \in \mathbb{N}}$ on $E^{\mathbb{N}}$. All the following discussion deals with the coding of a block of size $N$. We consider a family of coding distributions $\{\mathbb{Q}_\theta : \theta \in \mathcal{M}_+^1(E)\}$ for the blocks of size $N$ defined by $\mathbb{Q}_\theta = \theta^{\otimes N}$. We are looking for a prior distribution on the parameters which produces a mixture of coding distributions close to the minimax estimator for the model $\{\mathbb{P}_\theta = \mathbb{Q}_\theta : \theta \in \Theta\}$. According to proposition 1.2.4, it is suitable to look for a prior distribution $\rho$ such that

$$\theta \longmapsto \mathcal{R}(\mathbb{Q}_\theta, \mathbb{Q}_\rho)$$

is approximately constant (at least on the support of $\rho$). Let us notice that

$$\begin{aligned}
\mathcal{R}(\mathbb{Q}_\theta, \mathbb{Q}_\rho) &= \mathbb{E}_{\mathbb{Q}_\theta(dX_1^N)}\left(\log_2\left(\frac{\exp(-N\mathcal{K}(\bar{P}_{X_1^N}, \theta))}{\mathbb{E}_{\rho(d\theta)}\exp(-N\mathcal{K}(\bar{P}_{X_1^N}, \theta))}\right)\right) \\
&= -N\mathbb{E}_{\mathbb{Q}_\theta(dX_1^N)}\big(\mathcal{R}(\bar{P}_{X_1^N}, \theta)\big) \\
&\quad - \mathbb{E}_{\mathbb{Q}_\theta(dX_1^N)}\Big(\log_2\Big(\mathbb{E}_{\rho(d\theta)}\exp(-N\mathcal{K}(\bar{P}_{X_1^N}, \theta))\Big)\Big),
\end{aligned}$$

where $\bar{P}_{x_1^N} \in \mathcal{M}_+^1(E)$ is the *empirical probability distribution* of the sample $x_1^N$, namely

$$\bar{P}_{x_1^N} = \frac{1}{N} \sum_{i=1}^{N} \delta_{x_i}.$$

It is easy to check that

$$\lim_{N \to +\infty} N \mathbb{E}_{\mathbb{Q}_\theta(dX_1^N)} \bigl( \mathcal{R}(\bar{P}_{X_1^N}, \theta) \bigr) = \frac{|E| - 1}{2 \log(2)},$$

noticing that

$$\mathcal{R}(\bar{P}_{X_1^N}, \theta) = \frac{1}{\log(2)} \mathbb{E}_{\theta(dx)} \left( \Phi \left( \frac{\bar{P}_{X_1^N}(x)}{\theta(x)} \right) \right)$$

where

$$\Phi(r) = 1 - r + r \log(r) \underset{r \to 1}{\sim} \frac{1}{2}(1 - r)^2.$$

Moreover, if $\rho(d\theta) = f(\theta)\lambda(d\theta)$, where $\lambda$ is the Lebesgue measure on the simplex $\Theta$ and where $f$ is a smooth function, it is easy to check [1], using the Laplace method, that for any $x_1^N$ such that $\bar{P}_{x_1^N}$ belongs to the interior of the simplex:

$$\mathbb{E}_{\rho(d\theta)} \Bigl( \exp(-N \mathcal{K}(\bar{P}_{x_1^N}, \theta)) \Bigr)$$

[1] The second equivalence can be obtained by choosing as coordinate system for the simplex the $d-1$ first coordinates of the ambient space $\mathbb{R}^d$, namely $(\theta_1, \dots, \theta_{d-1})$. (One assumes in this computation that $E = \{1, \dots, d\}$ and uses the notations $\theta(x) = \theta_x$, $\bar{P}_{x_1^N}(x) = \pi_x$.) The Lebesgue measure on the simplex is equal to the restriction of the exterior differential form $\sqrt{d}\, \partial\theta_1 \wedge \dots \wedge \partial\theta_{d-1}$. Indeed one checks that $\sqrt{d}\, \partial\theta_1 \wedge \dots \wedge \partial\theta_{d-1} \wedge \partial n$, where $\partial n = (d^{-\frac{1}{2}}, \dots, d^{-\frac{1}{2}})$ is the normal vector to the simplex, is the volume form on $\mathbb{R}^d$. Thus we have to compute an equivalent for

$$\int_\Theta \exp \left( -\frac{N}{2} \left( \sum_{i=1}^{d-1} \frac{(\theta_i - \pi_i)^2}{\pi_i} + \frac{(1 - \sum_{i=1}^{d-1} \theta_i - \pi_d)^2}{\pi_d} \right) \right) \sqrt{d}\, \partial\theta_1 \wedge \dots \wedge \partial\theta_{d-1}.$$

It is given by $\sqrt{d} \left( \frac{2\pi}{N} \right)^{\frac{d-1}{2}} (\mathrm{Det}(M))^{-1/2}$, where $M$ is the matrix:

$$M = \begin{pmatrix} \pi_1^{-1} + \pi_d^{-1} & \pi_d^{-1} & \pi_d^{-1} \dots & \pi_d^{-1} \\ \pi_d^{-1} & \pi_2^{-1} + \pi_d^{-1} & \pi_d^{-1} \dots & \pi_d^{-1} \\ \dots & \dots & \dots & \dots \\ \pi_d^{-1} & \pi_d^{-1} & \pi_d^{-1} \dots & \pi_{d-1}^{-1} + \pi_d^{-1} \end{pmatrix},$$

whose determinant is equal to $\prod_{i=1}^d \pi_i^{-1} \sum_{i=1}^d \pi_i = \prod_{i=1}^d \pi_i^{-1}$.

$$\underset{N\to+\infty}{\sim} \int_{\Theta} \exp\left( - \sum_{x\in E} \frac{N}{2} \frac{\left(\theta(x) - \bar{P}_{x_1^N}(x)\right)^2}{\bar{P}_{x_1^N}(x)} \right) f(\bar{P}_{x_1^N})\lambda(d\theta)$$

$$\underset{N\to+\infty}{\sim} \sqrt{|E|} f(\bar{P}_{X_1^N}) \left(\frac{2\pi}{N}\right)^{\frac{|E|-1}{2}} \prod_{x\in E} \bar{P}_{x_1^N}(x)^{1/2}.$$

This little asymptotic evaluation suggests to choose for $\rho(d\theta)$ the *Dirichlet distribution* with parameter $1/2$, defined by

$$\rho(d\theta) = \frac{1}{\sqrt{|E|}} \frac{\Gamma\left(\frac{|E|}{2}\right)}{\Gamma\left(\frac{1}{2}\right)^{|E|}} \prod_{x\in E} \theta(x)^{-\frac{1}{2}} \lambda(d\theta). \tag{1.4.1}$$

**Lemma 1.4.1.**

$$\mathbb{Q}_{\rho}(x_1^N) = \frac{\Gamma\left(\frac{|E|}{2}\right) \prod_{x\in E} \Gamma\left(N\bar{P}_{x_1^N}(x) + \frac{1}{2}\right)}{\Gamma\left(\frac{1}{2}\right)^{|E|} \Gamma\left(N + \frac{|E|}{2}\right)},$$

*where $\Gamma(z) = \int_{\mathbb{R}_+} t^{z-1}\exp(-t)dt$ stands for the Gamma function.*

*Proof.* Let us prove that for any $(\alpha_i)_{i=1}^d \in ]-1, +\infty[^d$,

$$\int_{\Theta} \prod_{i=1}^{d} \theta_i^{\alpha_i} \lambda(d\theta) = \sqrt{d} \frac{\prod_{i=1}^d \Gamma(\alpha_i + 1)}{\Gamma(\sum_{i=1}^d \alpha_i + d)}, \tag{1.4.2}$$

where $\Theta$ is the simplex of $\mathbb{R}^d$, and $\lambda$ is the Lebesque measure on the simplex. This identity is obtained by a change of variables:

$$\int_{\mathbb{R}_+^d} \prod_{i=1}^{d} t_i^{\alpha_i} e^{-t_i} dt_i = \int_{\mathbb{R}_+} \int_{\theta\in\Theta} \prod_{i=1}^{d} \theta_i^{\alpha_i} s^{\sum_{i=1}^d \alpha_i} e^{-s} s^{d-1} \lambda(d\theta) \frac{ds}{\sqrt{d}},$$

where we have put $s = \sum_{i=1}^d t_i$ et $\theta_i = t_i/s$. □

**Proposition 1.4.1 (Krichevski - Trofimov).** *For any $x_1^N \in E^N$,*

$$\mathbb{Q}_{\rho}(x_1^N) \geq \frac{2^{-NH(\bar{P}_{x_1^N})}}{|E| N^{\frac{|E|-1}{2}}} = \frac{1}{|E| N^{\frac{|E|-1}{2}}} \sup_{\theta\in\Theta} \mathbb{Q}_{\theta}(x_1^N).$$

*Proof.* Let us put for any $a = (a_i)_{i=1}^d \in \mathbb{N}^d$,

$$\Delta(a) = \frac{\Gamma\left(\frac{d}{2}\right)}{\Gamma\left(\frac{1}{2}\right)^d} \frac{\prod_{i=1}^d \Gamma\left(a_i + \frac{1}{2}\right)}{\Gamma\left(\sum_{i=1}^d a_i + \frac{d}{2}\right)} \frac{\left(\sum_{i=1}^d a_i\right)^{\sum_i a_i + \frac{d-1}{2}}}{\prod_{i=1}^d a_i^{a_i}}.$$

We have to show that $\Delta(a) \geq \frac{1}{d}$. To this purpose, let us notice that $\Delta((1,0,\dots,0)) = \frac{1}{d}$, that $\Delta(a)$ is invariant under any permutation of the $a_i$s. It is therefore enough to check that $\Delta(a) \geq \Delta((a_1 - 1, a_2, \dots, a_d))$. Let us put $s = \sum_{i=1}^{d} a_i$ and $t = a_1$. With these simplified notations,

$$\Delta(a) = \Delta(a_1 - 1, a_2^d) \frac{(t-\frac{1}{2})(t-1)^{t-1}}{t^t} \frac{s^{s+\frac{d-1}{2}}}{(s+\frac{d}{2}-1)(s-1)^{s-1+\frac{d-1}{2}}}.$$

We have to check that

$$\frac{(t-\frac{1}{2})(t-1)^{t-1}}{t^t} \frac{s^{s+\frac{d-1}{2}}}{(s+\frac{d}{2}-1)(s-1)^{s-1+\frac{d-1}{2}}} \geq 1, \qquad t \geq 1, s \geq 2.$$

We are going to show that

$$f(t) = \log\left(\frac{(t-\frac{1}{2})(t-1)^{t-1}}{t^t}\right) \geq -1,$$

$$g(s) = \log\left(\frac{s^{s+\frac{d-1}{2}}}{(s-1)^{s-1+\frac{d-1}{2}}(s+\frac{d}{2}-1)}\right) \geq 1.$$

Let us notice first that

$$\lim_{t\to+\infty} f(t) = \lim_{t\to+\infty} \log\left(1-\frac{1}{2t}\right) + (t-1)\log\left(1-\frac{1}{t}\right) = -1,$$

$$\lim_{s\to+\infty} g(s) = \lim_{s\to+\infty} -\left(s-1+\frac{d-1}{2}\right)\log\left(1-\frac{1}{s}\right) - \log\left(1+\frac{d-2}{2s}\right) = 1.$$

It is enough to conclude to show that $f'(t) \leq 0$ and $g'(s) \leq 0$. But

$$\begin{aligned} f'(t) &= \frac{1}{t-\frac{1}{2}} + \log\left(\frac{t-1}{t}\right), \\ \lim_{t\to+\infty} f'(t) &= 0, \\ f''(t) &= -\frac{1}{\left(t-\frac{1}{2}\right)^2} + \frac{1}{t(t-1)} \\ &= -\frac{1}{\left(t-\frac{1}{2}\right)^2} + \frac{1}{\left(t-\frac{1}{2}\right)^2 - \frac{1}{4}} \geq 0, \end{aligned}$$

showing that $f'(t) \leq 0$ for any $t > 1$. In the same way [2]

$$g'(s) = \log\left(\frac{s}{s-1}\right) + \frac{d-1}{2}\left(\frac{1}{s} - \frac{1}{s-1}\right) - \frac{1}{s+\frac{d}{2}-1},$$

$$\lim_{s\to+\infty} g'(s) = 0,$$

$$g''(s) \geq 0.$$

It follows that $g'(s) \leq 0$ for any $s \geq 2$. □

The previous proposition can be made more precise in the following way:

**Proposition 1.4.2 (Barron and Xie).** *For any $x_1^N \in E^N$,*

$$\exp\left\{-\sum_{y\in\mathcal{X}}\left[\frac{\mathbb{1}(\bar{P}_{x_1^N}(y)\neq 0)}{23N\bar{P}_{x_1^N}(y)} + \frac{\mathbb{1}(\bar{P}_{x_1^N}(y)=0)}{2}\log(2)\right] - \frac{(|E|-1)^2}{4N}\right\}$$

---

2

$$\begin{aligned} g''(s) &= \frac{1}{s} - \frac{1}{s-1} + \frac{d-1}{2}\left(\frac{1}{(s-1)^2} - \frac{1}{s^2}\right) + \frac{1}{\left(s+\frac{d}{2}-1\right)^2} \\ &= \frac{(d-1)\left(s-\frac{1}{2}\right)}{(s-1)^2 s^2} - \frac{1}{(s-1)s} + \frac{1}{\left(s+\frac{d}{2}-1\right)^2} \\ &= \frac{(d-1)\left(s-\frac{1}{2}\right)}{(s-1)^2 s^2} - \frac{(d-1)s + \frac{(d-2)^2}{4}}{s(s-1)\left(s+\frac{d}{2}-1\right)^2} \\ &= \frac{(d-1)\left(s-\frac{1}{2}\right)\left(s+\frac{d}{2}-1\right)^2 - \left((d-1)s + \frac{(d-2)^2}{4}\right)s(s-1)}{s^2(s-1)^2\left(s+\frac{d}{2}-1\right)^2} \\ &= (d-1)\frac{\left(d-\frac{3}{2}-\frac{(d-2)^2}{4(d-1)}\right)s^2 + \left(\frac{(d-2)^2}{4} - \frac{d-2}{2} + \frac{(d-2)^2}{4(d-1)}\right)s - \frac{(d-2)^2}{8}}{s^2(s-1)^2\left(s+\frac{d}{2}-1\right)^2} \end{aligned}$$

It is then easily checked (by taking derivatives) that the numerator of this last expression is a non decreasing function of $s$ for any $s \geq 2$ and $d \geq 2$. It follows that

$$\begin{aligned} g''(s) &\geq (d-1)\frac{4d-6-\frac{(d-2)^2}{d-1} + \frac{(d-2)^2}{2} - d + 2 + \frac{(d-2)^2}{2(d-1)} - \frac{(d-2)^2}{8}}{s^2(s-1)^2\left(s+\frac{d}{2}-1\right)^2} \\ &= (d-1)\frac{3d-4+\frac{3(d-2)^2}{8} - \frac{(d-2)^2}{2(d-1)}}{s^2(s-1)^2\left(s+\frac{d}{2}-1\right)^2} \geq (d-1)\frac{3d-4+\frac{3(d-2)^2}{8} - \frac{(d-2)}{2}}{s^2(s-1)^2\left(s+\frac{d}{2}-1\right)^2} \\ &= (d-1)\frac{\frac{5}{2}d - 3 + \frac{3(d-2)^2}{8}}{s^2(s-1)^2\left(s+\frac{d}{2}-1\right)^2} \geq \frac{2(d-1)}{s^2(s-1)^2\left(s+\frac{d}{2}-1\right)^2} \geq 0, \end{aligned}$$

$$\leq \left(\frac{N}{2\pi}\right)^{\frac{|E|-1}{2}} \frac{\Gamma\left(\frac{1}{2}\right)^{|E|}}{\Gamma\left(\frac{|E|}{2}\right)} \frac{\mathbb{Q}_\rho(x_1^N)}{\sup_{\theta\in\Theta} \mathbb{Q}_\theta(x_1^N)}$$

$$\leq \exp\left\{ - \sum_{y\in\mathcal{X}} \left[ \frac{\mathbb{1}(\bar{P}_{x_1^N}(y) \neq 0)}{29N\bar{P}_{x_1^N}(y)} + \frac{\mathbb{1}(\bar{P}_{x_1^N}(y) = 0)}{2} \log(2) \right] \right\}. \quad (1.4.3)$$

*In particular*

$$\exp\left[ -\tfrac{|E|}{2}\log(2) - \frac{(|E|-1)^2}{4N} \right]$$

$$\leq \left(\frac{N}{2\pi}\right)^{\frac{|E|-1}{2}} \frac{\Gamma\left(\frac{1}{2}\right)^{|E|}}{\Gamma\left(\frac{|E|}{2}\right)} \frac{\mathbb{Q}_\rho(x_1^N)}{\sup_{\theta\in\Theta} \mathbb{Q}_\theta(x_1^N)} \leq \exp\left(-\frac{|E|}{29N}\right) \leq 1. \quad (1.4.4)$$

*Proof.* Let us recall first Stirling's formula (see e.g. [39, tome 1] page 54):

$$\sqrt{2\pi} n^{n+\frac{1}{2}} \exp\left(-n + \tfrac{1}{12n+1}\right) < n! < \sqrt{2\pi} n^{n+\frac{1}{2}} \exp\left(-n + \tfrac{1}{12n}\right).$$

To simplify computations, let us use the same notations as in the previous proof and put $d = |E|$ and $(a_i)_{i=1}^d = \left(N\bar{P}_{x_1^N}(y)\right)_{y\in\mathcal{X}}$, so that $(a_i)_{i=1}^d \in \mathbb{N}^d$. With these notations, let us put moreover

$$\Delta \overset{\text{def}}{=} \left(\frac{N}{2\pi}\right)^{\frac{|E|-1}{2}} \frac{\Gamma\left(\frac{1}{2}\right)^{|E|}}{\Gamma\left(\frac{|E|}{2}\right)} \frac{\mathbb{Q}_\rho(x_1^N)}{\sup_{\theta\in\Theta} \mathbb{Q}_\theta(x_1^N)} = (2\pi)^{-\frac{d-1}{2}} \frac{\prod_{i=1}^d \Gamma(a_i + \frac{1}{2})}{\Gamma(N + \frac{d}{2})} \frac{N^{N+\frac{d-1}{2}}}{\prod_{i=1}^d a_i^{a_i}}.$$

Let us remark now that for any integer $a$

$$\Gamma(a + \tfrac{1}{2}) = \Gamma\left(\tfrac{1}{2}\right) \prod_{k=1}^{a} \left(a - k + \tfrac{1}{2}\right) = \frac{(2a)!}{2^{2a} a!} \sqrt{\pi}.$$

Thus, from Stirling's formula,

$$\frac{(2a)^{2a+\frac{1}{2}} \exp\left(-2a + \frac{1}{24a+1}\right)}{2^{2a} a^{a+\frac{1}{2}} \exp\left(-a + \frac{1}{12a}\right)} \sqrt{\pi} \leq \Gamma\left(a + \tfrac{1}{2}\right)$$

$$\leq \frac{(2a)^{2a+\frac{1}{2}} \exp\left(-2a + \frac{1}{24a}\right)}{2^{2a} a^{a+\frac{1}{2}} \exp\left(-a + \frac{1}{12a+1}\right)} \sqrt{\pi},$$

or putting it otherwise

$$\exp\left(-\frac{1}{23a}\right)$$

$$\leq \exp\left(\frac{1}{24a+1} - \frac{1}{12a}\right) \leq \frac{\Gamma\left(a+\frac{1}{2}\right)}{\sqrt{2\pi}a^a \exp(-a)} \leq \exp\left(\frac{1}{24a} - \frac{1}{12a+1}\right)$$
$$\leq \exp\left(-\frac{1}{29a}\right).$$

Let us put $I = \{i : a_i > 0\}$. It follows that

$$\prod_{i\in I} \exp\left(-\tfrac{1}{23a_i}\right) \leq \frac{\Delta 2^{\frac{d-|I|}{2}} \exp(N)\Gamma\left(N+\frac{d}{2}\right)}{\sqrt{2\pi}N^{N+\frac{d-1}{2}}} \leq \prod_{i\in I} \exp\left(-\tfrac{1}{29a_i}\right). \quad (1.4.5)$$

When $d$ is even, one gets that

$$\prod_{i\in I} \exp\left(-\tfrac{1}{23a_i}\right)$$
$$\times \exp\left(\tfrac{d-2}{2} - \tfrac{(d-|I|)\log(2)}{2} - \tfrac{1}{12N+6d-12} - (N+\tfrac{d-1}{2})\log(1+\tfrac{d-2}{2N})\right)$$
$$\leq \Delta \leq \prod_{i\in I} \exp\left(-\tfrac{1}{29a_i}\right)$$
$$\times \exp\left(\tfrac{d-2}{2} - \tfrac{(d-|I|)\log(2)}{2} - \tfrac{1}{12N+6d-11} - (N+\tfrac{d-1}{2})\log(1+\tfrac{d-2}{2N})\right).$$

Let us notice that $\sup_{r>0} r - (1+r)\log(1+r) < 0$, it shows that

$$\exp\left\{-\sum_{i\in I} \tfrac{1}{23a_i} - \tfrac{(d-|I|)\log(2)}{2} - \tfrac{(d-1)^2}{4N}\right\}$$
$$\leq \Delta \leq \exp\left\{-\sum_{i\in I} \tfrac{1}{29a_i} - \tfrac{(d-|I|)\log(2)}{2}\right\}.$$

In the case when $d$ is odd,

$$\Gamma(N+\tfrac{d}{2}) = \Gamma(N+\tfrac{d-1}{2}+\tfrac{1}{2}),$$

where $N + \frac{d-1}{2}$ is an integer, whence

$$\exp\left(-\tfrac{1}{23(N+\frac{d-1}{2})}\right) \leq \frac{\Gamma(N+\frac{d}{2})\exp(N+\frac{d-1}{2})}{\sqrt{2\pi}(N+\frac{d-1}{2})^{(N+\frac{d-1}{2})}} \leq \exp\left(-\tfrac{1}{29(N+\frac{d-1}{2})}\right).$$

Coming back to (1.4.5), we conclude that

$$\exp\left\{-\sum_{i\in I} \tfrac{1}{23a_i} - \tfrac{(d-|I|)\log(2)}{2} - \tfrac{(d-1)^2}{4N}\right\} \leq$$
$$\exp\left\{-\sum_{i\in I} \tfrac{1}{23a_i} + \tfrac{d-1}{2} - \tfrac{(d-|I|)\log(2)}{2}\right.$$

$$+ \frac{1}{29(N+\frac{d-1}{2})} - \left(N + \tfrac{d-1}{2}\right) \log\left(1 + \tfrac{d-1}{2N}\right)\Bigg\}$$

$$\leq \Delta \leq \exp\Bigg\{ - \sum_{i \in I} \frac{1}{29 a_i} + \frac{d-1}{2} - \frac{(d-|I|)\log(2)}{2}$$

$$+ \frac{1}{23(N+\frac{d-1}{2})} - \left(N + \tfrac{d-1}{2}\right) \log\left(1 + \tfrac{d-1}{2N}\right)\Bigg\}$$

$$\leq \exp\Bigg\{ - \sum_{i \in I} \frac{1}{29 a_i} - \frac{(d-|I|)\log(2)}{2}\Bigg\}.$$

The last of these inequalities is established by noticing that

$$\begin{aligned} r - (1+r)\log(1+r) &= \int_0^r -\log(1+s)ds \\ &\leq \int_0^{r \wedge 1} (-s + \tfrac{s^2}{2})ds \leq -\tfrac{(r\wedge 1)^2}{2} + \tfrac{(r \wedge 1)^3}{6} \leq -\tfrac{(r\wedge 1)^2}{3}, \end{aligned}$$

and therefore that

$$\begin{aligned} N\left[\tfrac{d-1}{2N} - (1 + \tfrac{d-1}{2N})\log(1 + \tfrac{d-1}{2N})\right] &+ \frac{1}{23(N+\frac{d-1}{2})} \\ &\leq -\tfrac{1}{3} N \left(\tfrac{d-1}{2N} \wedge 1\right)^2 + \frac{1}{23(N+\frac{d-1}{2})} \leq 0. \end{aligned}$$

Let us remark also that we could have used instead of Stirling's formula for $n!$ the bounds for the Gamma function given in [81, page 253] : for any real number $r > 0$

$$-\tfrac{1}{360 r^3} < \log[\Gamma(r)] - (r - \tfrac{1}{2})\log(r) + r - \tfrac{1}{2}\log(2\pi) - \tfrac{1}{12r} < 0.$$

□

**Corollary 1.4.1.** *The Krichevski - Trofimov mixture of coding distributions is such that for* any *probability distribution* $\mathbb{P} \in \mathcal{M}_+^1(E^N)$

$$\mathcal{R}(\mathbb{P}, \mathbb{Q}_\rho) \leq \inf_{\theta \in \Theta} \mathcal{R}(\mathbb{P}, \mathbb{Q}_\theta) + \frac{|E|-1}{2} \log_2(N) + \log_2|E|.$$

*More precisely, for any* $d > 2$*, equation* (1.4.4) *implies that*

$$\mathcal{R}(\mathbb{P}, \mathbb{Q}_\rho) \leq \inf_{\theta \in \Theta} \mathcal{R}(\mathbb{P}, \mathbb{Q}_\theta) + \frac{|E|-1}{2} \log_2\left(\tfrac{2N}{|E|-2}\right) + \tfrac{|E|}{2\log(2)} + \tfrac{(|E|-1)^2}{4N\log(2)}.$$

*Remark* 1.4.1. This is an "oracle" inequality, comparing the performances of a family of coding distributions for an *arbitrary* source. It applies in particular to the case when the distribution of the source is one of the coding distributions used in the mixture, namely in the case when $\mathbb{P} \in \{\mathbb{Q}_\theta : \theta \in \Theta\}$. Rissanen's theorem 1.3.2 then shows that the quantity $\frac{|E|-1}{2}\log_2(N)$ is asymptotically optimal.

## 1.5 Double mixtures and adaptive compression

### 1.5.1 General principle

Let us assume that we have built a sequence of parametric models $\{\mathbb{P}_\theta \in \mathcal{M}_+^1(\mathcal{X}) : \theta \in \Theta_i\}$, $i \in \mathbb{N}$, where $\mathcal{X}$ is, as previously, a finite set. Let us assume that we know for each $i \in \mathbb{N}$ of some probability measure $\rho_i \in \mathcal{M}_+^1(\Theta_i)$ on the parameter space, which is "almost minimax" in the sense that

$$\frac{\sup_{\theta \in \Theta_i} \mathcal{R}(\mathbb{P}_\theta, \mathbb{P}_{\rho_i})}{\mathbb{E}_{\rho_i(d\theta)}\big(\mathcal{R}(\mathbb{P}_\theta, \mathbb{P}_{\rho_i})\big)} \leq (1+\epsilon_i).$$

Let us consider an (arbitrary) probability measure $\pi \in \mathcal{M}_+^1(\mathbb{N})$ on the integers, and let us define on $\Theta = \bigsqcup_{i \in \mathbb{N}} \Theta_i$ (the disjoint union of $\Theta_i$, $i \in \mathbb{N}$) the "double mixture" $\alpha \in \mathcal{M}_+^1(\Theta)$ defined as

$$\alpha(A) = \sum_{i \in \mathbb{N}} \pi(i) \rho_i(A \cap \Theta_i).$$

**Proposition 1.5.1.** *Under the above hypotheses and with the above notations, the coding distribution $\mathbb{P}_\alpha$ is adaptively of minimax order in the sense that*

$$\sup_{\theta_i \in \Theta_i} \mathcal{R}(\mathbb{P}_{\theta_i}, \mathbb{P}_\alpha) \leq (1+\epsilon_i) \inf_{\mathbb{Q} \in \mathcal{M}_+^1(\mathcal{X})} \sup_{\theta_i \in \Theta_i} \mathcal{R}(\mathbb{P}_{\theta_i}, \mathbb{Q}) - \log\big(\pi(i)\big), \qquad i \in \mathbb{N}.$$

*Remark* 1.5.1. If $\mathcal{X} = E^N$ is the set of blocks of length $N$ and if the sets $\Theta_i$ are parametric sets of dimension $d_i$ in the sense of Rissanen's theorem 1.3.2, the minimax redundancies of the left member are of order $d_i \log_2(N)/2$. If moreover the prior distribution $\pi$ on the integer indices is chosen independently of $N$, then the quantity $-\log\big(\pi(i)\big)$ is of a smaller order of magnitude than the minimax redundancy of model number $i$ when $N$ tends to infinity.

*Proof.* As $\mathbb{P}_\alpha(x) \geq \pi(i)\mathbb{P}_{\rho_i}(x)$, for any $\theta_i \in \Theta_i$,

$$\mathcal{R}(\mathbb{P}_{\theta_i}, \mathbb{P}_\alpha) \leq \mathcal{R}(\mathbb{P}_{\theta_i}, \mathbb{P}_{\rho_i}) - \log\big(\pi(i)\big).$$

The conclusion comes from proposition 1.2.4. □

### 1.5.2 The context tree weighting coding distribution

We are going to describe in some details the case of context tree models. We will show for these models an oracle inequality of the same type as for the Krichevski-Trofimov estimator.

In this section we will deal with a source using a finite alphabet $E$, which we will represent by the canonical process $(X_n)_{n \in \mathbb{Z}}$ on $E^{\mathbb{Z}}$.

**Definition 1.5.1.** A source is said to be a finite memory source, if there is a finite set $\mathcal{E}$ and a function

$$\begin{aligned} f : \mathbb{E}^{\mathbb{Z}_-} &\longrightarrow \mathcal{E} \\ x_{-\infty}^{-1} &\longmapsto e \end{aligned}$$

such that for any $x_{-\infty}^{n-1} \in E^{]-\infty,n-1]}$, any $x_n \in E$,

$$\mathbb{P}\big(X_n = x_n \,|\, X_{-\infty}^{n-1} = x_{-\infty}^{n-1}\big) = \mathbb{P}\big(X_1 = x_n \,|\, f(X_{-\infty}^{0}) = f\big(x_{-\infty}^{n-1}\big)\big).$$

The function $f$ is called the *context function.*

**Definition 1.5.2.** A *bounded memory* source is a finite memory source whose context function $f$ depends only on a finite number of coordinates. (A bounded memory source is a homogenous Markov chain).

*Remark* 1.5.2. This definition may seem to coincide with the definition of a homogeneous Markov chain of rank $k$. Anyhow, introducing the context function $f$ affords a distinction, among the chains of order $k$, between submodels defined by a restricted number of parameters — equal to $|\mathcal{E}| \times (|E| - 1)$ — instead of $|E|^k(|E| - 1)$ for a "generic" Markov source of rank $k$.

**Definition 1.5.3.** A context tree source is a bounded memory source whose context function $f$ is defined with the help of some complete suffix dictionary [3] $\mathcal{D}$ by the identity:

$$f_{\mathcal{D}}(x_{-\infty}^{-1}) = x_{-k}^{-1} \in \mathcal{D}.$$

Let us consider a coding application in which the receiver knows the initial context $X_{-D}^{0}$ and wants to transmit the block of data $X_1^N$. Let $\ell(\mathcal{D})$ be the maximal length of the words of the suffix dictionary $\mathcal{D}$. Our aim will be to build an adaptive coding distribution satisfying an oracle inequality involving all the the context tree coding distributions based on dictionaries of length $\ell(\mathcal{D})$ not geater than $D$.

A context tree coding distribution is parametrized by a complete suffix dictionary $\mathcal{D}$ and by a family $\Theta_{\mathcal{D}} = \{\theta_m \in \mathcal{M}_+^1(E) : m \in \mathcal{D}\}$ of conditional probablity measures. One interpretation of this situation is to consider that the dictionary $\mathcal{D}$ defines a "model" of the source, and that $\Theta_{\mathcal{D}}$ is the parameter space for this source model.

For each model $\mathcal{D}$, let $\rho_{\mathcal{D}}$ be the prior distribution on the parameter space defined by

$$\rho_{\mathcal{D}}(d\theta) = \prod_{m \in \mathcal{D}} \rho(d\theta_m),$$

[3] By definition, a complete suffix dictionary $\mathcal{D}$ is a finite set of finite words such that the set of reversed words $\{(x_1, \dots, x_k) : (x_k, \dots, x_1) \in \mathcal{D}\}$ forms a complete prefix dictionary. This property ensures that for any sequence of past symbols $x_{-\infty}^{-1} \in E^{\mathbb{Z}_-}$, there is a unique index $k$, which may change from one sequence to the other, such that $x_{-k}^{-1} \in \mathcal{D}$.

where $\rho$ is the Krichevski-Trofimov distribution defined by equation (1.4.1). Let $\mathcal{F}_D = \{\mathcal{D} : \ell(\mathcal{D}) \leq D\}$ be the set of complete suffix dictionaries (which may be void) of length not greater than $D$. By convention, we consider that the void dictionary contains a single word "of null length", noted $\varnothing$. With this convention, i.i.d. coding distributions are a special case of context tree coding model, corresponding to the void context dictionary. Each dictionary $\mathcal{D}$ can be seen as the genealogy tree of a realization of a branching process whose survival probability at generation number $D$ is null. More precisely let us consider a branching process in which each individual has a probability $\alpha$ to give birth to an offspring exactly equal to $E$ and a probablity $1-\alpha$ to have no offspring at all, this being valid up to the $D$th generation for which the probability to have an offspring is null. This branching process defines on $\mathcal{F}_D$ the probability distribution

$$\pi(\mathcal{D}) = \prod_{m \in \mathring{\mathcal{D}}} \alpha \prod_{\substack{m \in \mathcal{D};\\ \ell(m) < D}} (1-\alpha), \tag{1.5.1}$$

where $\mathring{\mathcal{D}}$ is the interior of $\mathcal{D}$, that is the set of words $m \in E^* \cup \{\varnothing\}$ which are strict suffixes of words of $\mathcal{D}$. It is easy to check by induction that

$$|\mathcal{D}| = |\mathring{\mathcal{D}}|(|E|-1)+1.$$

This allows to write

$$\pi(\mathcal{D}) = \alpha^{\frac{|\mathcal{D}|-1}{|E|-1}}(1-\alpha)^{|\{m \in \mathcal{D} : \ell(m) < D\}|}.$$

Consider on the set of parameters $\Theta = \bigsqcup_{\mathcal{D} \in \mathcal{F}_D} \Theta_{\mathcal{D}}$ the double mixture distribution $\mu \in \mathcal{M}_+^1(\Theta)$ defined by

$$\mu(A) = \sum_{\mathcal{D} \in \mathcal{F}_D} \pi(\mathcal{D}) \rho_{\mathcal{D}}(A \cap \Theta_{\mathcal{D}}).$$

For any couple $(\mathcal{D}, \theta)$ where $\theta \in \Theta_{\mathcal{D}}$, let us define the coding distribution

$$\mathbb{Q}_{\mathcal{D},\theta}(x_1^N \mid x_{-D+1}^0) = \prod_{i=1}^N \theta_{\mathcal{D}(x_{i-D}^{i-1})}(x_i),$$

where we have used the short notation $f_{\mathcal{D}}(x_{-\infty}^{i-1}) = \mathcal{D}(x_{i-D}^{i-1})$. Let us consider the mixture

$$\mathbb{Q}_\mu = \int_\Theta \mathbb{Q}_{\mathcal{D},\theta} d\mu.$$

**Proposition 1.5.2.** *The computation of* $\mathbb{Q}_\mu(x_1^N \mid x_{1-D}^0)$ *from the counters*

$$a_m^y(x_1^N) = \sum_{k=1}^N \mathbb{1}\big(x_{k-\ell(m)}^{k-1} = m,\ x_k = y\big), \qquad m \in E^* \cup \varnothing,\ y \in E,$$

$$b_m(x_1^N) = \sum_{y \in E} a_m^y(x_1^N);$$

$$K_m(x_1^N) = \frac{\Gamma\left(\frac{|E|}{2}\right) \prod_{y \in E} \Gamma\left(a_m^y(x_1^N) + \frac{1}{2}\right)}{\Gamma\left(\frac{1}{2}\right)^{|E|} \Gamma\left(b_m(x_1^N) + \frac{|E|}{2}\right)};$$

*is given by the following identity:*

$$\mathbb{Q}_\mu(x_1^N \mid x_{1-D}^0) = \sum_{\mathcal{D} \in \mathcal{F}_D} \pi(\mathcal{D}) \prod_{m \in \mathcal{D}} K_m(x_1^N).$$

*This expression may itself be computed by the following induction on words of length not greater than D:*

$$\Upsilon_m(x_1^N) = K_m(x_1^N), \qquad m \in E^D,$$

$$\vdots$$

$$\Upsilon_m(x_1^N) = (1-\alpha) K_m(x_1^N) + \alpha \prod_{y \in E} \Upsilon_{(y,m)}(x_1^N), \quad m \in E^* \cup \{\varnothing\}, \ell(m) < D,$$

$$\vdots$$

$$\Upsilon_\varnothing(x_1^N) = \mathbb{Q}_\mu(x_1^N \mid x_{1-D}^0).$$

*The counters may moreover be updated on the fly while the $x_n$ are being read, according to the following update rules:*

$$a_m^y(\varnothing) = 0;$$
$$b_m(\varnothing) = 0;$$
$$K_m(\varnothing) = 1;$$
$$\Upsilon_m(\varnothing) = 1;$$

$$\vdots$$

$$a_m^y(x_1^n) = \begin{cases} a_m^y(x_1^{n-1}) + 1 & \textit{if } m = x_{n-\ell(m)}^{n-1} \textit{ and } y = x_n, \\ a_m^y(x_1^{n-1}) & \textit{otherwise;} \end{cases}$$

$$b_m(x_1^n) = \begin{cases} b_m(x_1^{n-1}) + 1 & \textit{if } m = x_{n-\ell(m)}^{n-1}, \\ b_m(x_1^{n-1}) & \textit{otherwise;} \end{cases}$$

$$K_m(x_1^n) = \begin{cases} K_m(x_1^{n-1}) \frac{a_m^{x_n}(x_1^n) - \frac{1}{2}}{b_m(x_1^n) + \frac{|E|}{2} - 1} & \textit{if } m = x_{n-\ell(m)}^{n-1}, \\ K_m(x_1^{n-1}) & \textit{otherwise;} \end{cases}$$

$$\Upsilon_m(x_1^n) = \begin{cases} \Upsilon_m(x_1^{n-1}) & \text{if } x_{n-\ell(m)}^{n-1} \neq m, \\ K_m(x_1^n) & \text{if } x_{n-D}^{n-1} = m, \\ (1-\alpha)K_m(x_1^n) & \\ \quad + \alpha \Upsilon_{x_{n-\ell(m)-1}^{n-1}}(x_1^n) & \text{if } x_{n-\ell(m)}^{n-1} = m, \\ \quad \times \prod_{y \in E \setminus \{x_{n-\ell(m)-1}\}} \Upsilon_{(y,m)}(x_1^{n-1}) & \text{and } \ell(m) < D. \end{cases}$$

*To perform this computation, one can choose to store in memory only the visited nodes, namely those for which $b_m(x_1^N) > 0$. Nodes can be created dynamically as the $x_n$s are being read. When a block $x_1^N$ of length $N$ is processed, at most $DN$ nodes are thus created (allocated into memory). The induction formulas also show that the number of operations needed to compute $\Upsilon_\varnothing(x_1^N)$ and therefore $\mathbb{Q}_\mu(x_1^N \,|\, x_{1-D}^0)$, increases with $N$ and $D$ as $DN$. A modified algorithm where the dependence with $D$ is cancelled will be described subsequently.*

*Proof.* Let us show first that

$$\int_{\theta \in \Theta_\mathcal{D}} \mathbb{Q}_{\mathcal{D},\theta}(x_1^N \,|\, x_{1-D}^0)\rho_\mathcal{D}(d\theta) = \prod_{m \in \mathcal{D}} K_m(x_1^N).$$

It is only a matter of remembering the definitions and using equation (1.4.2):

$$\mathbb{Q}_{\mathcal{D},\theta}(x_1^N \,|\, x_{1-D}^0) = \prod_{m \in \mathcal{D}} \prod_{y \in E} \theta_m(y)^{a_m^y(x_1^N)},$$

$$\begin{aligned} &\int_{\theta \in \Theta_\mathcal{D}} \mathbb{Q}_{\mathcal{D},\theta}(x_1^N \,|\, x_{1-D}^0)\rho_\mathcal{D}(d\theta) \\ &\quad = \prod_{m \in \mathcal{D}} \left( \frac{\Gamma\left(\frac{|E|}{2}\right)}{\sqrt{|E|}\Gamma\left(\frac{1}{2}\right)^{|E|}} \int_{\theta \in \mathcal{M}_+^1(E)} \prod_{y \in E} \theta(y)^{a_m^y(x_1^N) - \frac{1}{2}} \lambda(d\theta) \right) \\ &\qquad = \prod_{m \in \mathcal{D}} \frac{\Gamma\left(\frac{|E|}{2}\right)\Gamma\left(a_m^y(x_1^N) + \frac{1}{2}\right)}{\Gamma\left(\frac{1}{2}\right)^{|E|}\Gamma\left(b_m(x_1^N) + \frac{|E|}{2}\right)}, \end{aligned}$$

as announced. This shows that

$$\mathbb{Q}_\mu(x_1^N \,|\, x_{1-D}^0) = \sum_{\mathcal{D} \in \mathcal{F}_D} \pi(\mathcal{D}) \prod_{m \in \mathcal{D}} K_m(x_1^N).$$

Let us prove by induction with respect to $\ell(m) = D, \ldots, 0$ that

$$\Upsilon_m(x_1^N) = \sum_{\mathcal{D} \in \mathcal{F}_{D-\ell(m)}} \pi_{D-\ell(m)}(\mathcal{D}) \prod_{s \in \mathcal{D}} K_{(s,m)}(x_1^N), \tag{1.5.2}$$

where $\pi_{D-\ell(m)}$ is the distribution of the genealogy tree of a branching process stopped at depth $D - \ell(m)$.

Let us notice first that

$$\mathcal{F}_{D-\ell(m)} = \Big\{ \{(s, y) : y \in E,\ s \in \mathcal{D}_y\} : \mathcal{D}_y \in \mathcal{F}_{D-\ell(m)-1} \Big\} \cup \Big\{ \{\varnothing\} \Big\},$$

with

$$\pi_{D-\ell(m)} \big( \{(s, y) : y \in E,\ s \in \mathcal{D}_y\} \big) = \alpha \prod_{y \in E} \pi_{D-\ell(m)-1}(\mathcal{D}_y),$$

$$\pi_{D-\ell(m)}(\{\varnothing\}) = 1 - \alpha.$$

Let us assume that (1.5.2) has been proved for words of length $d$, which is easily checked in the case when $d = D$. Let $m$ be a word of length $d - 1$.

$$\begin{aligned}
\Upsilon_m(x_1^N) &= (1-\alpha) K_m(x_1^N) + \alpha \prod_{y \in E} \sum_{\mathcal{D} \in \mathcal{F}_{D-d}} \pi_{D-d}(\mathcal{D}) \prod_{s \in \mathcal{D}} K_{(s,y,m)}(x_1^N) \\
&= (1-\alpha) K_m(x_1^N) + \sum_{\mathcal{D}_y \in \mathcal{F}_{D-d}} \alpha \prod_{y \in E} \pi_{D-d}(\mathcal{D}_y) \prod_{s \in \mathcal{D}_y} K_{(s,y,m)}(x_1^N) \\
&= \pi_{D-\ell(m)}(\varnothing) K_m(x_1^N) \\
&\qquad + \sum_{\mathcal{D}_y \in \mathcal{F}_{D-d}} \pi_{D-\ell(m)} \big( \{(s, y) : s \in \mathcal{D}_y\} \big) \prod_{(s,y) \in \mathcal{D}_y \times E} K_{(s,y,m)}(x_1^N) \\
&= \sum_{\mathcal{D} \in \mathcal{F}_{D-\ell(m)}} \pi_{D-\ell(m)}(\mathcal{D}) \prod_{s \in \mathcal{D}} K_{(s,m)}(x_1^N).
\end{aligned}$$

This shows (1.5.2) for $m$ and thus completes the proof by induction. It is thus a proved fact that

$$\Upsilon_\varnothing(x_1^N) = \sum_{\mathcal{D} \in \mathcal{F}_D} \pi(\mathcal{D}) \prod_{m \in \mathcal{D}} K_m(x_1^N) = \mathbb{Q}_\mu(x_1^N \mid x_{1-D}^0).$$

The fact that $\Upsilon_m(x_1^N) = 1$ when $K_m(x_1^N) = 1$ is a direct consequence of equation (1.5.2), given that it implies $K_{sm}(x_1^N) = 1$ for any $s \in E^*$.

Alternatively, a more probabilistic proof of the induction formula for $\Upsilon_m(x_1^N)$ can be provided. Let us write $w \geq m$ when the word $m \in E^* \cup \{\varnothing\}$ is a suffix of the word $w \in E^* \cup \{\varnothing\}$ and $\mathcal{D} \geq m$ when the suffix dictionary $\mathcal{D}$ contains a word $w$ such that $w \geq m$.

Let us also for any word $m \in E^* \cup \{\varnothing\}$ and any dictionary $\mathcal{D} \in \mathcal{F}_D$ such that $\mathcal{D} \geq m$ define

$$\mathcal{D}_m \stackrel{\text{def}}{=} \{w \in \mathcal{D} : w \geq m\}$$

and

$$\mathcal{F}_D(m) \stackrel{\text{def}}{=} \{\mathcal{D} \in \mathcal{F}_D : \mathcal{D} \geq m\}.$$

Let us notice that $\mathcal{F}_D(m)$ is a product set :

$$\mathcal{F}_D(m) \simeq \{\mathcal{D}_m : \mathcal{D} \in \mathcal{F}_D(m)\} \times \{\mathcal{D} \setminus \mathcal{D}_m : \mathcal{D} \in \mathcal{F}_D(m)\}.$$

The branching property of $\pi$ can be expressed as

$$\pi(\mathcal{D} \,|\, \mathcal{D} \geq m) = \pi(\mathcal{D}_m \,|\, \mathcal{D} \geq m)\pi(\mathcal{D} \setminus \mathcal{D}_m \,|\, \mathcal{D} \geq m). \tag{1.5.3}$$

The quantities $\Upsilon_m$ can then be *defined* as

$$\Upsilon_m \stackrel{\text{def}}{=} \mathbb{E}_{\pi(d\mathcal{D} | \mathcal{D} \geq m)} \left( \prod_{w \in \mathcal{D}_m} K_w \right),$$

and decomposed into

$$\Upsilon_m = \pi(m \in \mathcal{D} \,|\, \mathcal{D} \geq m) K_m \\ + \pi(m \notin \mathcal{D} \,|\, \mathcal{D} \geq m) \mathbb{E}_{\pi(d\mathcal{D} | \mathcal{D} \geq m, m \notin \mathcal{D})} \left( \prod_{w \in \mathcal{D}_m} K_w \right).$$

Noticing the identity of events

$$\{\mathcal{D} \in \mathcal{F}_D(m) : m \notin \mathcal{D}\} = \{\mathcal{D} \in \mathcal{F}_D(m) : \mathcal{D} \geq (x, m)\}, \qquad x \in E,$$

we can repeatedly apply the branching property (1.5.3) to the words $(x, m)$, $x \in E$, to show that

$$\pi(\mathcal{D} \,|\, \mathcal{D} \geq m, m \notin \mathcal{D}) = \prod_{x \in E} \pi\big(\mathcal{D}_{(x,m)} \,|\, \mathcal{D} \geq (x, m)\big),$$

and thus that

$$\Upsilon_m = \pi(m \in \mathcal{D} \,|\, \mathcal{D} \geq m) K_m + \pi(m \notin \mathcal{D} \,|\, \mathcal{D} \geq m) \prod_{x \in E} \Upsilon_{(x,m)}.$$

As $\pi(m \in \mathcal{D} \,|\, \mathcal{D} \geq m) = (1 - \alpha)$, this ends the proof of proposition 1.5.2.

□

The following theorem gives an upper bound for the redundancy of the coding distribution $\mathbb{Q}_\mu$ :

**Theorem 1.5.1.** *Let* $\gamma : \mathbb{R}_+ \to \mathbb{R}_+$ *be the smallest* concave *function upper bounding on* $\mathbb{R}_+$ *the map*

$$x \mapsto \begin{cases} 0 & \text{if } x = 0 \\ \frac{|E|-1}{2} \log_2(x) + \min\Big\{ \log_2(|E|), \\ \qquad - \frac{|E|-1}{2} \log_2\left(\frac{|E|-2}{2}\right) + \frac{(|E|-1)^2}{4x \log(2)} + \frac{|E|}{2\log(2)} \Big\} & \text{if } x > 0. \end{cases}$$

*Let us notice that when $|E| \geq 6$ and $x$ is large enough*

$$\gamma(x) = \frac{|E|-1}{2} \log_2\left(\frac{2x}{|E|-2}\right) + \frac{|E|}{2\log(2)} + \frac{(|E|-1)^2}{4x\log(2)}.$$

*Let us also notice that*

$$\gamma(x) \leq \begin{cases} x^{\frac{|E|-1}{2}} \exp\left(\frac{2\log(|E|)}{|E|-1} - 1\right) & \text{if } x \leq \exp\left(1 - \frac{2\log(|E|)}{|E|-1}\right) \\ \frac{|E|-1}{2} \log_2(x) + \log_2(|E|) & \text{otherwise.} \end{cases}$$

*With these notations*

$$-\log_2\Big(\mathbb{Q}_\mu(x_1^N \,|\, x_{1-D}^0)\Big) \leq \inf_{\mathcal{D}\in\mathcal{F}_D} \Bigg\{ \inf_{\theta\in\Theta_{\mathcal{D}}} -\log_2\Big(\mathbb{Q}_{\mathcal{D},\theta}(x_1^N \,|\, x_{1-D}^0)\Big)$$
$$+ |\mathcal{D}|\gamma\left(\frac{N}{|\mathcal{D}|}\right) - \frac{|\mathcal{D}|-1}{|E|-1}\log_2(\alpha) - |\mathcal{D}|\log_2(1-\alpha)\Bigg\}.$$

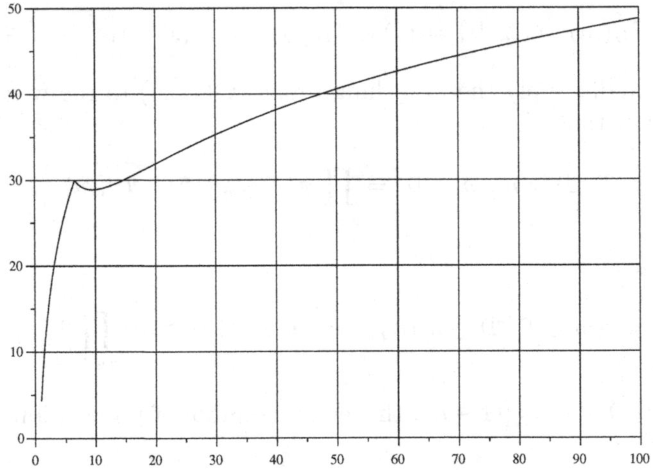


**Fig. 1.1.** *This diagram represents the function of which $\gamma$ is the concave upper envelope in the case when $|E| = 20$.*

*Proof.* From proposition 1.4.1

$$-\log_2\left(\int_{\Theta_{\mathcal{D}}} \mathbb{Q}_{\mathcal{D},\theta}(x_1^N \,|\, x_{1-D}^0)\rho_{\mathcal{D}}(d\theta)\right) \leq \inf_{\theta\in\Theta_{\mathcal{D}}} -\log_2\left(\mathbb{Q}_{\mathcal{D},\theta}(x_1^N \,|\, x_{1-D}^0)\right)$$
$$+ \sum_{m\in\mathcal{D}} \gamma\big[b_m(x_1^N)\big]$$

$$= \inf_{\theta \in \Theta_{\mathcal{D}}} -\log_2\left(\mathbb{Q}_{\mathcal{D},\theta}(x_1^N \mid x_{1-D}^0)\right) + |\mathcal{D}| \sum_{m \in \mathcal{D}} \frac{1}{|\mathcal{D}|} \gamma\left[b_m(x_1^N)\right]$$

$$\leq \inf_{\theta \in \Theta_{\mathcal{D}}} -\log_2\left(\mathbb{Q}_{\mathcal{D},\theta}(x_1^N \mid x_{1-D}^0)\right) + |\mathcal{D}| \gamma\left(\frac{N}{|\mathcal{D}|}\right),$$

where it is needed to use the property that function $\gamma$ is concave to obtain the last inequality. It implies that

$$-\log_2\left(\mathbb{Q}_\mu(x_1^N \mid x_{1-D}^0)\right) \leq \inf_{\mathcal{D} \in \mathcal{F}_D} \left( \inf_{\theta \in \Theta_D} -\log_2\left(\mathbb{Q}_{\mathcal{D},\theta}(x_1^N \mid x_{1-D}^0)\right) + |\mathcal{D}| \gamma\left(\frac{N}{|\mathcal{D}|}\right) - \log_2\left(\pi(\mathcal{D})\right) \right).$$

□

### 1.5.3 Context tree weighting without initial context

It is easy to adapt the double mixture of context trees to the case when the initial context $X_{1-D}^0$ is unavailable. The idea is to restraint from coding, or more precisely to code according to the uniform coding distribution the $x_n$s for which the context is not observed. One way to put it is to add to $E$ an extra letter $\epsilon$ and to define consequently $\tilde{E} = E \cup \{\epsilon\}$. The sequence $x_1^N$ may then be extended to $\tilde{x}_{-\infty}^n$ defined by

$$\tilde{x}_n = \begin{cases} x_n & \text{if } n > 0, \\ \epsilon & \text{if } n \leq 0. \end{cases}$$

Any complete suffix dictionary $\mathcal{D}$ of $E^*$ may be extended to the smallest complete suffix dictionary $\tilde{\mathcal{D}}$ of $\tilde{E}^*$ which contains it: it is obtained by adding to $\mathcal{D}$ all the strict suffixes of the words of $\mathcal{D}$, preceded by the letter $\epsilon$. The parameter vector $\theta \in \Theta_{\mathcal{D}}$ may then be extended to the parameter vector $\tilde{\theta} \in \Theta_{\tilde{\mathcal{D}}}$, defined by $\tilde{\theta}_m(\epsilon) = 0$ and

$$\tilde{\theta}_m(y) = \begin{cases} \theta_m(y) & \text{if } y \in E, m \in \mathcal{D}, \\ |E|^{-1} & \text{if } y \in E, m \in \tilde{\mathcal{D}} \setminus \mathcal{D}. \end{cases}$$

(Letters with a non fully observed context are not coded.)

It is then possible to modify the definition of the coding distributions $\mathbb{Q}_{\mathcal{D},\theta}$ putting for any word $m \in \tilde{E}^*$:

$$\tilde{a}_m^y(x_1^N) = \sum_{k=1}^N \mathbb{1}\left(\tilde{x}_{k-\ell(m)}^k = (m, y)\right),$$

$$\tilde{b}_m(x_1^N) = \sum_{y \in E} \tilde{a}_m^y \left( = \sum_{y \in \tilde{E}} \tilde{a}_m^y \right),$$

$$\begin{aligned}\tilde{\mathbb{Q}}_{\mathcal{D},\theta}(x_1^N) &= \prod_{k=1}^{N} \theta_{\tilde{\mathcal{D}}(\tilde{x}_{-\infty}^{n-1})}(\tilde{x}_n) \\ &= \prod_{m \in \tilde{\mathcal{D}}} \prod_{y \in E} \left(\tilde{\theta}_m(y)\right)^{\tilde{a}_m^y(x_1^N)}.\end{aligned}$$

Let us use the same prior distribution $\mu$ as before. We get as above

$$\tilde{\mathbb{Q}}_{\rho_{\mathcal{D}}}(x_1^N) = \prod_{m \in \tilde{\mathcal{D}}} \tilde{K}_m(x_1^N),$$

where

$$\tilde{K}_m(x_1^N) = \begin{cases} \dfrac{\Gamma\left(\frac{|E|}{2}\right) \prod_{y \in E} \Gamma\left(\tilde{a}_m^y(x_1^N) + \frac{1}{2}\right)}{\Gamma\left(\frac{1}{2}\right)^{|E|} \Gamma\left(\tilde{b}_m(x_1^N) + \frac{|E|}{2}\right)} & \text{if } m \in \mathcal{D}, \\ |E|^{-\tilde{b}_m(x_1^N)} = |E|^{-1} & \text{if } m \in \tilde{\mathcal{D}} \setminus \mathcal{D}. \end{cases}$$

Let us also notice that

$$0 \leq \sum_{m \in \tilde{\mathcal{D}} \setminus \mathcal{D}} \tilde{b}_m(x_1^N) \leq \max_{m \in \mathcal{D}} \ell(m).$$

Moreover

$$\tilde{\mathbb{Q}}_\mu(x_1^N) = \sum_{\mathcal{D} \in \mathcal{F}_D} \pi(\mathcal{D}) \prod_{m \in \mathcal{D}} \tilde{K}_m(x_1^N).$$

It can therefore be computed along the following induction scheme:

$$\tilde{\mathbb{Q}}_\mu(x_1^N) = \Upsilon_\varnothing(x_1^N),$$

$$\Upsilon_m(x_1^N) = \begin{cases} \tilde{K}_m(x_1^N) & \text{if } \ell(m) = D, \\ (1-\alpha)\tilde{K}_m(x_1^N) + \alpha \prod_{y \in E} \Upsilon_{(y,m)}(x_1^N) & \text{otherwise.} \end{cases}$$

This induction may as in the previous case be updated sequentially as the $x_n$s are being read.

**Proposition 1.5.3.** *With the preceding notations and definitions, for any block $x_{1-D}^N \in E^{N+D}$,*

$$-\log_2\left(\tilde{\mathbb{Q}}_\mu(x_1^N)\right) \leq \inf_{\mathcal{D} \in \mathcal{F}_D} \left\{ \inf_{\theta \in \Theta_{\mathcal{D}}} -\log_2\left(\mathbb{Q}_{\mathcal{D},\theta}(x_1^N \mid x_{1-D}^0)\right) + \ell(\mathcal{D}) \log_2(|E|) \right.$$

$$+ |\mathcal{D}|\gamma\left(\tfrac{N}{|\mathcal{D}|}\right) - \tfrac{|\mathcal{D}|-1}{|E|-1}\log_2(\alpha) - |\mathcal{D}|\log_2(1-\alpha)\Bigg\},$$

*where* $\ell(\mathcal{D}) = \max_{m\in\mathcal{D}} \ell(m)$.

*Proof.* The same computation as in the proof of theorem 1.5.1 shows that

$$-\log_2(\tilde{\mathbb{Q}}_\mu(x_1^N)) \leq \inf_{\mathcal{D}\in\mathcal{F}_D}\Bigg\{\inf_{\theta\in\Theta_{\mathcal{D}}} -\log_2\left(\tilde{\mathbb{Q}}_{\mathcal{D},\theta}(x_1^N)\right)$$
$$+ |\mathcal{D}|\gamma\left(\tfrac{\sum_{m\in\mathcal{D}}\tilde{b}_m(x_1^N)}{|\mathcal{D}|}\right) - \tfrac{|\mathcal{D}|-1}{|E|-1}\log_2(\alpha) - |\mathcal{D}|\log_2(1-\alpha)\Bigg\}.$$

To complete the proof, it is then enough to remark that $\sum_{m\in\mathcal{D}}\tilde{b}_m(x_1^N) \leq N$ and that

$$\inf_{\theta\in\Theta_{\mathcal{D}}}\left\{-\log_2\left[\tilde{Q}_{\mathcal{D},\theta}(x_1^N)\right]\right\}$$
$$\leq \inf_{\theta\in\Theta_{\mathcal{D}}}\left\{-\log_2\left[\mathbb{Q}_{\mathcal{D},\theta}(x_1^N \mid x_{1-D}^0)\right]\right\} + \log_2(|E|)\max_{m\in\mathcal{D}}\ell(m).$$

□

### 1.5.4 Weighting of trees of arbitrary depth

In this section, we will extend the context tree weighting method to the set $\mathcal{F}_\infty$ of all complete suffix dictionaries (of arbitrary length). From a theoretical point of view, this generalization is straightforward: it is just a matter of considering on $\mathcal{F}_\infty$ the distribution of a branching process without stopping rule at generation $D$. We will therefore put here, for any $\mathcal{D}\in\mathcal{F}_\infty$

$$\pi(\mathcal{D}) = \alpha^{\frac{|\mathcal{D}|-1}{|E|-1}}(1-\alpha)^{|\mathcal{D}|}.$$

(When $\alpha$ is super-critical, this is only a sub-probability distribution on $\mathcal{F}_\infty$).

It remains to show how to perform the computation in practice.

Let us assume for a while that $D$ is finite, but that it is larger than $N$, and let us show that the computation of $\Upsilon_m(x_1^N)$ does not depend on the choice of $D$ in this case. This will show that the same computation is still valid when $D$ tends to infinity.

Let us notice first that in the case when $\tilde{b}_m(x_1^N) = 0$, namely when the context $m$ has not been observed yet, $\Upsilon_m(x_1^N) = 1$, as it was established above. In the same way, when $\tilde{b}_m(x_1^N) = 1$, $\tilde{K}_m(x_1^N) = |E|^{-1}$, and it is easy to check from the inductive definition of $\Upsilon$ that $\Upsilon_m(x_1^N) = |E|^{-1}$ in this case. Eventually in the case when $\tilde{b}_{(m,s)}(x_1^N) = \tilde{b}_s(x_1^N)$, namely in the case when no branching has occured in the trees of observed contexts between nodes $s$ and $(m,s)$,

$$\Upsilon_s(x_1^N) = \big(1 - \alpha^{\ell(m)}\big)\tilde{K}_s(x_1^N) + \alpha^{\ell(m)}\Upsilon_{(m,s)}(x_1^N).$$

It is thus possible to choose the following finite representation of the infinite context tree of $\tilde{x}_{-\infty}^N$ :

- Let us decide first to represent only the nodes $m \in \tilde{E}^*$ which have been visited, that is such that $\tilde{b}_m(x_1^N) > 0$. Let us code these nodes by a couple of integers $(t, n)$ such that $m = \tilde{x}_t^n$.
- Among these visited nodes, let us represent only those for which

$$\tilde{b}_{m^0_{2-\ell(m)}}(x_1^N) > \tilde{b}_m(x_1^N).$$

Thus we only represent in the computer memory the context nodes following a branching. As all the infinite contexts $\big(\tilde{x}_{-\infty}^n\big)_{n=1}^N$ are distinct, each of them contributes for exactly one more branching. If this branching is made at a point where a previous branching already occured, it leads to represent into memory one more node. If it occurs at a point where there was no branching before, it leads to represent into memory two more nodes. The number of nodes stored in memory is thus bounded by $2N$. The number of operations needed to compute $\tilde{\mathbb{Q}}_\mu(x_1^N)$ grows in the worst case as $N^2$, if the tree of visited contexts is very ill-balanced (it will for instance be the case when the observed sequence $x_1^N$ is constant).

Weighting trees of arbitrary depths provides a way to compress with the best possible asymptotic compression rate any stationary source:

**Proposition 1.5.4.** *The mixture of coding distributions $\tilde{\mathbb{Q}}_\mu$ obtained by the context tree weighting method applied to trees of arbitrary depths without knowledge of the initial context is such that for any stationary source distributed according to $\mathbb{P}\big(d(X_n)_{n\in\mathbb{Z}}\big)$*

$$\lim_{N\to+\infty} -\frac{1}{N}\mathbb{E}\Big(\log_2\big(\tilde{\mathbb{Q}}_\mu(X_1^N)\big)\Big) = \bar{H}\big(\mathbb{P}(dX)\big).$$

*Moreover, if $\mathbb{P}(dX)$ is ergodic,*

$$\lim_{N\to+\infty} -\frac{1}{N}\log_2\big[\tilde{\mathbb{Q}}_\mu(X_1^N)\big] = \bar{H}\big[\mathbb{P}(dX)\big], \qquad \textit{in } \mathbb{L}^1\big(\mathbb{P}(dX)\big). \tag{1.5.4}$$

*The random variables $\frac{1}{N}\log_2\big[\tilde{\mathbb{Q}}_\mu(X_1^N)\big]$ being uniformly bounded, convergence in $\mathbb{L}^1$ implies convergence in all the $\mathbb{L}^p(P(dX))$ spaces. Almost sure convergence also holds:*

$$\lim_{N\to+\infty} -\frac{1}{N}\log_2\big(\tilde{\mathbb{Q}}_\mu(X_1^N)\big) = \bar{H}\big(\mathbb{P}(dX)\big), \qquad \mathbb{P}\ \textit{p.s.}.$$

*Proof.* Let us apply proposition 1.5.3 to the void dictionary $\mathcal{D} = \{\varnothing\}$ and to the uniform distribution on $E^N$, we get that

$$0 \leq -\frac{1}{N}\log_2\bigl[\tilde{\mathbb{Q}}_\mu(X_1^N)\bigr] \leq \log_2(|E|) + \frac{1}{N}\biggl\{\frac{(|E|-1)}{2}\log_2(N) + \log_2(|E|) - \log_2(1-\alpha)\biggr\},$$

which proves that the random variables $-\frac{1}{N}\log_2\bigl[\tilde{\mathbb{Q}}_\mu(X_1^N)\bigr]$ are uniformly bounded. Let us consider now the dictionary $\mathcal{D} = E^d$. Proposition 1.5.3 shows that for $N$ large enough,

$$-\frac{1}{N}\log_2\bigl(\tilde{\mathbb{Q}}_\mu(X_1^N)\bigr) \leq -\frac{1}{N}\sum_{n=1}^{N}\log_2\bigl(\mathbb{P}(X_n \,|\, X_{n-d}^{n-1})\bigr) + \frac{1}{N}\biggl\{d\log_2(|E|) + \frac{(|E|-1)|E|^d}{2}\biggl[\log_2\biggl(\frac{N}{|E|^d}\biggr) + \frac{2\log_2(|E|)}{|E|-1}\biggr] - \frac{|E|^d-1}{|E|-1}\log_2(\alpha) - |E|^d\log_2(1-\alpha)\biggr\}. \tag{1.5.5}$$

Let us introduce now the conditional Shannon entropies

$$H_d(\mathbb{P}) = -\mathbb{E}_{\mathbb{P}(dX_{1-d}^1)}\Bigl(\log_2\bigl(\mathbb{P}(X_1 \,|\, X_{1-d}^0)\bigr)\Bigr).$$

The concavity of Shannon's entropy implies that the sequence $H_d(\mathbb{P})$ is non increasing with respect to $d$, and therefore has a limit when $d$ tends to $+\infty$. Moreover

$$\frac{1}{D+1}\sum_{d=0}^{D}H_d(\mathbb{P}) = \frac{1}{D+1}\sum_{d=0}^{D}\mathbb{E}_{\mathbb{P}}\Bigl(\log_2\bigl(\mathbb{P}(X_d \,|\, X_0^{d-1})\bigr)\Bigr) = \frac{H\bigl(\mathbb{P}(dX_0^D)\bigr)}{D+1}.$$

A converging sequence having the same limit as its Cesaro mean, we get

$$\lim_{d\to+\infty} H_d(\mathbb{P}) = \bar{H}(\mathbb{P}). \tag{1.5.6}$$

Integrating (1.5.5) gives

$$\limsup_{N\to+\infty} -\frac{1}{N}\mathbb{E}_{\mathbb{P}(dX_1^N)}\Bigl(\log_2\bigl(\tilde{\mathbb{Q}}_\mu(X_1^N)\bigr)\Bigr) \leq \bar{H}(\mathbb{P}).$$

As

$$-\frac{1}{N}\mathbb{E}_{\mathbb{P}(dX_1^N)}\Bigl(\log_2\bigl(\tilde{\mathbb{Q}}_\mu(X_1^N)\bigr)\Bigr) \geq \frac{H\bigl(\mathbb{P}(dX_1^N)\bigr)}{N},$$

this implies that

$$-\lim_{N\to+\infty}\frac{1}{N}\mathbb{E}_{\mathbb{P}(dX_1^N)}\Bigl(\log_2\bigl(\tilde{\mathbb{Q}}_\mu(X_1^N)\bigr)\Bigr) = \bar{H}(\mathbb{P}).$$

In the case when the source is ergodic, the ergodic theorem shows that que

$$\lim_{N \to +\infty} -\frac{1}{N} \sum_{n=1}^{N} \log_2 \bigl( \mathbb{P}(X_n \,|\, X_{n-d}^{n-1}) \bigr) = H_d(\mathbb{P}), \qquad \mathbb{P} \text{ a.s. and in } \mathbb{L}^1. \quad (1.5.7)$$

(Let us notice that the random variable $\log_2 \bigl[ \mathbb{P}(X_1 \,|\, X_{1-d}^0) \bigr]$ takes only a finite number of finite values $\mathbb{P}$ a.s. and therefore belongs to $\mathbb{L}^\infty(\mathbb{P})$, implying that convergence in all the spaces $\mathbb{L}^p(\mathbb{P})$, $1 \leq p < \infty$ are equivalent.)

Let us put $S_d(N) = -\frac{1}{N} \sum_{n=1}^N \log_2 \mathbb{P}(X_n \,|\, X_{n-d}^{n-1})$. Equation (1.5.5) shows that

$$-\tfrac{1}{N} \log_2 \bigl[ \tilde{\mathbb{Q}}_\mu(X_1^N) \bigr] \leq S_d(N) + \epsilon_d(N),$$

where $\lim_{N \to +\infty} \epsilon_d(N) = 0$ for any $d$. Hence

$$\begin{aligned} \mathbb{E} \Bigl( \bigl| \tfrac{1}{N} \log_2 \bigl[ \tilde{\mathbb{Q}}_\mu(X_1^N) \bigr] + S_d(N) \bigr| \Bigr) &\leq \mathbb{E} \Bigl( \tfrac{1}{N} \log_2 \bigl[ \tilde{\mathbb{Q}}_\mu(X_1^N) \bigr] + S_d(N) \Bigr) + 2\epsilon_d(N) \\ &\leq \mathbb{E} \Bigl( \tfrac{1}{N} \log_2 \bigl[ \tilde{\mathbb{Q}}_\mu(X_1^N) \bigr] \Bigr) + H_d(\mathbb{P}) + 2\epsilon_d(N) \\ &\leq 2\epsilon_d(N) + H_d(\mathbb{P}) - \tfrac{1}{N} H \bigl[ \mathbb{P}(dX_1^N) \bigr], \end{aligned}$$

and therefore

$$\begin{aligned} &\mathbb{E} \Bigl( \bigl| \tfrac{1}{N} \log_2 \bigl[ \tilde{\mathbb{Q}}_\mu(X_1^N) \bigr] + \bar{H} \bigr| \Bigr) \\ &\qquad \leq 2\epsilon_d(N) + H_d(\mathbb{P}) - \frac{1}{N} H \bigl[ \mathbb{P}(dX_1^N) \bigr] \\ &\qquad\qquad + \mathbb{E} \Bigl[ \bigl| S_d(N) - H_d(\mathbb{P}) \bigr| \Bigr] + H_d(\mathbb{P}) - \bar{H}. \end{aligned}$$

Convergence in $\mathbb{L}^1(\mathbb{P})$ (1.5.4) is obtained by letting $N$ and then $d$ tend to $+\infty$.

Let us consider now the issue of almost sure convergence. Equation (1.5.5) shows that

$$\limsup_{N \to +\infty} -\frac{1}{N} \log_2 \bigl( \tilde{\mathbb{Q}}_\mu(X_1^N) \bigr) \leq H_d(\mathbb{P}), \qquad \mathbb{P} \text{ a.s..}$$

A countable union of events of null measure being of null measure, we get

$$\limsup_{N \to +\infty} -\frac{1}{N} \log_2 \bigl( \tilde{\mathbb{Q}}_\mu(X_1^N) \bigr) \leq \lim_{d \to +\infty} H_d(\mathbb{P}) = \bar{H}(\mathbb{P}), \qquad \mathbb{P} \text{ a.s..}$$

Moreover if we had

$$\mathbb{P} \Bigl( \limsup_{N \to +\infty} -\frac{1}{N} \log_2 \bigl( \tilde{\mathbb{Q}}_\mu(X_1^N) \bigr) < \bar{H}(\mathbb{P}) \Bigr) > 0,$$

this would imply that

$$\limsup_{N \to +\infty} \mathbb{E} \Bigl( -\frac{1}{N} \log_2 \bigl( \tilde{\mathbb{Q}}_\mu(X_1^N) \bigr) \Bigr) \leq \mathbb{E} \Bigl( \limsup_{N \to +\infty} -\frac{1}{N} \log_2 \bigl( \tilde{\mathbb{Q}}_\mu(X_1^N) \bigr) \Bigr) < \bar{H}(\mathbb{P}).$$

As this strict inequality is in contradiction with the first part of the proposition, which has already been proved, it implies

$$\limsup_{N\to+\infty} -\frac{1}{N}\log_2(\tilde{\mathbb{Q}}_\mu(X_1^N)) = \bar{H}(\mathbb{P}), \qquad \mathbb{P} \text{ a.s..}$$

The lower bound of the inferior limit is valid for any probability distribution $Q \in \mathcal{M}_+^1(E^{\mathbb{N}})$, and is a consequence of the Shannon-McMillan Breiman theorem. More precisely,

$$\begin{aligned}\mathbb{P}\Big(\frac{1}{N}\log_2\big[Q(X_1^N)\big] \geq \frac{1}{N}\log_2\big[\mathbb{P}(X_1^N)\big] + \epsilon\Big) &= \mathbb{P}\Big(\frac{Q(X_1^N)}{\mathbb{P}(X_1^N)} \geq e^{N\epsilon}\Big)\\ &\leq e^{-N\epsilon}\mathbb{E}_{\mathbb{P}}\left(\frac{Q(X_1^N)}{\mathbb{P}(X_1^N)}\right)\\ &= e^{-N\epsilon}.\end{aligned}$$

This last upper bound being summable with respect to $N$, it follows that

$$\liminf_{N\to+\infty} -\frac{1}{N}\log_2\big[Q(X_1^N)\big] \geq \liminf_{N\to+\infty} -\frac{1}{N}\log_2\big[\mathbb{P}(X_1^N)\big] - \epsilon, \qquad \mathbb{P} \text{ a.s.}$$

It is enough then to consider a sequence of values of $\epsilon$ decreasing to 0, to obtain

$$\liminf_{N\to+\infty} -\frac{1}{N}\log_2\big[Q(X_1^N)\big] \geq \liminf_{N\to+\infty} -\frac{1}{N}\log_2\big[\mathbb{P}(X_1^N)\big], \qquad \mathbb{P} \text{ a.s.}$$

According to the Shannon-McMillan-Breiman theorem,

$$\lim_{N\to+\infty} -\frac{1}{N}\log_2\big[\mathbb{P}(X_1^N)\big] = \bar{H}, \qquad \mathbb{P} \text{ a.s.,}$$

showing consequently that

$$\liminf_{N\to+\infty} -\frac{1}{N}\log_2\big[Q(X_1^N)\big] \geq \bar{H} \qquad \mathbb{P} \text{ a.s.}$$

Let us notice that we have confined in this proof the use of the Shannon-McMillan-Breiman theorem to a minimum, but that a shorter proof would result from showing first a.s. convergence: as the concerned variables are uniformly bounded, it implies convergence in all the $\mathbb{L}^p(\mathbb{P})$ spaces, for $1 \leq p < \infty$. The proof of equality for the almost sure superior limit also becomes superfluous once the lower bound for the inferior limit is proved. □

# Appendix

## 1.6 Fano's lemma

It is interesting to compare theorem 1.3.1 with Fano's lemma as it is for instance stated in [8, page 198] :

**Lemma 1.6.1.** *Let $\{P_\theta \in \mathcal{M}_+^1(\mathcal{X}) : \theta \in \Theta\}$ be a regular family of probability distributions and $d$ a distance defined on $\Theta$.*

*If there exists a finite subset $S \subset \Theta$ such that $d(\theta, \theta') \geq J$ for any $\theta \neq \theta' \in S$ and such that*

$$\sup_{\theta, \theta' \in S} \mathcal{K}(P_\theta, P_{\theta'}) \leq \gamma \log(|S|) - \log(2),$$

*then for any estimator $\hat{\theta}(X)$ of $\theta$*

$$\frac{1}{|S|} \sum_{\theta \in S} \mathbb{E}_{P_\theta} \bigl[ d(\hat{\theta}, \theta)^q \bigr] \geq (1 - \gamma) 2^{-q} J^q.$$

*Proof.* Let us consider on $S \times \mathcal{X}$ the probability distribution

$$P(\{\theta\} \times A) = \frac{1}{|S|} P_\theta(A)$$

and let $\mu$ denote the uniform distribution on $S$. We have seen in the proof of theorem 1.3.1 that

$$\mathcal{I}(\theta, X) \geq H(\mu) - 1 - H(\mu) P\bigl[ d(\hat{\theta}, \theta) \geq J/2 \bigr].$$

On the other hand, $\theta'$ being some fixed point in $S$,

$$\begin{aligned} \mathcal{I}(\theta, X) &= \mathbb{E}_{\mu(d\theta)} \bigl[ \mathcal{K}(P_\theta, P_\mu) \bigr] \\ &\leq \mathbb{E}_{\mu(d\theta)} \bigl[ \mathcal{K}(P_\theta, P_{\theta'}) \bigr] \end{aligned}$$

$$\leq \gamma \frac{\log(|S|)}{\log(2)} - 1$$
$$= \gamma H(\mu) - 1.$$

It follows that $P\big[d(\hat{\theta}, \theta) \geq J/2\big] \geq 1 - \gamma$, and therefore that Fano's lemma is satisfied. □

## 1.7 Decomposition of the Kullback divergence function

**Proposition 1.7.1.** *Let $\rho$ and $\mu$ be two probability distributions defined on some measurable set $(\Theta, \mathcal{T})$. Let $\{P_\theta \in \mathcal{M}_+^1(\mathcal{X}, \mathcal{F}) : \theta \in \Theta\}$ and $\{Q_\theta \in \mathcal{M}_+^1(\mathcal{X}, \mathcal{F}) : \theta \in \Theta\}$ be two families of probability distributions defined on some other measurable set $(\mathcal{X}, \mathcal{F})$. Let us assume that for any measurable subset $B \in \mathcal{F}$ the functions*

$$\begin{aligned} &\theta \mapsto Q_\theta(B) : && \Theta \to \mathbb{R}_+ \\ \text{and} \quad &\theta \mapsto P_\theta(B) : && \Theta \to \mathbb{R}_+ \quad \text{are measurable.} \end{aligned}$$

*For any subset $F \subset \Theta \times \mathcal{X}$ which is measurable with respect to the product $\sigma$-algebra $\mathcal{T} \otimes \mathcal{F}$, we let $F_\theta = \{x \in \mathcal{X} : (\theta, x) \in F\}$ be the trace of $F$ with respect to $\theta$. In this case $F_\theta \in \mathcal{F}$, and the functions*

$$\begin{aligned} P : \mathcal{T} \otimes \mathcal{F} &\to \mathbb{R}_+ & \qquad Q : \mathcal{T} \otimes \mathcal{F} &\to \mathbb{R}_+ \\ F &\mapsto \int_\Theta P_\theta(F_\theta)\rho & F &\mapsto \int_\Theta Q_\theta(F_\theta)\mu \end{aligned}$$

*are probability distributions on $(\Theta \times \mathcal{X}, \mathcal{T} \otimes \mathcal{X})$. Let us assume moreover that one of the following conditions are satisfied:*

1. *$\mathcal{F} = \sigma(\mathcal{G})$, where $\mathcal{G}$ is a countable family of subsets of $\mathcal{X}$.*
2. *$\Theta$ is countable.*

*In this case, the function*

$$\theta \mapsto \mathcal{K}\big(P_\theta, Q_\theta\big) : \qquad \Theta \to [0, \infty]$$

*is "pseudo-integrable" with respect to $\rho$, in the sense that its upper integral is equal to its lower integral: letting $\mathcal{L}_+(\Theta, \mathcal{T})$ denote the set of measurable functions with values ranging in $[0, \infty]$,*

$$\begin{aligned} \sup \left\{ \int_\Theta f\rho : f \in \mathcal{L}_+(\Theta, \mathcal{T}) \text{ and } f(\theta) \leq \mathcal{K}(P_\theta, Q_\theta) \text{ for any } \theta \in \Theta \right\} \\ = \inf \left\{ \int_\Theta f\rho : f \in \mathcal{L}_+(\Theta, \mathcal{T}) \text{ and } f(\theta) \geq \mathcal{K}(P_\theta, Q_\theta) \text{ for any } \theta \in \Theta \right\}. \end{aligned}$$

*(With this definition, any function equal to $+\infty$ on a set of positive measure is pseudo-integrable. Moreover a pseudo-integrable function with a finite integral is measurable with respect to the completion of the $\sigma$-algebra $\mathcal{T}$ with respect to the measure $\rho$.)*

*Under these hypotheses, the Kullback divergence function satisfies the following decomposition formula:*

$$\mathcal{K}(P,Q) = \mathcal{K}(\rho,\mu) + \mathbb{E}_{\rho(d\theta)}\Big[\mathcal{K}\big(P_\theta, Q_\theta\big)\Big].$$

*Proof.* The fact that $F_\theta$ is measurable is a classical preliminary to the proof of Fubini's theorem: the sets $F \in \mathcal{T} \otimes \mathcal{F}$ which satisfy this property form a $\sigma$-algebra containing the rectangles. It can then be established that for any $F \in \mathcal{T} \otimes \mathcal{F}$

$$\theta \mapsto P_\theta(F_\theta)$$

is measurable. Indeed, this property holds for disjoint unions of rectangles, it is also stable when increasing and decreasing limits of subsets are taken (a pointwise limit of measurable functions being measurable), therefore it holds for any set of the product $\sigma$-algebra, according to the monotone class theorem.

The fact that $P$ and $Q$ are measures on $(\Theta \times \mathcal{X}, \mathcal{T} \otimes \mathcal{F})$ is then a consequence of the monotone convergence theorem.

Another consequence is that for any function $h(\theta, x) \in \mathcal{L}_+(\mathcal{T} \otimes \mathcal{X})$,

$$\theta \mapsto \int_{\mathcal{X}} h(\theta, x) P_\theta(dx)$$

is measurable. Indeed, in the case when $h$ is bounded, it can be uniformly approximated by a sequence of simple functions (that is by finite linear combinations of indicator functions of measurable sets). The property is then extended to non bounded functions $h$ : any such function is the increasing limit of a sequence of bounded functions, to which the monotone convergence theorem can be applied for each value of $P_\theta$.

Let us also notice that for any function $h(\theta, x) \in \mathcal{L}_+(\Theta \times \mathcal{X}, \mathcal{T} \otimes \mathcal{F})$

$$\int hP = \int_\Theta \left[ \int_{\mathcal{X}} h(\theta, x) P_\theta(dx) \right] \rho(d\theta),$$

and that the same holds for $Q$. Indeed, in the case when $h$ is bounded, it can be uniformly approximated by simple functions. The case when $h$ is not bounded is dealt with the help of the monotone convergence theorem.

Let us discuss first the case when $P \ll Q$. As $\rho(A) = P(A \times \mathcal{X})$ and $\mu(A) = Q(A \times \mathcal{X})$, it follows that $\rho \ll \mu$. Let $\frac{P}{Q}$ and $\frac{\rho}{\mu}$ be some versions of the Radon Nikodym density of the corresponding measures. Let $f(\theta, x) = \frac{P}{Q}\left[\frac{\rho}{\mu}\right]^{-1} \mathbb{1}\left(\frac{\rho}{\mu} \neq 0\right)$. We are going to show for $\rho$ almost all $\theta$ that $P_\theta \ll Q_\theta$ and that $x \mapsto f(\theta, x)$ is a version of the density of $P_\theta$ with respect to $Q_\theta$. Indeed, for any $A \in \mathcal{T}$ and any $B \in \mathcal{F}$,

$$\begin{aligned}
\int_A \Big[ \int_B f(\theta,x) Q_\theta(dx) \Big] \rho(d\theta) \\
&= \int_A \Big\{ \int_B \tfrac{P}{Q}(\theta,x) \left[\tfrac{\rho}{\mu}(\theta)\right]^{-1} \mathbb{1}\left(\tfrac{\rho}{\mu}(\theta) \neq 0\right) Q_\theta(dx) \Big\} \tfrac{\rho}{\mu}(\theta)\mu(d\theta) \\
&= \int_A \Big\{ \int_B \tfrac{P}{Q}(\theta,x) \mathbb{1}\left(\tfrac{\rho}{\mu} \neq 0\right) Q_\theta(dx) \Big\} \mu(d\theta) \\
&= \int_{A\times B} \tfrac{P}{Q} \mathbb{1}\left(\tfrac{\rho}{\mu} \neq 0\right) Q.
\end{aligned}$$

Moreover

$$\begin{aligned}
\int_{A\times B} \tfrac{P}{Q} \mathbb{1}\big(\tfrac{\rho}{\mu} = 0\big) Q &= \int_{A\times B} \mathbb{1}\big(\tfrac{\rho}{\mu}(\theta) = 0\big) P(d(\theta,x)) \\
&= \int_A \mathbb{1}\big(\tfrac{\rho}{\mu}(\theta) = 0\big) P_\theta(B)\rho(d\theta) \\
&= \int_A \tfrac{\rho}{\mu}(\theta) \mathbb{1}\big(\tfrac{\rho}{\mu}(\theta) = 0\big) P_\theta(B)\mu(d\theta) \\
&= 0.
\end{aligned}$$

This shows that

$$\int_A \Big[ \int_B f(\theta,x) Q_\theta(dx) \Big] \rho(d\theta) = \int_{A\times B} \tfrac{P}{Q} Q = P(A\times B) = \int_A P_\theta(B)\rho(d\theta).$$

Consequently, for any $B \in \mathcal{F}$

$$\rho\Big( \int_B f(\theta,x) Q_\theta(dx) \neq P_\theta(B) \Big) = 0. \tag{1.7.1}$$

When $\Theta$ is countable, putting $N = \big\{\theta \in \Theta : \rho(\{\theta\}) = 0\big\}$, one concludes that for any $B \in \mathcal{F}$ and any $\theta \in \Theta \setminus N$

$$\int_B f(\theta,x) Q_\theta(dx) = P_\theta(B),$$

and therefore that for all these values of $\theta$, $x \mapsto f(\theta,x)$ is a version of $\frac{P_\theta}{Q_\theta}$.

On the other hand, if it is assumed that $\mathcal{F} = \sigma(\mathcal{G})$ where $\mathcal{G}$ is countable, the following reasoning can be carried. Let $\mathcal{G}' = \{B \in \mathcal{F} : \mathcal{X} \setminus B \in \mathcal{G}\} \cup \mathcal{G}$, let $\mathcal{G}''$ be the set of finite intersections of sets of $\mathcal{G}'$ and $\mathcal{G}'''$ the finite unions of sets of $\mathcal{G}''$. It is easily seen that these three sets are countable, because a countable union of countable sets is countable. Moreover, $\mathcal{G}'''$ is the algebra generated by $\mathcal{G}$: it is closed with respect to taking finite intersections, finite unions and complements (because any Boolean expression can be put in disjunctive normal form). Identity (1.7.1) is true for any $B \in \mathcal{G}'''$. As $\mathcal{G}'''$ is countable, the union of these events is still of null probability with respect to $\rho$:

$$\rho\left(\int_B f(\theta, x) Q_\theta(dx) = P_\theta(B) \text{ for any } B \in \mathcal{G}'''\right) = 1.$$

But for any fixed value of $\theta$, $\left\{C \in \mathcal{F} : \int_C f Q_\theta = P_\theta(C)\right\}$ is a monotone class of $\mathcal{F}$ (from the monotone convergence theorem). It is therefore, according to the monotone class theorem, equal to the whole $\sigma$-algebra $\mathcal{F}$, since it contains $\mathcal{G}'''$. This proves that for some $N \in \mathcal{T}$ such that $\rho(N) = 0$, for any $\theta \in \Theta \setminus N$,

$$\int_B f Q_\theta = P_\theta(B) \text{ for any } B \in \mathcal{F}.$$

Consequently, for any $\theta \in \Theta \setminus N$, $x \mapsto f(\theta, x)$ is a version of $\frac{P_\theta}{Q_\theta}$. Thus

$$\begin{aligned}\mathcal{K}(P_\theta, Q_\theta) &= \int \log\left[f(\theta, x)\right] P_\theta(dx) \\ &= \int \log\left[f(\theta, x)\right]_+ P_\theta(dx) - \int \log\left[f(\theta, x)\right]_- P_\theta(dx), \qquad \theta \in \Theta \setminus N,\end{aligned}$$

where $[r]_+ = \max\{r, 0\}$ and $[r]_- = -\min\{r, 0\}$. Let us notice that the last equality makes sense because $x \mapsto \log\left[f(\theta, x)\right]_-$ belongs to $\mathbb{L}^1(P_\theta)$:

$$\int \log\left[f(\theta, x)\right]_- P_\theta(dx) = \int f(\theta, x) \log\left[f(\theta, x)\right]_- Q_\theta(dx) \leq e^{-1}, \qquad \theta \in \Theta \setminus N,$$

(because $r \log[r]_- \leq e^{-1}$). This shows that the function $\theta \mapsto \mathcal{K}(P_\theta, Q_\theta)$ is measurable on $\Theta \setminus N$, and therefore that it is measurable on $\Theta$ with respect to the completion of the $\sigma$-algebra $\mathcal{T}$ with respect to the measure $\rho$.

Let us now prove that the function $[\log(f)]_-$ belongs to $\mathbb{L}^1(P)$. Indeed

$$\begin{aligned}\int [\log(f)]_- P &= \int_\Theta \left\{\int_\mathcal{X} [\log(f)]_- P_\theta\right\} \rho(d\theta) \\ &= \int_\Theta \left\{\int_\mathcal{X} \left[\log\left(\tfrac{P_\theta}{Q_\theta}\right)\right]_- P_\theta\right\} \rho(d\theta) \\ &= \int_\Theta \left\{\int_\mathcal{X} \tfrac{P_\theta}{Q_\theta} \left[\log\left(\tfrac{P_\theta}{Q_\theta}\right)\right]_- Q_\theta\right\} \rho(d\theta) \\ &\leq e^{-1}.\end{aligned}$$

In the same way $(\theta, x) \mapsto \log\left[\frac{\rho}{\mu}(\theta)\right]_-$ is in $\mathbb{L}^1(P)$. As $P(\frac{\rho}{\mu} = 0) = 0$, we can write

$$\begin{aligned}\mathcal{K}(P, Q) &= \int \mathbb{1}(\tfrac{\rho}{\mu} \neq 0) \log\left(\tfrac{P}{Q}\right) P \\ &= \int_{\{\frac{\rho}{\mu} \neq 0\}} \left[\log(f) + \log\left(\tfrac{\rho}{\mu}\right)\right] P \\ &= \int_{\{\frac{\rho}{\mu} \neq 0\}} \log(f) P + \int_{\{\frac{\rho}{\mu} \neq 0\}} \log\left(\tfrac{\rho}{\mu}\right) P\end{aligned}$$

$$= \int_{\{\frac{\rho}{\mu} \neq 0\}} \left\{ \int_{\mathcal{X}} \log[f(\theta, x)] P_\theta(dx) \right\} \rho(d\theta) + \int_\theta \log\left(\frac{\rho}{\mu}\right) \rho(d\theta)$$

$$= \int_\theta \mathcal{K}(P_\theta, Q_\theta)\rho(d\theta) + \mathcal{K}(\rho, \mu).$$

Let us discuss now the case when $P$ is not absolutely continuous with respect to $Q$. Let us introduce the auxiliary probability distribution

$$Q'(F) = \int_\Theta Q_\theta(F_\theta)\rho(d\theta).$$

In the case when $P \ll Q'$, then $\theta \mapsto \mathcal{K}(P_\theta, Q_\theta)$ is $\mathcal{T}_\rho$ measurable (where $\mathcal{T}_\rho$ is the completion of $\mathcal{T}$ with respect to the measure $\rho$). In the case when $P$ is not absolutely continuous with respect to $Q'$, there exists $F \in \mathcal{T} \otimes \mathcal{X}$ such that $P(F) > 0$ and $Q'(F) = 0$. As $P(F) = \int_\theta P_\theta(F_\theta)\rho(d\theta)$, there is some set $A \in \mathcal{T}$ such that $\rho(A) > 0$ and $P_\theta(F_\theta) > 0$ for any $\theta \in A$. On the other hand $Q'(F) = \int_\Theta Q_\theta(F_\theta)\rho(d\theta)$, which proves that $Q_\theta(F_\theta) = 0$ for any $\theta \in \Theta \setminus N$ where $N \in \mathcal{T}$ is some set such that $\rho(N) = 0$. It follows that $P_\theta$ is not absolutely continuous with respect to $Q_\theta$ at any point $\theta$ in the set $A \setminus N$ of measure $\rho(A \setminus N) > 0$. This implies that $\theta \mapsto \mathcal{K}(P_\theta, Q_\theta)$ is equal to $+\infty$ on the set $A \setminus N$ of positive measure $\rho(A \setminus N)$. It is therefore pseudo-integrable with respect to $\rho$ in the sense indicated in the proposition.

Now that measurability and integrability issues are settled, it remains to prove the identity $\mathcal{K}(P, Q) = \mathcal{K}(\rho, \mu) + \int_\Theta \mathcal{K}(P_\theta, Q_\theta)\rho(d\theta)$. The lefthand side being equal to $+\infty$, it is to be shown that one of the two terms on the righthand side is necessarily also equal to $+\infty$ (as they both take their values in $[0, +\infty]$, there is no ambiguity about the definition of the sum). In the case when $\mathcal{K}(\rho, \mu) = +\infty$, we are done, thus we can assume that $\mathcal{K}(\rho, \mu) < +\infty$. In this latter case, $\rho \ll \mu$, and as it was assumed that $P \not\ll Q$, there exists $F \in \mathcal{T} \otimes \mathcal{F}$ such that $P(F) > 0$ and $Q(F) = 0$. The identities $P(F) = \int_\Theta P_\theta(F_\theta)\rho(d\theta)$ and $Q(F) = \int_\Theta Q_\theta(F_\theta)\mu(d\theta)$ show that there are two $\mathcal{T}$ measurable sets $A$ and $N$ such that $\rho(A) > 0$, $\mu(N) = 0$, $P_\theta(F_\theta) > 0$ for any $\theta \in A$, and $Q_\theta(F_\theta) = 0$ for any $\theta \in \Theta \setminus N$. As $\rho \ll \mu$, we also have $\rho(N) = 0$, and therefore $\rho(A \setminus N) > 0$. On this last set $\mathcal{K}(P_\theta, Q_\theta) = +\infty$, because $P_\theta \not\ll Q_\theta$. The conclusion is that $\int_\Theta \mathcal{K}(P_\theta, Q_\theta)\rho(d\theta) = +\infty$. □

2

# Links between data compression and statistical estimation

The aim of this chapter is to show how lossless data compression theory provides statistical inference methods with a controled *cumulated risk* and a controled *asymptotic risk*. We will discuss three kinds of problems : density estimation with respect to the Kullback Leibler divergence, pattern recognition and least square regression.

## 2.1 Estimating a conditional probability distribution with respect to the Kullback divergence

As was mentioned about the Rissanen theorem, the decomposition

$$\mathcal{R}(P(dX_1^N), Q(dX_1^N)) = \sum_{k=1}^{N} \frac{1}{\log(2)} \mathbb{E}_{P(dX_1^{k-1})} \Big[ \mathcal{K}\big[P(dX_k \,|\, X_1^{k-1}), Q(dX_k \,|\, X_1^{k-1})\big] \Big]$$

shows how to infer from an upper bound on the redundancy of the coding distribution $Q(dX_1^N)$ of a block of length $N$ an upper bound for the *cumulated risk* of the estimator $Q(dX_k \,|\, X_1^{k-1})$ of the conditional probability distribution of $X_k$ knowing $X_1^{k-1}$. This allows to deal in the same framework with the sequential prediction of a sequence of dependent data and the estimation of a probability distribution from an i.i.d. sample.

From an upper bound of the form

$$\mathcal{K}(P(dX_1^N), Q(dX_1^N)) \leq \inf_{\theta \in \Theta} \mathcal{K}(P(dX_1^N), Q_\theta(X_1^N)) + A \log(N) + B$$

it is moreover possible to deduce an asymptotic speed of convergence of the type

$$\sup_{\theta \in \Theta} \liminf_{N \to +\infty} N \Big\{ \mathbb{E}_{P(dX_1^{N-1})} \Big[ \mathcal{K}(P(dX_N | X_1^{N-1}), Q(dX_N | X_1^{N-1}))$$

$$- \mathcal{K}(P(dX_N|X_1^{N-1}), Q_\theta(dX_N|X_1^{N-1}))\Big]\Big\} \leq A.$$

Let us notice that this type of result is not completely satisfactory for more than one reason:

- This is an asymptotic result, in which the influence of constant $B$ is not felt.
- It would be more satisfactory if the inferior limit could be replace with a true limit. It is only proved that for any $\epsilon > 0$, there is an increasing sequence of sample sizes for which the risk is upper bounded by the bias term augmented with $\frac{A+\epsilon}{N}$. Although it is natural to conjecture that this property will be true for any large enough sample sizes in numerous situations, it does not provide a proof of it.
- This is a result for a fixed value of $\theta$: a result about $\lim_N \sup_{\theta \in \Theta}$ of the same quantity would be more uniform and therefore more informative about the adaptivity of the estimator for a fixed size of problem.

However, the bounds obtained from compression theory have the priceless advantage to be universally robust with respect to what the true distribution of the source $P(dX_1^N)$ may be, because they are valid for individual sequences, and therefore do not require to make *any* hypothesis about the source.

## 2.2 Least square regression

The least square regression problem is concerned with the observation of an i.i.d. sequence $(X_i, Y_i)_{i=1}^N \in (\mathcal{X} \times \mathbb{R})^N$, where $(\mathcal{X}, \mathcal{F})$ is a measurable space and $(\mathbb{R}, \mathcal{B})$ is the real line with the Borel sigma algebra. Its aim is to estimate solely the conditional expectation $\mathbb{E}(Y|X)$ and not the full conditional distribution of $Y$ knowing $X$ In this framework, an estimator is described as a *regression function* $f : \mathcal{X} \to \mathbb{R}$ (depending on the observation $(X_i, Y_i)_{i=1}^N$), and its quality is measured by the quadratic risk

$$\mathcal{R}(f) = \mathbb{E}\Big[\big(Y - f(X)\big)^2\Big].$$

A countable family of estimators

$$\Big\{\widehat{f}_\theta : \mathcal{X} \times \bigsqcup_{n \in \mathbb{N}} (\mathcal{X} \times \mathbb{R})^n \to \mathbb{R} \, ; \, \theta \in \Theta\Big\}$$

being given, data compression theory shows how to build the following adaptive estimator. Let us choose a constant $\lambda > 0$ (we will set its value later), and let us define for any $i = 1, \dots, N$ the Gaussian conditional probability density (with respect to the Lebesgue measure on the real line)

$$q_\theta(Y_i \,|\, X_1^N, Y_1^{i-1}) = \sqrt{\frac{\lambda}{2\pi}} \exp\Big[-\tfrac{\lambda}{2}\big(Y_i - \widehat{f}_\theta(X_i, Z_1^{i-1})\big)^2\Big],$$

where we have put $Z_i = (X_i, Y_i)$ to shorten notations. Let us notice that this density does not depend on $X_{i+1}^N$, meaning that $Y_i$ is conditionally independent from $X_{i+1}^N$ knowning $(X_i, Z_1^{i-1})$.

We can then consider the global conditional probability density:

$$q_\theta(Y_1^N \mid X_1^N) = \prod_{i=1}^N q_\theta(Y_i \mid X_1^N, Y_1^{i-1}) = \prod_{i=1}^N q_\theta(Y_i \mid X_i, Z_1^{i-1}).$$

Coding theory invites us to introduce a prior distribution $\pi \in \mathcal{M}_+^1(\Theta)$ and to put

$$q(Y_1^N \mid X_1^N) = \sum_{\theta \in \Theta} \pi(\theta) q_\theta(Y_1^N \mid X_1^N).$$

Let us notice that the conditional independence property mentioned above still holds for the mixture $q$:

$$\begin{aligned} q(Y_i \mid X_1^N, Y_1^{i-1}) &= \frac{\mathbb{E}_{\pi(d\theta)}\big[q_\theta(Y_1^i \mid X_1^N)\big]}{\mathbb{E}_{\pi(d\theta)}\big[q(Y_1^{i-1} \mid X_1^N)\big]} \\ &= \frac{\mathbb{E}_{\pi(d\theta)}\big[q_\theta(Y_1^i \mid X_1^i)\big]}{\mathbb{E}_{\pi(d\theta)}\big[q(Y_1^{i-1} \mid X_1^{i-1})\big]} \\ &= q(Y_i \mid X_i, Z_1^{i-1}). \end{aligned}$$

Obviously, for any sequence $(X_i, Y_i)_{i=1}^N \in (\mathcal{X} \times \mathbb{R})^N$,

$$-\log\big[q(Y_1^N \mid X_1^N)\big] \leq \inf_{\theta \in \Theta} - \log\big[q_\theta(Y_1^N \mid X_1^N)\big] - \log\big[\pi(\theta)\big].$$

Let us notice that this also can be written as

$$-\sum_{i=1}^N \log\big[q(Y_i \mid X_i, Z_1^{i-1})\big] \leq \inf_{\theta \in \Theta} \left[-\sum_{i=1}^N \log\big[q_\theta(Y_i \mid X_i, Z_1^{i-1})\big]\right] - \log\big[\pi(\theta)\big].$$

To make this formula more explicit, we can introduce the posterior distributions

$$\rho_i(\theta) = \frac{q_\theta(Y_1^{i-1} \mid X_1^{i-1})\pi(\theta)}{\mathbb{E}_{\pi(d\theta')}\big[q_{\theta'}(Y_1^{i-1} \mid X_1^{i-1})\big]}.$$

With this new notations, the previous inequality can also be written as

$$\begin{aligned} -\sum_{i=1}^N \log\Big\{\mathbb{E}_{\rho_i(d\theta)}\Big[q_\theta(Y_i \mid X_i, Z_1^{i-1})\Big]\Big\} \\ \leq \inf_{\theta \in \Theta} - \sum_{i=1}^N \Big[\log\big[q_\theta(Y_i \mid X_i, Z_1^{i-1})\big]\Big] - \log\big[\pi(\theta)\big]. \end{aligned}$$

More explicitely, we obtain

$$- \sum_{i=1}^{N} \log \mathbb{E}_{\rho_i(d\theta)} \Bigl[ \exp\Bigl\{ -\frac{\lambda}{2} (Y_i - \widehat{f}_\theta(X_i, Z_1^{i-1}))^2 \Bigr\} \Bigr]$$
$$\leq \inf_{\theta \in \Theta} \frac{\lambda}{2} \sum_{i=1}^{N} (Y_i - \widehat{f}_\theta(X_i, Z_1^{i-1}))^2 - \log[\pi(\theta)].$$

Let us assume then that the random variable $Y_i$ is almost surely bounded. Namely, let us assume that for some positive constant $B$, $P(|Y_i| > B) = 0$. In this case, we can also assume without loss of generality that $|\widehat{f}_\theta(X_i, Z_1^{i-1})| \leq B$, truncating if necessary the regression functions. Due to the fact that the Gaussian function $x \mapsto \exp(-x^2/2)$ is concave on the interval $[-1, 1]$, the inequality

$$\mathbb{E}_{\rho_i(d\theta)} \Bigl[ \exp\Bigl\{ -\frac{\lambda}{2} (Y_i - \widehat{f}_\theta(X_i, Z_1^{i-1}))^2 \Bigr\} \Bigr]$$
$$\leq \exp\Bigl[ -\frac{\lambda}{2} \Bigl\{ Y_i - \mathbb{E}_{\pi(d\theta)}\bigl(\widehat{f}_\theta(X_i, Z_1^{i-1})\bigr) \Bigr\}^2 \Bigr].$$

is satisfied when $\lambda = \frac{1}{4B^2}$. We have proved the following proposition.

**Proposition 2.2.1.** *In the case when* $\lambda = \frac{1}{4B^2}$, *the estimators* $\widehat{f}_\theta$ *being truncated to the interval* $[-B, B]$, *for any probability distribution* $P \in \mathcal{M}_+^1\Bigl((\mathcal{X} \times \mathbb{R})^N, (\mathcal{F} \otimes \mathcal{B})^{\otimes N}\Bigr)$ *such that* $P(|Y_i| > B) = 0, \quad i = 1, \dots, N$

$$\frac{1}{N} \sum_{i=1}^{N} \mathbb{E}_{P(dZ_1^{i-1}, dX_i)} \Bigl\{ \bigl[\mathbb{E}_P(Y_i \,|\, X_i, Z_1^{i-1}) - \mathbb{E}_{\rho_i(d\theta)}\bigl[\widehat{f}_\theta\bigr](X_i, Z_1^{i-1})\bigr]^2 \Bigr\}$$
$$\leq \inf_{\theta \in \Theta} \Biggl\{ \frac{1}{N} \sum_{i=1}^{N} \mathbb{E}_{P(dZ_1^{i-1}, dX_i)} \Bigl\{ \bigl[\mathbb{E}_P(Y_i \,|\, X_i, Z_1^{i-1}) - \widehat{f}_\theta(X_i, Z_1^{i-1})\bigr]^2 \Bigr\}$$
$$- \frac{8B^2}{N} \log\bigl[\pi(\theta)\bigr] \Biggr\}.$$

*Remark* 2.2.1. This result is given as an example of what can be obtained. We will see further on how to deal with the case when the noise is unbounded. One distinctive feature of this kind of results is that the hypotheses about $P$ are very weak. In particular the preceding proposition covers the *autoregressive* case when $X_i = (Y_j)_{j=i-D}^{i-1}$, $P$ a.s..

## 2.3 Pattern recognition

This framework is close to the regression one, with the difference that $\mathcal{Y}$ is now a finite set and that the considered risk is the error rate. More precisely, let us consider a measurable space $(\mathcal{X}, \mathcal{F})$ and a finite set $(\mathcal{Y}, \mathcal{P})$ where $\mathcal{P}$ is the trivial sigma algebra consisting in all the subsets of $\mathcal{Y}$. Let $\{f_\theta : \mathcal{X} \to \mathcal{Y}\,;\, \theta \in \Theta\}$ be a family of classification rules, and let the risk to be minimized be $P(f_\theta(X) \neq Y)$ where $P \in \mathcal{M}_+^1(\mathcal{X} \times \mathcal{Y}, \mathcal{F} \otimes \mathcal{P})$ is known through an i.i.d. sample $Z_1^N = (X_i, Y_i)_{i=1}^N$ distributed according to $P^{\otimes N}$. Universal data compression theory invites us to consider the slightly more general framework, where a probability distribution $P \in \mathcal{M}_+^1\bigl((\mathcal{X} \times \mathcal{Y})^N, (\mathcal{F} \otimes \mathcal{P})^{\otimes N}\bigr)$ is considered and the *cumulated risk* of any classification estimator

$$\widehat{f} : \mathcal{X} \times \bigsqcup_{n \in \mathbb{N}} (\mathcal{X} \times \mathcal{Y})^n \to \mathcal{Y},$$

is defined to be $$\mathcal{R}_N(\widehat{f}) = \frac{1}{N} \sum_{i=1}^N P\bigl(\widehat{f}(X_i, Z_1^{i-1}) \neq Y_i\bigr).$$

For a parameter $\lambda \in \mathbb{R}_+$ to be chosen afterwards, let us consider the exponential models

$$Q_\theta(Y_i \,|\, X_i) = \frac{1}{W(\lambda)} \exp\Bigl\{-\lambda \mathbb{1}\bigl(f_\theta(X) \neq Y\bigr)\Bigr\},$$

where $W(\lambda) = \bigl(|\mathcal{Y}| - 1\bigr)\exp(-\lambda) + 1$. We will use the notation $Q_\theta(Y_1^N \,|\, X_1^N) = \prod_{i=1}^N Q_\theta(Y_i \,|\, X_i)$ and consider the mixture

$$Q(Y_1^N \,|\, X_1^N) = \mathbb{E}_{\pi(d\theta)}\Bigl[Q_\theta(Y_1^N \,|\, X_1^N)\Bigr],$$

where $\pi \in \mathcal{M}_+^1(\Theta, \mathcal{T})$ is a prior distribution on the parameter space. At this stage, some difference between least square regression and classification is already perceptible: whatever the nature of the parameter space $(\Theta, \mathcal{T})$ may be, (countable or continuous), the set $\{Q_\theta(Y_1^N \,|\, X_1^N) : \theta \in \Theta\}$ of values taken by the likelihood when $\theta$ ranges in $\Theta$ is always finite and therefore corresponds to a finite partition of $\Theta$. In order to get a partition not depending on the values of $Y$, let us consider the partition defined by the *traces* (i.e. the restrictions)

$$t(f_\theta) = f_{\theta \{X_i : i = 1, \dots, N\}} \in \mathcal{Y}^N$$

of the functions $f_\theta$ to the finite set $\{X_i : i = 1, \dots, N\}$. It is immediately seen that $\{\theta' : t(f_{\theta'}) = t(f_\theta)\} \subset \{\theta' : Q_{\theta'}(Y_1^N \,|\, X_1^N) = Q_\theta(Y_1^N \,|\, X_1^N)\}$. This implies that

$$-\log\bigl[Q(Y_1^N \,|\, X_1^N)\bigr] \leq \inf_{\theta \in \Theta} -\log\bigl[Q_\theta(Y_1^N \,|\, X_1^N)\bigr] - \log\bigl[\pi(\{\theta' : t(f_{\theta'}) = t(f_\theta)\})\bigr].$$

Let us put

$$\rho_i(d\theta) = \frac{Q_\theta(Y_1^{i-1} \,|\, X_1^{i-1}) \pi(d\theta)}{\mathbb{E}_{\pi(d\theta')}\bigl[Q_{\theta'}(Y_1^{i-1} \,|\, X_1^{i-1})\bigr]}.$$

We get

$$-\frac{1}{N}\sum_{i=1}^{N}\log\Bigl\{\mathbb{E}_{\rho_i(d\theta)}\bigl[Q_\theta(Y_i\,|\,X_i)\bigr]\Bigr\}$$
$$\leq \inf_{\theta\in\Theta} -\frac{1}{N}\sum_{i=1}^{N}\log\bigl[Q_\theta(Y_i\,|\,X_i)\bigr] - \frac{\log\bigl[\pi\bigl(\{\theta' : t(f_{\theta'}) = t(f_\theta)\}\bigr)\bigr]}{N}.$$

We can then notice that

$$\log\Bigl\{\mathbb{E}_{\rho_i(d\theta)}\bigl[Q_\theta(Y_i\,|\,X_i)\bigr]\Bigr\} + \log\bigl[W(\lambda)\bigr]$$
$$= \log\Bigl\{\mathbb{E}_{\rho_i(d\theta)}\Bigl[\exp\Bigl(-\lambda\mathbb{1}\bigl(Y_i \neq f_\theta(X_i)\bigr)\Bigr)\Bigr]\Bigr\}$$
$$= \log\Bigl\{1 + (e^{-\lambda} - 1)\rho_i\bigl(Y_i \neq f_\theta(X_i)\bigr)\Bigr\}$$
$$\leq -(1 - e^{-\lambda})\rho_i\bigl(Y_i \neq f_\theta(X_i)\bigr),$$

hence

$$\frac{1}{N}\sum_{i=1}^{N}\rho_i\bigl(Y_i \neq f_\theta(X_i)\bigr)$$
$$\leq \inf_{\theta\in\Theta}\frac{\lambda}{1-e^{-\lambda}}\frac{1}{N}\sum_{i=1}^{N}\mathbb{1}\bigl(Y_i \neq f_\theta(X_i)\bigr)$$
$$- \frac{\log\bigl[\pi\bigl(\{\theta' : t(f_\theta) = t(f_{\theta'})\}\bigr)\bigr]}{N(1-e^{-\lambda})}.$$

**Proposition 2.3.1.**
*For any probability distribution* $P \in \mathcal{M}_+^1\bigl((\mathcal{X}\times\mathcal{Y})^N, (\mathcal{F}\otimes\mathcal{P})^{\otimes N}\bigr)$,

$$\frac{1}{N}\sum_{i=1}^{N}\mathbb{E}_{P(dZ_1^{i-1})}\mathbb{E}_{\rho_i(d\theta)}\Bigl[P\bigl(Y_i \neq f_\theta(X_i)\,|\,Z_1^{i-1}\bigr)\Bigr]$$
$$\leq \inf_{\theta\in\Theta}\frac{\lambda}{1-e^{-\lambda}}\frac{1}{N}\sum_{i=1}^{N}P\bigl(Y_i \neq f_\theta(X_i)\bigr)$$
$$- \frac{\mathbb{E}_{P(dX_1^N)}\Bigl\{\log\bigl[\pi\bigl(\{\theta' : t(f_{\theta'}) = t(f_\theta)\}\bigr)\bigr]\Bigr\}}{N(1-e^{-\lambda})}.$$

*The quantity* $\mathbb{E}_{P(dX_1^N)}\Bigl\{\log\bigl[\pi\bigl(\{\theta' : t(f_{\theta'}) = t(f_\theta)\}\bigr)\bigr]\Bigr\}$ *measures the* mean margin *of* $f_\theta$.

*Remark* 2.3.1. If it is known that for some value $\overline{\theta} \in \Theta$, $Y_i = f_{\overline{\theta}}(X_i)$ $P$ almost surely, then it is advisable to take $\lambda = +\infty$, and thus to use the maximum likelihood estimator (randomized to chose between ties), if it is known that $\inf_{\theta\in\Theta}\frac{1}{N}\sum_{i=1}^{N}P\bigl(Y_i \neq f_\theta(X_i)\bigr)$ is of order $1/N$, then it is advisable to take $\lambda$ of order 1, and if this same quantity is of order 1, then it is advisable to take $\lambda$ of order $1/\sqrt{N}$.

A possible variant to this approach consists in making use of the Vapnik entropy of the family $\{f_\theta : \theta \in \Theta\}$, considering a measure $\pi$ depending on the "design" $X_1^N$. Let us assume that the space of parameters is decomposed into the countable disjoint union of several models $\Theta = \bigsqcup_{j\in\mathbb{N}} \Theta_j$. We have in mind here a partition of $\Theta$ made on a quite arbitrary basis, reflecting the way the statistician has chosen to introduce some "structure" in his set of parameters. For any $j \in \mathbb{N}$, let us consider the quotient set $\Theta_j/\tau$, where $\tau$ is the relation $t(f_\theta) = t(f_{\theta'})$. This quotient set is finite and can be equiped with the uniform probability measure $\pi_j(\overline{\theta}) = |\Theta_j/\tau|^{-1}$. Let us moreover consider some probability measure $\alpha \in \mathcal{M}_+^1(\mathbb{N})$ and define the prior distribution $\pi \in \mathcal{M}_+^1\Bigl(\bigsqcup_{j\in\mathbb{N}}(\Theta_j/\tau)\Bigr)$ by the formula $\pi(A) = \sum_{j\in\mathbb{N}} \alpha(j)\pi_j\Bigl[A \cap (\Theta_j/\tau)\Bigr]$. A computation similar to the one precedingly carried shows that the posterior distributions $\rho_i(d\overline{\theta}) \propto Q_{\overline{\theta}}(Y_1^{i-1} \,|\, X_1^{i-1})\pi(d\overline{\theta})$ are such that

**Proposition 2.3.2.** *For any probability distribution* $P \in \mathcal{M}_+^1\bigl((\mathcal{X}\times\mathcal{Y})^N, (\mathcal{F}\otimes\mathcal{P})^{\otimes N}\bigr)$,

$$\frac{1}{N}\sum_{i=1}^N \mathbb{E}_{P(dX_1^N, dY_1^{i-1})}\mathbb{E}_{\rho_i(d\overline{\theta})}\Bigl[P(Y_i \neq f_{\overline{\theta}}(X_i) \,|\, X_1^N, Y_1^{i-1})\Bigr]$$

$$\leq \inf_{j\in\mathbb{N}} \inf_{\theta\in\Theta_j} \frac{\lambda}{1-e^{-\lambda}} \frac{1}{N}\sum_{i=1}^N P(Y_i \neq f_\theta(X_i) \,|\, X_1^N) + \frac{\mathbb{E}_{P(dX_1^N)}\{\log[|\Theta_j/\tau|]\} - \log[\alpha(j)]}{N(1-e^{-\lambda})}.$$

*In the binary classification case when* $\mathcal{Y} = \{0,1\}$, *the quantity* $\mathbb{E}_{P(dX_1^N)}\Bigl\{\log\bigl[|\Theta_j/\tau|\bigr]\Bigr\}$ *is called the* Vapnik entropy *of the model* $\{f_\theta : \theta \in \Theta_j\}$; *it measures its* complexity *for the classification problems where the patterns to be recognized are distributed according to* $P(dX_1^N)$. *Moreover when the* Vapnik Cervonenkis dimension $d_j$ *of* $\Theta_j$ *is finite, for any* $(X_1, \dots, X_N) \in \mathcal{X}^N$, $\log\bigl[|\Theta_j/\tau|\bigr] \leq d_j\Bigl(\log(N/d_j) + 1\Bigr)$.

*Remark* 2.3.2. More details about Vapnik's entropy and the Vapnik Cervonenkis dimension can be found in [76]. It is defined as

$$d_j = \max\Bigl\{N : \max_{x_1^N\in\mathcal{X}^N} |\Theta_j/\tau(x_1^N)| = 2^N\Bigr\},$$

where $\tau(x_1^N) = \{(\theta, \theta') \in \Theta^2 : f_\theta(x_i) = f_{\theta'}(x_i), i = 1, \dots, N\}$. For example, the family of half spaces in dimension $d$ has a Vapnik Cervonenkis dimension equal to $d+1$.

It is to be noticed that the posterior distribution $\rho_i(d\overline{\theta})$ here depends on $(X_1, \dots, X_N)$.

### 2.3.1 Prediction from classification trees

In the previous chapter about data compression, we have seen how to build a coding distribution which mixes context tree models in an asymptotically optimal way when the quantity to be maximized is the log-likelihood. In a classification problem, it is often the case that a pattern $x$ is to be recognized through the successive answers to a series of questions. If the order in which the questions are asked is fixed, it is natural (after some change of variables) to consider that $\mathcal{X} = E^{\mathbb{Z}_-}$ where $E$ is the finite set of the possible answers to the questions (most of the time it will be $\{0,1\}$, generalization to the case when $E$ depends on the question is also possible and left as an exercise). Let $\mathcal{Y}$ be a finite set of labels. Let us consider the classification functions of the type

$$f_{\mathcal{D},\theta}(x_{-\infty}^0) = \theta\big[\mathcal{D}(x_{-\infty}^0)\big], \qquad x_{-\infty}^0 \in \mathcal{X} = E^{\mathbb{Z}_-},$$

where $\mathcal{D} \in \mathcal{F}_D$ is a complete suffix tree of depth not greater than $D$, where $\mathcal{D}(x_{-\infty}^0)$ is the (unique) suffix of $x_{-\infty}^0$ belonging to $\mathcal{D}$, and where $\theta \in \mathcal{Y}^{\mathcal{D}} \overset{\text{def}}{=} \Theta_{\mathcal{D}}$ is a function from $\mathcal{D}$ to $\mathcal{Y}$. Let us notice also that this framework covers the autoregressive case when a sequence $(Z_i)_{i\in\mathbb{Z}} \in E^{\mathbb{Z}}$ is being considered and when the observation is made of the couples of random variables $(X_i, Y_i)_{i=1}^N = (Z_{-\infty}^{i-1}, Z_i)_{i=1}^N$. Let us define the parameter space $\Theta = \bigsqcup_{\mathcal{D}\in\mathcal{F}_D} \Theta_{\mathcal{D}}$ and the conditional distribution models

$$Q_{\mathcal{D},\theta}(y\,|\,x) = W(\lambda)^{-1}\exp\Big\{-\lambda\mathbb{1}\big[y \neq f_{\mathcal{D},\theta}(x)\big]\Big\}, \qquad \theta \in \Theta_{\mathcal{D}}, (x,y) \in \mathcal{X}\times\mathcal{Y},$$

where $W(\lambda) = (|\mathcal{Y}|-1)\exp(-\lambda)+1$. Let us consider for any $\mathcal{D} \in \mathcal{F}_D$ the uniform distribution $\nu_{\mathcal{D}}$ on $\Theta_{\mathcal{D}}$, defined by $\nu_{\mathcal{D}}(\theta) = |\mathcal{Y}|^{-|\mathcal{D}|}$ for any $\theta \in \Theta_{\mathcal{D}}$. Let us also consider the branching distribution $\pi \in \mathcal{M}_+^1(\mathcal{F}_D)$ defined by equation (1.5.1), and let us build the "double mixture" prior distribution on $\Theta$

$$\mu(\theta) = \sum_{\mathcal{D}\in\mathcal{F}_D} \mathbb{1}(\theta \in \Theta_{\mathcal{D}})\pi(\mathcal{D})\nu_{\mathcal{D}}(\theta).$$

**Proposition 2.3.3.** *For any sequence of observations $\big(x(i), y(i)\big) \in (\mathcal{X}\times\mathcal{Y})^N$, the posterior distributions $\rho_i(\theta) \propto \prod_{j=1}^{i-1} Q_{\mathcal{D},\theta}\big(y(j)\,|\,x(j)\big)\mu(\theta)$ are such that*

$$\frac{1}{N}\sum_{i=1}^N \rho_i\big[f_\theta\big(x(i)\big) \neq y(i)\big]$$
$$\leq \inf_{\mathcal{D}\in\mathcal{F}_D}\inf_{\theta\in\Theta_{\mathcal{D}}} \frac{\lambda}{N(1-e^{-\lambda})}\sum_{i=1}^N \mathbb{1}\big[f_\theta\big(x(i)\big) \neq y(i)\big] + \frac{|\mathcal{D}|\log(|\mathcal{Y}|)}{N(1-e^{-\lambda})}$$
$$- \frac{(|\mathcal{D}|-1)}{N(1-e^{-\lambda})(|E|-1)}\log(\alpha) - \frac{|\mathcal{D}|}{N(1-e^{-\lambda})}\log(1-\alpha).$$

To implement this randomized classification scheme, we have to compute the random choice of $\theta$ according to the distribution $\rho_i(d\theta)$. More precisely, we have to draw at random $f_\theta[x(i)]$, which depends on two quantities: $\mathcal{D}[x(i)]$ and $\theta_{\mathcal{D}}\big(\mathcal{D}[x(i)]\big)$, because $f_\theta[x(i)] = \theta_{\mathcal{D}}\big(\mathcal{D}[x(i)]\big)$. Therefore, we can simulate in a first step the random variable $\tau = \ell\big[\mathcal{D}\big(x(i)\big)\big]$ according to the distribution $\rho_i(\tau)$, (because $\tau$ defines $\mathcal{D}\big(x(i)\big) = x(i)_{1-\tau}^0$), and once $\tau$ has been chosen, we can simulate in a second step $\theta_{\mathcal{D}}\big[x(i)_{1-\tau}^0\big]$ according to the distribution $\rho_i(d\theta \,|\, \tau)$.

To express $\rho_i$, it is useful to introduce the counters

$$a_m^y(i) = \sum_{j=1}^{i-1} \mathbb{1}\Big[\big(x(j)_{1-\ell(m)}^0, y(j)\big) = (m, y)\big)\Big],$$

$$b_m(i) = \sum_{y \in \mathcal{Y}} a_m^y(i).$$

We have

$$\prod_{j=1}^{i-1} Q_{\mathcal{D},\theta}\big(y(j) \,|\, x(j)\big) = \prod_{m \in \mathcal{D}} W(\lambda)^{-b_m(i)} \exp\Big[-\lambda\big(b_m(i) - a_m^{\theta(m)}(i)\big)\Big].$$

This shows immediately that

$$\rho_i(\theta \,|\, \mathcal{D}) \propto \prod_{m \in \mathcal{D}} \exp\Big[\lambda a_m^{\theta(m)}(i)\Big],$$

and therefore that

$$\rho_i\Big[f_\theta\big[x(i)\big] = y \,|\, \tau\Big] = \left(\sum_{z \in \mathcal{Y}} \exp\Big[\lambda a_{x(i)_{1-\tau}^0}^z\Big]\right)^{-1} \exp\Big[\lambda a_{x(i)_{1-\tau}^0}^y\Big].$$

Moreover

$$\rho_i(\mathcal{D}) \propto \pi(\mathcal{D}) \mathbb{E}_{\nu_{\mathcal{D}}(d\theta)} \left\{ \prod_{j=1}^{i-1} Q_{\mathcal{D},\theta}\big[y(j) \,|\, x(j)\big] \right\}$$

which can also be written as

$$\rho_i(\mathcal{D}) \propto \pi(\mathcal{D}) \prod_{m \in \mathcal{D}} K_m(i),$$

with

$$K_m(i) = W(\lambda)^{-b_m(i)} |\mathcal{Y}|^{-1} \sum_{y \in \mathcal{Y}} \exp\Big\{-\lambda\big[b_m(i) - a_m^y(i)\big]\Big\}.$$

We thus see that $\rho_i(\mathcal{D})$ is a branching process. More precisely, with the notations of the proof of proposition 1.5.2 of chapter 1,

$$\rho_i(\mathcal{D} \,|\, \mathcal{D} \geq m) \propto \pi(\mathcal{D}_m \,|\, \mathcal{D} \geq m) \prod_{w \in \mathcal{D}_m} K_w(i)$$

$$\times\, \pi(\mathcal{D} \setminus \mathcal{D}_m \,|\, \mathcal{D} \geq m) \prod_{w \in \mathcal{D} \setminus \mathcal{D}_m} K_w(i).$$

Therefore for any $m \in E^* \cup \{\varnothing\}$,

$$\rho_i(m \in \mathcal{D} \,|\, \mathcal{D} \geq m) = \frac{\pi(m \in \mathcal{D} \,|\, \mathcal{D} \geq m) K_m(i)}{\mathbb{E}_{\pi(d\mathcal{D}|\mathcal{D}\geq m)}\left(\prod_{w \in \mathcal{D}_m} K_w(i)\right)},$$

where $\pi(m \in \mathcal{D} \,|\, \mathcal{D} \geq m) = 1 - \alpha$ and, as in chapter 1, the quantities $\Upsilon_m = \mathbb{E}_{\pi(d\mathcal{D} \,|\, \mathcal{D} \geq m)}\left(\prod_{w \in \mathcal{D}_m} K_w(i)\right)$ can be computed by induction as

$$\begin{aligned} \Upsilon_m(i) &= K_m(i), && \ell(m) = D, \\ \Upsilon_m(i) &= (1-\alpha) K_m(i) + \alpha \prod_{x \in E} \Upsilon_{(x,m)}(i), && \ell(m) < D. \end{aligned}$$

As a special case,

$$\rho_i(\tau = k \,|\, \tau \geq k) = \rho_i(x(i)_{1-k}^0 \in \mathcal{D} \,|\, \mathcal{D} \geq x(i)_{1-k}^0) = \frac{(1-\alpha) K_{x(i)_{1-k}^0}}{\Upsilon_{x(i)_{1-k}^0}}.$$

Using this last formula, we can simulate the distribution of $\tau$ in no more than $D$ steps.

### 2.3.2 Noisy non ambiguous classification

Propositions 2.3.1, 2.3.2 and 2.3.3 can be improved in the case when the sample of labeled patterns is noisy (more precisely in the case when $P(Y \mid X)$ is far from being concentrated on one label), but when there exists a classification rule whose error rate is almost everywhere the best among the considered rules.

Let us consider here again the general case of an observation $(X_i, Y_i)_{i=1}^N \in (\mathcal{X} \times \mathcal{Y})^N$, where $\mathcal{Y}$ is a finite set and $(\mathcal{X}, \mathcal{F})$ is a measurable space. Let us assume that we observe an arbitrary sequence $X_1^N$ and that the conditional distribution of $Y_1^N$ knowing $X_1^N$ is $P(dY_1^N \mid X_1^N)$. Let us consider a measurable parameter space $(\Theta, \mathcal{T})$ and a family $\{f_\theta : \mathcal{X} \to \mathcal{Y} : \theta \in \Theta\}$ of classification rules. Let us consider some fixed distinguished value $\tilde{\theta}$ of $\theta$ the *margin at time i*

$$\alpha(i) \overset{\text{def}}{=} \min\Big\{ P\big(Y_i \neq f_\theta(X_i) \,|\, X_1^N, Y_1^{i-1}\big) - P\big(Y_i \neq f_{\tilde{\theta}}(X_i) \,|\, X_1^N, Y_1^{i-1}\big) \\ : \theta \in \Theta, f_\theta(X_i) \neq f_{\tilde{\theta}}(X_i) \Big\}.$$

Let us start then from inequality

$$- \mathbb{E}_{P(dY_1^N | X_1^N)} \Biggl\{ \sum_{i=1}^N \log \biggl[ \mathbb{E}_{\rho_i(d\theta)} \Bigl\{ \exp \Bigl[ -\lambda \bigl\{ \mathbb{1}\bigl[Y_i \neq f_\theta(X_i)\bigr] - \mathbb{1}\bigl[Y_i \neq f_{\tilde{\theta}}(X_i)\bigr] \bigr\} \Bigr] \Bigr\} \biggr] \Biggr\}$$

$$\leq \inf_{\theta \in \Theta} \lambda \sum_{i=1}^N \Bigl( P\bigl[Y_i \neq f_\theta(X_i) \,|\, X_1^N\bigr] - P\bigl[Y_i \neq f_{\tilde{\theta}}(X_i) \,|\, X_1^N\bigr] \Bigr) - \log\Bigl[\pi\Bigl(\bigl\{\theta' \in \Theta : t(f_{\theta'}) = t(f_\theta)\bigr\}\Bigr)\Bigr],$$

which is nothing else than proposition 2.3.1, where we have added the same constant to the left-hand and right-hand sides.

In the case when $\alpha(i) \geq \alpha$, where $\alpha$ is a strictly positive constant, it is to be noticed that

$$\begin{aligned}
\log\Bigl[\mathbb{E}_{\rho_i(d\theta)}\Bigl\{\exp\Bigl[-\lambda\bigl\{\mathbb{1}\bigl[Y_i \neq f_\theta(X_i)\bigr] - \mathbb{1}\bigl[Y_i \neq f_{\tilde{\theta}}(X_i)\bigr]\bigr\}\Bigr]\Bigr\}\Bigr]& \\
\leq \mathbb{E}_{\rho_i(d\theta)}\Bigl\{(e^{-\lambda} - 1)\mathbb{1}\bigl[Y_i = f_{\tilde{\theta}}(X_i) \neq f_\theta(X_i)\bigr]& \\
+ (e^{\lambda} - 1)\mathbb{1}\bigl[Y_i = f_\theta(X_i) \neq f_{\tilde{\theta}}(X_i)\bigr]\Bigr\}& \\
= \mathbb{E}_{\rho_i(d\theta)}\Bigl\{(e^{-\lambda} - 1)\Bigl[\mathbb{1}\bigl[Y_i \neq f_\theta(X_i)\bigr] - \mathbb{1}\bigl[Y_i \neq f_{\tilde{\theta}}(X_i)\bigr]\Bigr]& \\
+ (e^{\lambda} - 1)(1 - e^{-\lambda})\mathbb{1}\bigl[Y_i = f_\theta(X_i) \neq f_{\tilde{\theta}}(X_i)\bigr]\Bigr\}&.
\end{aligned}$$

Moreover, again in the case when $\alpha(i) \geq \alpha$

$$\begin{aligned}
P\bigl(Y_i = f_\theta(X_i) \neq f_{\tilde{\theta}}(X_i) \,|\, X_1^N, Y_1^{i-1}\bigr)& \\
= \mathbb{1}\bigl(f_\theta(X_i) \neq f_{\tilde{\theta}}(X_i)\bigr)P\bigl(Y_i = f_\theta(X_i) \,|\, X_1^N, Y_1^{i-1}\bigr)& \\
\leq \frac{1-\alpha}{2}\mathbb{1}\bigl(f_\theta(X_i) \neq f_{\tilde{\theta}}(X_i)\bigr)& \\
\leq \frac{1-\alpha}{2\alpha}\Bigl[P\bigl(Y_i \neq f_\theta(X_i) \,|\, X_1^N, Y_1^{i-1}\bigr) - P\bigl(Y_i \neq f_{\tilde{\theta}}(X_i) \,|\, X_1^N, Y_1^{i-1}\bigr)\Bigr]&.
\end{aligned}$$

When $\alpha(i) < \alpha$, one can use the upper bound given in proposition 2.3.1:

$$\begin{aligned}
\log\Bigl\{\mathbb{E}_{\rho_i(d\theta)}\Bigl[\exp\Bigl\{-\lambda\Bigl[\mathbb{1}\bigl[Y_i \neq f_\theta(X_i)\bigr] - \mathbb{1}\bigl[Y_i \neq f_{\tilde{\theta}}(X_i)\bigr]\Bigr]\Bigr\}\Bigr]\Bigr\}& \\
\leq (e^{-\lambda} - 1)\mathbb{E}_{\rho_i(d\theta)}\Bigl\{\mathbb{1}\bigl[Y_i \neq f_\theta(X_i)\bigr]\Bigr\} + \lambda\mathbb{1}\bigl[Y_i \neq f_{\tilde{\theta}}(X_i)\bigr]& \\
= (e^{-\lambda} - 1)\mathbb{E}_{\rho_i(d\theta)}\Bigl\{\mathbb{1}\bigl[Y_i \neq f_\theta(X_i)\bigr] - \mathbb{1}\bigl[Y_i \neq f_{\tilde{\theta}}(X_i)\bigr]\Bigr\}& \\
+ (e^{-\lambda} - 1 + \lambda)\,\mathbb{1}\bigl[Y_i \neq f_{\tilde{\theta}}(X_i)\bigr]&.
\end{aligned}$$

Consequently

$$
\begin{aligned}
(1-e^{-\lambda})\mathbb{E}_{\rho_i(d\theta)}\Bigl\{P\bigl(Y_i \neq f_\theta(X_i)\,|\,X_1^N, Y_1^{i-1}\bigr) - P\bigl(Y_i \neq f_{\tilde\theta}(X_i)\,|\,X_1^N, Y_1^{i-1}\bigr)\Bigr\} \\
- \mathbb{1}\bigl[\alpha(i) \geq \alpha\bigr](1-e^{-\lambda})(e^{\lambda}-1)\frac{1-\alpha}{2\alpha} \\
\times \mathbb{E}_{\rho_i(d\theta)}\Bigl\{P\bigl(Y_i \neq f_\theta(X_i)\,|\,X_1^N, Y_1^{i-1}\bigr) - P\bigl(Y_i \neq f_{\tilde\theta}(X_i)\,|\,X_1^N, Y_1^{i-1}\bigr)\Bigr\} \\
- \mathbb{1}\bigl[\alpha(i) < \alpha\bigr](e^{-\lambda}-1+\lambda)P\bigl(Y_i \neq f_{\tilde\theta}(X_i)\,|\,X_1^N, Y_1^{i-1}\bigr) \\
\leq -\mathbb{E}_{P(dY_i | X_1^N, Y_1^{i-1})}\biggl\{\log\biggl[\mathbb{E}_{\rho_i(d\theta)}\biggl\{\exp\biggl[-\lambda\Bigl\{\mathbb{1}\bigl[Y_i \neq f_\theta(X_i)\bigr] \\
- \mathbb{1}\bigl[Y_i \neq f_{\tilde\theta}(X_i)\bigr]\Bigr\}\biggr]\biggr\}\biggr]\biggr\}.
\end{aligned}
$$

We therefore have proved the following.

**Proposition 2.3.4.** *With the notations introduced previously,*

$$
\begin{aligned}
\mathbb{E}_{P(dY_1^N | X_1^N)}\biggl\{\frac{1}{N}\sum_{i=1}^N \mathbb{E}_{\rho_i(d\theta)}\Bigl[P\bigl(Y_i \neq f_\theta(X_i)\,|\,X_1^N, Y_1^{i-1}\bigr)\Bigr]\biggr\} \\
\leq \frac{1}{N}\sum_{i=1}^N P\bigl(Y_i \neq f_{\tilde\theta}(X_i)\,|\,X_1^N\bigr) \\
+ C_1(\lambda)\frac{1}{N}\sum_{i=1}^N P\bigl[\alpha(i) < \alpha\,|\,X_1^N\bigr] + C_2(\lambda)\frac{1}{N}\sum_{i=1}^N P\bigl[\alpha(i) < 0\,|\,X_1^N\bigr] \\
+ \inf_{\theta\in\Theta}\biggl\{C_3(\lambda)\frac{1}{N}\sum_{i=1}^N\Bigl\{P\bigl(Y_i \neq f_\theta(X_i)\,|\,X_1^N\bigr) - P\bigl(Y_i \neq f_{\tilde\theta}(X_i)\,|\,X_1^N\bigr)\Bigr\} \\
- \frac{C_4(\lambda)}{N}\log\Bigl\{\pi\Bigl[\bigl\{\theta' \in \Theta : t(f_{\theta'}) = t(f_\theta)\bigr\}\Bigr]\Bigr\}\biggr\},
\end{aligned}
$$

*where*

$$
\begin{aligned}
C_1(\lambda) &= \left[1-(e^{\lambda}-1)\frac{1-\alpha}{2\alpha}\right]^{-1}\frac{(e^{-\lambda}-1+\lambda)}{(1-e^{-\lambda})} \underset{\lambda\downarrow 0}{\sim} \frac{\lambda}{2} \\
C_2(\lambda) &= \left[1-(e^{\lambda}-1)\frac{1-\alpha}{2\alpha}\right]^{-1}(e^{\lambda}-1)\frac{1-\alpha}{2\alpha} \underset{\lambda\downarrow 0}{\sim} \frac{1-\alpha}{2\alpha}\lambda \\
C_3(\lambda) &= \left[1-(e^{\lambda}-1)\frac{1-\alpha}{2\alpha}\right]^{-1}\frac{\lambda}{(1-e^{-\lambda})} \underset{\lambda\downarrow 0}{\sim} 1 \\
C_4(\lambda) &= \left[1-(e^{\lambda}-1)\frac{1-\alpha}{2\alpha}\right]^{-1}\frac{1}{(1-e^{-\lambda})} \underset{\lambda\downarrow 0}{\sim} \frac{1}{\lambda}
\end{aligned}
$$

*Remark* 2.3.3. It is possible to integrate with respect to $X_1^N$ the inequality stated in this proposition. As the posterior distribution $\rho_i(d\theta)$ depends only on $(X_1^{i-1}, Y_1^{i-1})$, and not on $(X_i^N, Y_i^N)$, we get an upper bound of the quantity

$$\mathbb{E}_{P(dX_1^N, dY_1^N)} \left\{ \frac{1}{N} \sum_{i=1}^{N} \mathbb{E}_{\rho_i(d\theta)} \Big[ P\big(Y_i \neq f_\theta(X_i) \,|\, X_1^{i-1}, Y_1^{i-1}\big) \Big] \right\}$$

which remains valid in the autoregressive case when $X_i = Y_{-\infty}^{i-1}$.

*Remark* 2.3.4. Propositions 2.3.2 and 2.3.3 can be transposed in the same way to adapt to the case of noisy non ambiguous classification problems.

### 2.3.3 Non randomized classification rules

In the case when a randomized classification rule is not wanted, it is possible to build from the posterior distributions $\rho_i(d\theta)$ a deterministic classification rule obtained by voting:

$$\widehat{f}_i(X_i) = \arg\max_{y \in \mathcal{Y}} \rho_i\big[f_\theta(X_i) = y\big],$$

and to remark that

$$\rho_i\big[Y_i \neq f_\theta(X_i)\big] \geq \frac{1}{2} \mathbb{1}\big[Y_i \neq \widehat{f}_i(X_i)\big].$$

(Indeed the distribution of $f_\theta(X_i)$ under $\rho_i(d\theta)$ can put a weight larger than $1/2$ to at most one point of $E$). Upper bounds for the mean cumulated risk of randomized estimators obtained in propositions 2.3.1, 2.3.2 and 2.3.3 are therefore still valid for $\widehat{f}_i$, with this difference that the righthand side member has to be multiplied by a factor 2.

In the noisy non ambiguous case, the following reasonning can be made: as soon as $\alpha(i) \geq \alpha$,

$$\begin{aligned}
\mathbb{E}_{\rho_i(d\theta)} \Big\{ P\big(Y_i \neq f_\theta(X_i) \,|\, X_1^N, Y_1^{i-1}\big) - P\big(Y_i \neq f_{\tilde{\theta}}(X_i) \,|\, X_1^N, Y_1^{i-1}\big) \Big\} \\
\geq \frac{\alpha}{2} \mathbb{1}\big[\hat{f}_i(X_i) \neq f_{\tilde{\theta}}(X_i)\big] \\
\geq \frac{\alpha}{2} \Big[ P\big(Y_i \neq \hat{f}_i(X_i) \,|\, X_1^N, Y_1^{i-1}\big) - P\big(Y_i \neq f_{\tilde{\theta}}(X_i) \,|\, X_1^N, Y_1^{i-1}\big) \Big]
\end{aligned}$$

On the other hand, when $\alpha(i) < \alpha$

$$\mathbb{E}_{\rho_i(d\theta)} \big[ P\big(Y_i \neq f_\theta(X_i) \,|\, X_1^N, Y_1^{i-1}\big) \big] \geq \frac{1}{2} P\big(Y_i \neq \hat{f}_i(X_i) \,|\, X_1^N, Y_1^{i-1}\big).$$

Therefore

$$
\begin{aligned}
- \mathbb{E}_{P(dY_i | X_1^N, Y_1^{i-1})} & \Biggl\{ \log \Biggl[ \mathbb{E}_{\rho_i(d\theta)} \Biggl\{ \exp \Biggl[ -\lambda \Bigl\{ \mathbb{1}\bigl[Y_i \neq f_\theta(X_i)\bigr] \\
& \qquad - \mathbb{1}\bigl[Y_i \neq f_{\tilde{\theta}}(X_i)\bigr] \Bigr\} \Biggr] \Biggr\} \Biggr] \Biggr\} \\
& \geq \mathbb{1}\bigl[\alpha(i) \geq \alpha\bigr] \frac{\alpha}{2}(1 - e^{-\lambda}) \Bigl[ 1 - (e^{\lambda} - 1) \frac{(1-\alpha)}{2\alpha} \Bigr] \\
& \times \Bigl[ P\bigl(Y_i \neq \hat{f}_i(X_i) \,|\, X_1^N, Y_1^{i-1}\bigr) - P\bigl(Y_i \neq f_{\tilde{\theta}}(X_i) \,|\, X_1^N, Y_1^{i-1}\bigr) \Bigr] \\
& + \mathbb{1}\bigl[\alpha(i) < \alpha\bigr] \Biggl\{ \\
& \frac{1}{2}(1 - e^{-\lambda}) \Bigl[ P\bigl(Y_i \neq \hat{f}_i(X_i) \,|\, X_1^N, Y_1^{i-1}\bigr) - P\bigl(Y_i \neq f_{\tilde{\theta}}(X_i) \,|\, X_1^N, Y_1^{i-1}\bigr) \Bigr] \\
& - \Bigl[ \frac{1}{2}(1 - e^{-\lambda}) + (e^{-\lambda} - 1 + \lambda) \Bigr] P\bigl(Y_i \neq f_{\tilde{\theta}}(X_i) \,|\, X_1^N, Y_1^{i-1}\bigr) \Biggr\} \\
& \geq \frac{\alpha}{2}(1 - e^{-\lambda}) \Bigl[ 1 - (e^{\lambda} - 1) \frac{(1-\alpha)}{2\alpha} \Bigr] \\
& \times \Bigl[ P\bigl(Y_i \neq \hat{f}_i(X_i) \,|\, X_1^N, Y_1^{i-1}\bigr) - P\bigl(Y_i \neq f_{\tilde{\theta}}(X_i) \,|\, X_1^N, Y_1^{i-1}\bigr) \Bigr] \\
& + \mathbb{1}\bigl[\alpha(i) < \alpha\bigr] \Biggl\{ \\
& \frac{1}{2}(1 - e^{-\lambda}) \Bigl\{ 1 - \alpha \Bigl[ 1 - (e^{\lambda} - 1) \frac{(1-\alpha)}{2\alpha} \Bigr] \Bigr\} \\
& \times \Bigl[ P\bigl(Y_i \neq \hat{f}_i(X_i) \,|\, X_1^N, Y_1^{i-1}\bigr) - P\bigl(Y_i \neq f_{\tilde{\theta}}(X_i) \,|\, X_1^N, Y_1^{i-1}\bigr) \Bigr] \\
& - \Bigl[ \frac{1}{2}(1 - e^{-\lambda}) + (e^{\lambda} - 1 + \lambda) \Bigr] \Biggr\}
\end{aligned}
$$

This proves the following proposition.

**Proposition 2.3.5.** *With the previous notations,*

$$
\begin{aligned}
\frac{1}{N} \sum_{i=1}^N P\bigl(Y_i \neq \hat{f}_i(X_i) \,|\, X_1^N\bigr) & \leq \frac{1}{N} \sum_{i=1}^N P\bigl(Y_i \neq f_{\tilde{\theta}}(X_i) \,|\, X_1^N\bigr) \\
& + C_1(\lambda) \frac{1}{N} \sum_{i=1}^N P\bigl[\alpha(i) < \alpha \,|\, X_1^N\bigr] + C_2(\lambda) \frac{1}{N} \sum_{i=1}^N P\bigl[\alpha(i) < 0 \,|\, X_1^N\bigr] \\
& + \inf_{\theta \in \Theta} \Biggl\{ C_3(\lambda) \frac{1}{N} \Bigl[ P\bigl(Y_i \neq f_\theta(X_i) \,|\, X_1^N\bigr) - P\bigl(Y_i \neq f_{\tilde{\theta}}(X_i) \,|\, X_1^N\bigr) \Bigr]
\end{aligned}
$$

$$- \frac{C_4(\lambda)}{N} \log\bigl[\pi\bigl(\{\theta' \in \Theta : t(f_{\theta'}) = t(f_\theta)\}\bigr)\bigr] \Biggr\},$$

*where*

$$C_1(\lambda) = \frac{1 + 2\dfrac{e^\lambda - 1 + \lambda}{1 - e^{-\lambda}}}{\alpha\Bigl[1 - (e^\lambda - 1)\dfrac{1-\alpha}{2\alpha}\Bigr]} \underset{\lambda \searrow 0}{\sim} \frac{1}{\alpha}$$

$$C_2(\lambda) = \alpha^{-1}\Bigl[1 - (e^\lambda - 1)\frac{1-\alpha}{2\alpha}\Bigr]^{-1} - 1 \underset{\lambda \searrow 0}{\sim} \frac{1}{\alpha} - 1$$

$$C_3(\lambda) = \frac{2\lambda}{\alpha}(1 - e^{-\lambda})^{-1}\Bigl[1 - (e^\lambda - 1)\frac{1-\alpha}{2\alpha}\Bigr]^{-1} \underset{\lambda \searrow 0}{\sim} \frac{2}{\alpha}$$

$$C_4(\lambda) = \frac{2}{\alpha}(1 - e^{-\lambda})^{-1}\Bigl[1 - (e^\lambda - 1)\frac{1-\alpha}{2\alpha}\Bigr]^{-1} \underset{\lambda \searrow 0}{\sim} \frac{2}{\alpha\lambda}.$$

# 3

# Oracle inequalities for the non-cumulated mean risk

The aim of this chapter and the following is to present two pseudo-Bayesian methods to mix estimators (or fixed probability distributions), for which it is possible to prove non asymptotic oracle inequalities for the non cumulated mean risk. As in the previous chapter, we will begin with estimating conditional probability distributions, and will afterwards deduce applications to mean square regression and classification. The first method to be presented, called here the progressive mixture rule, is a simple modification of the compression setting, which gives satisfactory results for countable families of estimators. The second method, called here the Gibbs estimator, introduces a temperature parameter and uses inequalities inspired by statistical mechanics, which give improved results for "continuous" families of estimators (or of fixed distributions). Whereas the results of previous chapters were proved for any sample distribution (and indeed for any individual sequence of observations), those of this chapter apply to an *exchangeable* sample $(X_i, Y_i)_{i=1}^N \in (\mathcal{X} \times \mathcal{Y})^N$ (i.e. any sample whose joint distribution is invariant under any permutation of the indices).

## 3.1 The progressive mixture rule

### 3.1.1 Estimation of a conditional density

Let $(\mathcal{X}, \mathcal{F})$ and $(\mathcal{Y}, \mathcal{B})$ be two measurable spaces and $(X_i, Y_i)_{i=1}^N \in (\mathcal{X} \times \mathcal{Y})^N$ a finite sequence of random variables distributed according to an *exchangeable* probability measure $\mathbb{P} \in \mathcal{M}_+^1\big((\mathcal{X} \times \mathcal{Y})^N, (\mathcal{F} \otimes \mathcal{B})^{\otimes N}\big)$. This means by definition that for any permutation $\sigma \in \mathfrak{S}_N$ and any measurable set $A \subset (\mathcal{F} \otimes \mathcal{B})^{\otimes N}$,

$$\mathbb{P}\big((X_i, Y_i)_{i=1}^N \in A\big) = \mathbb{P}\big((X_{\sigma(i)}, Y_{\sigma(i)})_{i=1}^N \in A\big).$$

In order to shorten notations, we will write $\mathcal{Z} = \mathcal{X} \times \mathcal{Y}$ and $Z_i = (X_i, Y_i)$.

Our aim will be to estimate the conditional distribution $\mathbb{P}\big(dY_N \,|\, X_1^N, Y_1^{N-1}\big)$ from the observed sample $Z_1^{N-1}$. To this purpose, we will use a countable family of estimators built from:

- A reference measure $\mu \in \mathcal{M}_+^1(\mathcal{Y}, \mathcal{B})$,
- A family of conditional probability density functions

$$\Big\{ q_\theta\big(y_N \,|\, x_N, z_1^K\big) : \theta \in \Theta \Big\},$$

where $(\Theta, \mathcal{T})$ is a measurable space of parameters and where

$$\begin{aligned} \Theta \times (\mathcal{X} \times \mathcal{Y})^{K+1} &\to \mathbb{R}_+ \\ (\theta, x_N, y_N, z_1^k) &\mapsto q_\theta\big(y_N \,|\, x_N, z_1^K\big) \end{aligned}$$

is a measurable function with respect to the sigma algebra $\mathcal{T} \otimes (\mathcal{F} \otimes \mathcal{B})^{K+1}$, such that

$$\int_{\mathcal{Y}} q_\theta(y_N \,|\, x_N, z_1^K)\mu(dy_N) = 1, \qquad \theta \in \Theta,\ x_N \in \mathcal{X},\ z_1^K \in \mathcal{Z}^K.$$

To avoid dealing with integrability issues, we will assume that there is a positive constant $\epsilon$ such that

$$\epsilon \leq q_\theta(y_N \,|\, x_N,\, z_1^K) \leq \epsilon^{-1}, \qquad (z_N,\, z_1^K) \in \mathcal{Z}^{K+1}.$$

This hypothesis is not really a restriction since the bounds we will prove will not depend on the value of $\epsilon$ and since the estimators $q_\theta$ are built by the statistician at his convenience (as opposed to the unknown distribution of observations $\mathbb{P}$).

The conditional likelihoods $\prod_{i=K+1}^{N-1} q_\theta\big(Y_i \,|\, X_i, Z_1^K\big)$ will be used to build a posterior distribution $\rho(d\theta)$ on the parameter space for which $\mathbb{E}_{\rho(d\theta)}\Big[q_\theta\big(Y_N \,|\, X_N, Z_1^K\big)\Big]\mu(dY_N)$ is almost as good an approximation of $\mathbb{P}\big(dY_N \,|\, X_N, Z_1^K\big)$ as the best estimator $q_\theta(Y_N \,|\, X_N, Z_1^K)\mu(dY_N)$, in the sense that

$$\begin{aligned} &- \mathbb{E}_{\mathbb{P}(dZ_1^N)}\Big\{\log\Big[q_\rho\big(Y_N \,|\, X_N, Z_1^K\big)\Big]\Big\} \\ &\qquad \leq \inf_{\theta \in \Theta} -\mathbb{E}_{\mathbb{P}(dZ_1^N)}\Big\{\log\Big[q_\theta\big(Y_N \,|\, X_N, Z_1^K\big)\Big]\Big\} + \gamma(\theta, N), \quad (3.1.1) \end{aligned}$$

where $\gamma(\theta, N)$ is a *universal* bound, that is a bound which does not depend on $\mathbb{P}$. Let us remark that equation (3.1.1) has a corollary expressed in terms of the Kullback divergence function:

$$\begin{aligned} &\mathbb{E}_{\mathbb{P}(dZ_1^{N-1}, dX_N)}\Big\{\mathcal{K}\big[\mathbb{P}\big(dY_N \,|\, X_N, Z_1^{N-1}\big), q_\rho\big(Y_N \,|\, X_N, Z_1^K\big)\mu(dY_N)\big]\Big\} \\ &\leq \inf_{\theta \in \Theta} -\mathbb{E}_{\mathbb{P}(dZ_1^{N-1}, dX_N)}\Big\{\mathcal{K}\big[\mathbb{P}\big(dY_N \,|\, X_N, Z_1^{N-1}\big), q_\theta\big(Y_N \,|\, X_N, Z_1^K\big)\mu(dY_N)\big]\Big\} \\ &\qquad\qquad + \gamma(\theta, N). \quad (3.1.2) \end{aligned}$$

However, this latter equation is slightly weaker than the former: indeed, in the case when $P\big(dY_N \,|\, X_N, Z_1^{N-1}\big)$ is not absolutely continuous with respect to the reference measure $\mu(dY_N)$ on a set of positive measure under $\mathbb{P}$, the two members of equation (3.1.2) are equal to $+\infty$, whereas equation (3.1.1) still makes sense. Moreover, to give a rigorous meaning to (3.1.2), one has to assume that $P\bigl(dY_N \,|\, X_N, Z_1^{N-1}\bigr)$ is a regular conditional probability distribution and that $\mathcal{B}$ is generated by a countable basis, these conditions being for instance satisfied if $(\mathcal{X}, \mathcal{F})$ and $(\mathcal{Y}, \mathcal{B})$ are Polish spaces with their Borel sigma algebras. We refer to the appendix 1.5.4.1.7 for more details on the existence and integrability of conditional Kullback divergence functions.

**Definition 3.1.1.** Let us consider a random probability measure depending on the observation points (often called the design) $(X_1, \dots, X_N)$ in a measurable and exchangeable way, namely a conditional probability $\pi(d\theta \,|\, X_1^N)$ such that

- For any $x_1^N \in \mathcal{X}^N$,

$$\begin{aligned} \mathcal{T} &\to \mathbb{R}_+ \\ A &\mapsto \pi(A \,|\, x_1^N) \end{aligned}$$

  is a probability measure on $(\Theta, \mathcal{B})$;
- For any $A \in \mathcal{T}$

$$\begin{aligned} \mathcal{X}^N &\to \mathbb{R}_+ \\ (x_1, \dots, x_N) &\mapsto \pi(A \,|\, x_1^N) \end{aligned}$$

  is a measurable and exchangeable function.

Let us define the progressive mixture posterior distribution by the formula

$$\rho(d\theta) = \frac{1}{N-K} \sum_{M=K}^{N-1} \frac{\displaystyle\prod_{i=K+1}^{M} q_\theta(Y_i \,|\, X_i, Z_1^K)}{\mathbb{E}_{\pi(d\theta')}\left[\displaystyle\prod_{i=K+1}^{M} q_{\theta'}(Y_i \,|\, X_i, Z_1^K)\right]} \pi(d\theta \,|\, X_1^N),$$

where by convention $\prod_{i=K+1}^{K} q_\theta(\dots) = 1$. Although we did not make it explicit in the notations, $\rho(d\theta)$ is indeed a conditional probability distribution depending on $(Y_1^{N-1}, X_1^N)$, and it would be more accurate, although somehow cumbersome to write it as $\rho(d\theta \,|\, X_1^N, Y_1^{N-1})$.

In this definition of progressive mixture, we described a situation where the observations are gathered into two pools: a *training* set, on which the base estimators are computed, and a *validation* set, which is used to decide how to combine them.

The interest of the progressive mixture posterior comes from the following theorem, in which we put

$$q_\rho(Y_N \,|\, X_N, Z_1^K) \overset{\text{def}}{=} \mathbb{E}_{\rho(d\theta \,|\, Z_1^{N-1}, X_N)}\Bigl[q_\theta(Y_N \,|\, X_N, Z_1^K)\Bigr].$$

**Theorem 3.1.1 (Barron).** *Under the previous hypotheses, for any* exchangeable *probability measure* $\mathbb{P} \in \mathcal{M}_+^1((\mathcal{X} \times \mathcal{Y})^N)$,

$$- \mathbb{E}_{\mathbb{P}(dZ_1^N)}\Bigl\{\log\bigl[q_\rho(Y_N \,|\, X_N, Z_1^K)\bigr]\Bigr\} \\ \leq -\frac{1}{N-K}\mathbb{E}_{P(dZ_1^N)}\left\{\log\left[\mathbb{E}_{\pi(d\theta)}\left(\prod_{i=K+1}^{N} q_\theta(Y_i \,|\, X_i, Z_1^K)\right)\right]\right\}. \quad (3.1.3)$$

**Corollary 3.1.1.** *Due to this theorem, it is possible to apply to the progressive mixture estimator all the upper bounds established for mixtures of coding distributions in the previous chapters. For instance:*

$$- \mathbb{E}_{\mathbb{P}(dZ_1^N)}\Bigl\{\log\bigl[q_\rho(Y_N \,|\, X_N, Z_1^K)\bigr]\Bigr\} \\ \leq \inf_{\theta \in \Theta} -\mathbb{E}_{\mathbb{P}(dZ_1^N)}\Bigl\{\log\bigl[q_\theta(Y_N \,|\, X_N, Z_1^K)\bigr]\Bigr\} - \frac{\mathbb{E}_{\mathbb{P}(dX_1^N)}\bigl\{\log[\pi(\{\theta\})]\bigr\}}{N-K}.$$

*Proof.* To shorten notations, we have omitted in the following computations the dependence in $Z_1^K$. Concavity of the logarithm function and exchangeability yield

$$- \mathbb{E}_{\mathbb{P}(dZ_1^N)}\Bigl\{\log\bigl[q_\rho(Y_N \,|\, X_N)\bigr]\Bigr\} \\ \leq -\frac{1}{N-K}\sum_{M=K}^{N-1}\mathbb{E}_{\mathbb{P}(dZ_1^N)}\left\{\log\left[\frac{\mathbb{E}_{\pi(d\theta)}\left\{q_\theta(Y_N \,|\, X_N)\prod_{i=K+1}^{M} q_\theta(Y_i \,|\, X_i)\right\}}{\mathbb{E}_{\pi(d\theta)}\left\{\prod_{i=K+1}^{M} q_\theta(Y_i \,|\, X_i)\right\}}\right]\right\} \\ = -\frac{1}{N-K}\sum_{M=K}^{N-1}\mathbb{E}_{\mathbb{P}(dZ_1^N)}\left\{\log\left[\frac{\mathbb{E}_{\pi(d\theta)}\left\{\prod_{i=K+1}^{M+1} q_\theta(Y_i \,|\, X_i)\right\}}{\mathbb{E}_{\pi(d\theta)}\left\{\prod_{i=K+1}^{M} q_\theta(Y_i \,|\, X_i)\right\}}\right]\right\} \\ = -\frac{1}{N-K}\mathbb{E}_{\mathbb{P}(dZ_1^N)}\left\{\log\left[\mathbb{E}_{\pi(d\theta)}\left(\prod_{i=K+1}^{N} q_\theta(Y_i \,|\, X_i)\right)\right]\right\}.$$

□

### 3.1.2 Some variants of the progressive mixture estimator

The progressive mixture posterior distribution is heavy to compute, inasmuch as it requires the computation of a sum of $N-K$ terms. It is possible to avoid this inconvenience in two different ways: one is to use a dichotomic scheme.

**Definition 3.1.2 (Dichotomic progressive mixture).** Let us choose some integers $K$ and $N$ such that $N - K = 2^r$. Let us then define the dichotomic progressive mixture posterior distribution as

$$
\begin{aligned}
\alpha^r(d\theta) &= \pi(d\theta), \\
&\vdots \\
\alpha^j(d\theta) &= \frac{1}{2}\left[\frac{\displaystyle\prod_{i=K+2^j}^{K+2^{j+1}-1} q_\theta(Y_i \mid X_i)\alpha^{j+1}(d\theta)}{\mathbb{E}_{\alpha^{j+1}(d\theta')}\left(\displaystyle\prod_{i=K+2^j}^{K+2^{j+1}-1} q_{\theta'}(Y_i \mid X_i)\right)} + \alpha^{j+1}(d\theta)\right], \\
&\vdots \\
\alpha^0(d\theta) &= \frac{1}{2}\left[\frac{q_\theta(Y_{K+1} \mid X_{K+1})\alpha^1(d\theta)}{\mathbb{E}_{\pi(d\theta')}\Bigl(q_{\theta'}(Y_{K+1} \mid X_{K+1})\Bigr)} + \alpha^1(d\theta)\right], \\
\rho(d\theta) &\overset{\text{def}}{=} \alpha^0(d\theta).
\end{aligned}
$$

**Definition 3.1.3 (Randomized progressive mixture).** Let $\mu$ be the uniform measure on the set $[K, N-1] \cup \mathbb{N}$ of integers lying between $K$ and $N-1$. Let $(M_j)_{j=1}^r \times (Z_i)_{i=1}^N$ be the canonical process on $([K,\, N-1] \cap \mathbb{N})^r \times \mathcal{Z}^N$. Let us define the randomised progressive mixture posterior distribution as

$$
\rho(d\theta) = \frac{1}{r}\sum_{j=1}^{r} \frac{\displaystyle\prod_{i=K+1}^{M_j} q_\theta(Y_i \mid X_i)}{\mathbb{E}_{\pi(d\theta')}\left(\displaystyle\prod_{i=K+1}^{M_j} q_{\theta'}(Y_i \mid X_i)\right)} \pi(d\theta).
$$

The proof of the following proposition is left as an exercise.

**Proposition 3.1.1.** *The dichotomic progressive mixture posterior distribution satisfies equation* (3.1.3)*. The randomised progressive mixture posterior distribution satisfies*

$$
\begin{aligned}
&- \mathbb{E}_{\mu(dM_1^r)}\Bigl\{\mathbb{E}_{\mathbb{P}(dZ_1^N)}\Bigl\{\log\Bigl[q_\rho(Y_N \mid X_N, Z_1^K)\Bigr]\Bigr\}\Bigr\} \\
&\leq -\frac{1}{N-K}\mathbb{E}_{P(dZ_1^N)}\left\{\log\left[\mathbb{E}_{\pi(d\theta)}\left(\prod_{i=K+1}^{N} q_\theta(Y_i \mid X_i, Z_1^K)\right)\right]\right\}. \qquad (3.1.4)
\end{aligned}
$$

## 3.2 Estimating a Bernoulli random variable

In the previous section, we described a two step scheme which uses a pool of base estimators $q_\theta(Y_N \mid X_N, Z_1^K)$. In some cases, nearly optimal results can be derived for a one step scheme, which corresponds to choosing $K = 0$ and mixing fixed (conditional) distributions. In other situations, the one step scheme does not yield the optimal convergence speed. For instance, mixing context trees with a one step scheme would provide a convergence speed of the Kullback divergence off from the optimal by a $\log(N)$ factor. A two step scheme can mend this imperfection, as long as appropriate base estimators are available. These base estimators have to satisfy sharp enough oracle inequalities. We are going to provide some in the important case of the estimation of a distribution on a finite set (often called a Bernoulli distribution).

### 3.2.1 The Laplace estimator

**Definition 3.2.1.** Let $E$ be a finite set with $d+1$ elements. Let us use the notations of section 1.4 of chapter 1. The Laplace estimator is the Bayesian estimator built from the uniform distribution $\frac{1}{\sqrt{d+1}}\lambda(d\theta)$ on the simplex. It can be expressed in the two following ways:

$$\begin{aligned} Q_\lambda(x_N \mid x_1^{N-1}) &= \frac{(N-1)\bar{P}_{x_1^{N-1}}(x_N) + 1}{N+d} \\ &= \frac{N\bar{P}_{x_1^N}(x_N)}{N+d}. \end{aligned}$$

(These two expressions of the Laplace estimator have an interest of their own: the former uses the empirical distribution of the observed sample, whereas it is possible to guess from the latter the oracle inequality stated in the next proposition.)

**Proposition 3.2.1.** *For any exchangeable probability distribution* $\mathbb{P} \in \mathcal{M}_+^1(E^N)$,

$$-\mathbb{E}_{\mathbb{P}(dX_1^N)}\Big\{\log\big[Q_\lambda(X_N \mid X_1^{N-1})\big]\Big\} \leq \inf_{\theta \in \mathcal{M}_+^1(E)} -\mathbb{E}_{\mathbb{P}(dX_N)}\Big\{\log\big[\theta(X_N)\big]\Big\} + \frac{d}{N}.$$

*Proof.*

$$\begin{aligned} &-\mathbb{E}_{\mathbb{P}(dX_1^N)}\Big\{\log\big[Q_\lambda(X_N \mid X_1^{N-1})\big]\Big\} \\ &\qquad = -\mathbb{E}_{\mathbb{P}(dX_1^N)}\left\{\log \frac{N\bar{P}_{X_1^N}(X_N)}{N+d}\right\} \\ &\qquad = -\mathbb{E}_{\mathbb{P}(dX_1^N)}\Big\{\log\big[\bar{P}_{X_1^N}(X_N)\big]\Big\} + \log\left(1 + \frac{d}{N}\right) \end{aligned}$$

$$= -\mathbb{E}_{\mathbb{P}(dX_1^N)} \left\{ \frac{1}{N} \sum_{i=1}^N \log\bigl[\bar{P}_{X_1^N}(X_i)\bigr] \right\} + \log\left(1 + \frac{d}{N}\right)$$

$$= \mathbb{E}_{\mathbb{P}(dX_1^N)} \left\{ \inf_{\theta \in \mathcal{M}_+^1(E)} -\frac{1}{N} \sum_{i=1}^N \log\bigl[\theta(X_i)\bigr] \right\} + \log\left(1 + \frac{d}{N}\right)$$

$$\leq \inf_{\theta \in \mathcal{M}_+^1(E)} -\mathbb{E}_{\mathbb{P}(dX_1^N)} \left\{ \frac{1}{N} \sum_{i=1}^N \log\bigl[\theta(X_i)\bigr] \right\} + \frac{d}{N}$$

$$= \inf_{\theta \in \mathcal{M}_+^1(E)} -\mathbb{E}_{\mathbb{P}(dX_1^N)} \Bigl\{ \log\bigl[\theta(X_N)\bigr] \Bigr\} + \frac{d}{N}.$$

□

### 3.2.2 Some adaptive Laplace estimator

The Laplace estimator is not well suited to the case when the support of the distribution to be estimated is much smaller than the base space $E$ itself. It is then possible to get better results with a modified estimator which adapts to the size of the support which can be guessed from the observed data.

**Definition 3.2.2.** In the previously described context, let us define the adaptive Laplace estimator as

$$\widetilde{Q}(X_N \mid X_1^{N-1}) \overset{\text{def}}{=} \frac{(N-1)\bar{P}_{X_1^{N-1}}(X_N) + \dfrac{1}{d+1}}{N} = \bar{P}_{X_1^N}(X_N) - \frac{d}{(d+1)N}.$$

We will use the following notations:

$$\begin{aligned} \bar{\mathcal{S}} &= \operatorname{supp}\bigl[\bar{P}_{X_1^N}\bigr] \\ S &= |\bar{\mathcal{S}}|, \\ \bar{\mathcal{S}}' &= \bigl\{ y \in E : N\bar{P}_{X_1^N}(y) = 1 \bigr\}, \\ S' &= |\bar{\mathcal{S}}'|. \end{aligned}$$

**Proposition 3.2.2.** *For any exchangeable distribution* $\mathbb{P} \in \mathcal{M}_+^1(\mathcal{X}^N)$,

$$\mathbb{E}_{\mathbb{P}(dX_1^N)} \Bigl\{ -\log\bigl[\widetilde{Q}(X_N \mid X_1^{N-1})\bigr] \Bigr\} \leq \inf_{\theta \in \mathcal{M}_+^1(E)} -\mathbb{E}_{\mathbb{P}(dX_1^N)} \Bigl\{ \log\bigl[\theta(X_N)\bigr] \Bigr\} + \frac{\gamma(N, \mathbb{P})}{N},$$

*where*

$$\gamma(N,\mathbb{P}) = -\mathbb{E}_{\mathbb{P}(dX_1^N)}\left\{\sum_{x\in\bar{\mathcal{S}}} N\bar{P}_{X_1^N}(x)\log\left(1-\frac{d}{(d+1)N\bar{P}_{X_1^N}(x)}\right)\right\}$$

$$\leq \bigl|\operatorname{supp}\bigl[\mathbb{P}(dX_N)\bigr]\bigr|\frac{d}{d+1}\mathbb{E}_{\mathbb{P}(dX_1^N)}\left\{\frac{1}{|\bar{\mathcal{S}}|}\sum_{x\in\bar{\mathcal{S}}}\left(1-\frac{d}{(d+1)N\bar{P}_{X_1^N}(x)}\right)^{-1}\right\}.$$

*Remark* 3.2.1. The reminder term $\gamma(N,\mathbb{P})$ can also be upper bounded by

$$\gamma(N,\mathbb{P}) \leq \mathbb{E}_{\mathbb{P}(dX_1^N)}\Bigl\{S'\log(d+1) + 2(S-S')\Bigr\}.$$

Moreover, when $\mathbb{P}(dX_1^N) = \theta^{\otimes N}(dX_1^N)$ is a product distribution, it is easily seen from the law of large numbers and Lebesgue's dominated convergence theorem that (as $N\bar{P}_{X_1^N}(x) \geq 1$, $x \in \bar{\mathcal{S}}$,)

$$\lim_{N\to+\infty}\gamma(N,\theta^{\otimes N}) = \frac{d}{d+1}\bigl|\operatorname{supp}(\theta)\bigr|.$$

*Proof.* From the definition of the adaptive Laplace estimator and the exchangeability of the distribution $\mathbb{P}$ it is straightfoward to see that

$$\mathbb{E}_{\mathbb{P}(dX_1^N)}\Bigl\{-\log\bigl[\widetilde{Q}(X_N\,|\,X_1^{N-1})\bigr]\Bigr\}$$

$$= \mathbb{E}_{\mathbb{P}(dX_1^N)}\left\{\sum_{x\in\bar{\mathcal{S}}} -\bar{P}_{X_1^N}(x)\log\left[\bar{P}_{X_1^N}(x) - \frac{d}{(d+1)N}\right]\right\}$$

$$= \mathbb{E}_{\mathbb{P}(dX_1^N)}\left\{\sum_{x\in\bar{\mathcal{S}}} -\bar{P}_{X_1^N}(x)\log\bigl[\bar{P}_{X_1^N}(x)\bigr]\right\}$$

$$+ \mathbb{E}_{\mathbb{P}(dX_1^N)}\left\{\sum_{x\in\bar{\mathcal{S}}} -\bar{P}_{X_1^N}(x)\log\left[1-\frac{d}{(d+1)N\bar{P}_{X_1^N}(x)}\right]\right\}$$

$$\leq \mathbb{E}_{\mathbb{P}(dX_1^N)}\left\{\inf_{\theta\in\mathcal{M}_+^1(E)}\sum_{x\in\bar{\mathcal{S}}} -\bar{P}_{X_1^N}(x)\log\bigl[\theta(x)\bigr]\right\} + \frac{\gamma(N,\mathbb{P})}{N}$$

$$\leq \inf_{\theta\in\mathcal{M}_+^1(E)} -\mathbb{E}_{\mathbb{P}(dX_1^N)}\Bigl\{\log\bigl[\theta(X_N)\bigr]\Bigr\} + \frac{\gamma(N,\mathbb{P})}{N}.$$

The upper bounds for $\gamma(N,\mathbb{P})$ stated in the proposition and in the following remark can be deduced from the inequality $-\log(1-r) = \log\bigl[1+r/(1-r)\bigr] \leq r/(1-r)$. □

## 3.3 Adaptive histograms

We will here deal with the estimation of a probability measure on an arbitrary measurable space $(\mathcal{X},\mathcal{F})$ by histograms. In order to describe the family of models to be used, we need a reference measure $\mu \in \mathcal{M}_+^1(\mathcal{X})$ and a countable family $\mathcal{S}$ of partitions of $\mathcal{X}$. Let us assume that all these partitions are composed of measurable subsets of $\mathcal{X}$ with a strictly positive finite measure under $\mu$. For any partition $S \in \mathcal{S}$, let us consider the set of continuous parameters $\Theta_S \overset{\text{def}}{=} \mathcal{M}_+^1(S)$ and let us define the following probability densities with respect to $\mu$ :

$$q_{S,\theta}(x) \overset{\text{def}}{=} \sum_{I \in S} \frac{\theta(I)}{\mu(I)} \mathbb{1}(x \in I).$$

For a fixed partition $S \in \mathcal{S}$, it is possible to consider the Laplace estimator

$$q_S(X_N \,|\, X_1^K) \overset{\text{def}}{=} \sum_{I \in S} \left( \frac{K \bar{P}_{X_1^K}(I) + 1}{K + |S|} \right) \left( \frac{\mathbb{1}(X_N \in I)}{\mu(I)} \right).$$

As $\log\bigl[q_{S,\theta}(x)\bigr]$ is equal to $\log\bigl[\theta(I)\bigr]$ up to a constant term independent of $\theta$, we can apply the results proved for the Laplace estimator of a distribution on the finite sample $S$ to show that for any exchangeable probability measure $\mathbb{P} \in \mathcal{M}_+^1(\mathcal{X}^N)$,

$$-\mathbb{E}_{\mathbb{P}(dX_1^N)}\Bigl\{\log\bigl[q_S(X_N \,|\, X_1^K)\bigr]\Bigr\} \leq \inf_{\theta_S \in \Theta_S} -\mathbb{E}_{\mathbb{P}(dX_1^N)}\Bigl\{\log\bigl[q_{S,\theta}(X_N)\bigr]\Bigr\} + \frac{|S| - 1}{K + 1}.$$

For any prior distribution on the partitions $\pi \in \mathcal{M}_+^1(\mathcal{S})$ the progressive mixture estimator $q_\rho(X_N \,|\, X_1^{N-1})$ thus satisfies

$$\begin{aligned} &- \mathbb{E}_{\mathbb{P}}\Bigl\{\log\bigl[q_\rho(X_N \,|\, X_1^{N-1})\bigr]\Bigr\} \\ &\quad \leq \inf_{S \in \mathcal{S}} \left\{ \inf_{\theta \in \Theta_S} -\mathbb{E}_{\mathbb{P}}\Bigl\{\log\bigl[q_{S,\theta}(X_N)\bigr]\Bigr\} + \frac{|S| - 1}{K + 1} - \frac{1}{N - K} \log\bigl[\pi(S)\bigr] \right\}. \end{aligned}$$

For example, if the family of partitions $\mathcal{S}$ is "small enough", so that for some constant $\beta > 0$, $\sum_{S \in \mathcal{S}} \exp\bigl[-\beta(|S| - 1)\bigr] = Z(\beta) < +\infty$, it is possible to choose for a prior distribution $\pi(S) = Z(\beta)^{-1} \exp\bigl[-\beta(|S| - 1)\bigr]$ and to set $K = \frac{N - \beta}{1 + \beta}$ (increasing slightly if necessary the value of $\beta$ to obtain an integer).

**Proposition 3.3.1.** *With the previous notations and choice of $\pi$ and $K$,*

$$\begin{aligned} &- \mathbb{E}_{\mathbb{P}}\Bigl\{\log\bigl[q_\rho(X_N \,|\, X_1^{N-1})\bigr]\Bigr\} \\ &\quad \leq \inf_{S \in \mathcal{S}} \left\{ \inf_{\theta \in \Theta_S} -\mathbb{E}_{\mathbb{P}}\Bigl\{\log\bigl[q_{S,\theta}(X_N)\bigr]\Bigr\} \right. \\ &\qquad\qquad \left. + (1 + \beta) \frac{2(|S| - 1) + \frac{\log[Z(\beta)]}{\beta}}{N + 1} \right\}. \end{aligned}$$

## 3.4 Some remarks on approximate Monte-Carlo computations

In this section, we want to point out briefly how stochastic sampling and optimisation techniques can be used to compute the progressive mixture estimator. This will be useful in the case when no fast deterministic algorithm is available.

An approximate computation of $\mathbb{E}_{\rho(dm)} q_m^K(x_N \,|\, x_1^K)$ is legitimate as soon as the computed value $\mathbb{E}_{\tilde{\rho}(dm)} q_m^K(x_N \,|\, x_1^K)$ satisfies

$$\sup_{x_1^N \in \mathcal{X}^N} \log \frac{\mathbb{E}_{\rho(dm)} q_m^K(x_N \,|\, x_1^K)}{\mathbb{E}_{\tilde{\rho}(dm)} q_m^K(x_N \,|\, x_1^K)} \leq \frac{c}{N},$$

for some positive constant $c$ of "moderate" size. As we are discussing practical matters here, we will assume that the index set $\mathfrak{M}$ of the primary estimators is finite.

The progressive mixture estimator can be written as a mean of expectations with respect to Gibbs measures of decreasing temperatures: let us introduce some notations first. Let $R_{x_{K+1}^{K+M}} = \frac{1}{M} \sum_{n=K+1}^{K+M} \delta_{x_n}$ be the empirical distribution of $x_{K+1}^{K+M}$, let $L(R_{x_{K+1}^{K+M}}, q_m^K(\cdot \,|\, x_1^K))$ be the "Shannon" relative entropy $L(R,q) = E_R(-\log q)$, and let

$$\rho_M(m) = \frac{\pi(m) \exp\Big(-ML(R_{x_{K+1}^{K+M}}, q_m^K(\cdot \,|\, x_1^K))\Big)}{\sum_{m' \in \mathfrak{M}} \pi(m') \exp\Big(-ML(R_{x_{K+1}^{K+M}}, q_{m'}^K(\cdot \,|\, x_1^K))\Big)}$$

be the Gibbs distribution with reference measure $\pi$, energy function

$$m \longmapsto L(R_{x_{K+1}^{K+M}}, q_m^K(\cdot \,|\, x_1^K))$$

and temperature $1/M$. Then

$$\mathbb{E}_{\rho(dm)} q_m^K(x_N \,|\, x_1^{N-1}) = \frac{1}{N-K} \sum_{M=0}^{N-K-1} \sum_{m \in \mathfrak{M}} \rho_M(m) q_m^K(x_N \,|\, x_1^K),$$

where it should be understood as previously that $\rho_0 = \pi$. When the sample distribution is a product measure $\mathbb{P}_N = \mathbb{P}^{\otimes N}$ and $\log q_m^K(\cdot \,|\, x_1^K)$ is in $\mathbb{L}_1(\mathbb{P})$, the "empirical Shannon entropies" $L(R_{X_{K+1}^{K+M}}, q_m^K(\cdot \,|\, x_1^K))$ tends to $L(\mathbb{P}, q_m^K(\cdot \,|\, x_1^K))$ when $M$ tends to infinity. Therefore for large values of $M$, the Gibbs measures $\rho_M$ have almost the same energy function. This gives the idea to use some kind simulated annealing algorithm with stepwise constant temperature schedule to compute $\mathbb{E}_{\rho(dm)} q_m^K(x_N \,|\, x_1^K)$. For this,

- we need an irreducible reversible Markov matrix
$$\Upsilon : \mathfrak{M} \times \mathfrak{M} \mapsto [0,1]$$
with invariant measure $\pi$,
- we introduce the thermalized Markov matrices
$$\Upsilon_M(m,m') = \Upsilon(m,m') \exp\Bigl(-M\bigl(L_M(m') - L_M(m)\bigr)^+\Bigr), \qquad m \neq m' \in \mathfrak{M},$$
where we have used the abbreviation $L_M(m) = L\bigl(R_{x_{K+1}^{K+M}}, q_m^K(\cdot \,|\, x_1^K)\bigr)$,
- we define on the estimator index set $\mathfrak{M}$ the stepwise homogeneous Markov chain $(Y_k)_{k\in\mathbb{N}}$ with transitions
$$\mathbb{P}(Y_k = m' \,|\, Y_{k-1} = m) = \Upsilon_M(m,m'),$$
$$\tau M < k \leq \tau(M+1), 0 \leq M < N-K,$$
where $\tau$ is the step length,
- we compute the approximate Monte-Carlo posterior distribution
$$\tilde{\rho}(m) = \frac{1}{\tau(N-K)} \sum_{k=1}^{\tau(N-K)} \delta_{Y_k},$$
(where $\delta_{Y_k}$ is the Dirac mass at estimator index $Y_k \in \mathfrak{M}$).

It is easy to see that
$$\lim_{\tau \to +\infty} \mathbb{E}_{\tilde{\rho}(dm)} q_m^K(x_N \,|\, x_1^K) = \mathbb{E}_{\rho(dm)} q_m^K(x_N \,|\, x_1^K),$$
by a simple application of the ergodic theorem for homogeneous Markov chains. Discussing the speed of convergence of this limit would drive us too far from the main scope of this chapter. The reader can find some results about the convergence speed of simulated annealing algorithms with a piecewise constant temperature schedule in [28] and in [19], although these two studies are focused on the convergence in distribution of the marginals towards states of low energy and not on the ergodic properties of the process. We refer also to 7.7.7 and to [69] for the convergence of homogeneous Markov chains and to 7.7.8, [18] and [75] for a study of the optimal convergence rate of simulated annealing algorithms.

## 3.5 Selection and aggregation : a toy example pointing out some differences

We give this example to show that a selection rule, that selects a single primary estimator, cannot be substituted to the progressive mixture rule in theorem 3.1.1. Some supplementary assumptions on the structure of the family of estimators have to be added.

We consider a sample $(X_1^N) \in \{0,1\}^N$ of binary variables, and two simple models containing only one distribution. For any real number $\lambda \in [0,1]$, let $B_\lambda$ be the Bernoulli distribution with parameter $\lambda : B_\lambda(1) = 1 - B_\lambda(0) = \lambda$. Let us consider two models, the first one containing only the distribution $B_{1/4}$ and the second one only the distribution $B_{3/4}$. The best estimators for these two models are obviously the constant estimators

$$Q_{-1}^K(\cdot \mid x_1^K) = Q_{-1}(\cdot) = B_{1/4},$$
$$Q_{+1}^K(\cdot \mid x_1^K) = Q_{+1}(\cdot) = B_{3/4}.$$

Consider the two possible sample distributions

$$\mathbb{P}_i^{\otimes N} = B_{1/2+i\alpha/\sqrt{N}}^{\otimes N}, \quad i \in \{-1,+1\},$$

where $\alpha$ is some positive parameter to be chosen afterwards. We have

$$\mathcal{K}(\mathbb{P}_i, B_{1/4}) - \mathcal{K}(\mathbb{P}_i, B_{3/4}) = i\frac{2\alpha\log 3}{\sqrt{N}},$$

and, according to Pinsker's inequality (see [56, 29, 66]),

$$\begin{aligned}\frac{1}{2}\Big(\mathrm{Var}\left(\mathbb{P}_{+1}^{\otimes N}, \mathbb{P}_{-1}^{\otimes N}\right)\Big)^2 &\leq \mathcal{K}\left(\mathbb{P}_{+1}^{\otimes N}, \mathbb{P}_{-1}^{\otimes N}\right)\\ &= N\mathcal{K}\left(\mathbb{P}_{+1}, \mathbb{P}_{-1}\right)\\ &= 2\alpha\sqrt{N}\log\left(1 + \frac{2\alpha}{\sqrt{N}\left(1 - \frac{\alpha}{\sqrt{N}}\right)}\right)\\ &\leq \frac{4\alpha^2}{1 - \frac{\alpha}{\sqrt{N}}}\end{aligned}$$

Let $\hat{m}(X_1^N) \in \{-1,+1\}$ be any selection rule between $Q_{-1}$ and $Q_{+1}$. The following inequalities are satisfied :

$$\begin{aligned}\max_{i\in\{-1,+1\}}&\left(\mathbb{E}_{\mathbb{P}_i^{\otimes N}}\mathcal{K}(\mathbb{P}_i, Q_{\hat{m}(X_1^N)}) - \min_{j\in\{-1,+1\}}\mathcal{K}(\mathbb{P}_i, Q_j)\right)\\ &= \max_{i\in\{-1,+1\}}\mathbb{E}_{\mathbb{P}_i^{\otimes N}}\frac{\left(1 - i\,\hat{m}(X_1^N)\right)\alpha\log 3}{\sqrt{N}}\\ &\geq \frac{1}{2}\sum_{i\in\{-1,+1\}}\mathbb{E}_{\mathbb{P}_i^{\otimes N}}\frac{\left(1 - i\,\hat{m}(X_1^N)\right)\alpha\log 3}{\sqrt{N}}\\ &\geq \frac{\alpha\log 3}{\sqrt{N}}\sum_{x_1^N\in\{-1,+1\}^N}\min_{i\in\{-1,+1\}}\mathbb{P}_i^{\otimes N}(x_1^N)\end{aligned}$$

$$= \frac{\alpha \log 3}{\sqrt{N}} \left( 1 - \frac{1}{2} \operatorname{Var} \left( \mathbb{P}_{-1}^{\otimes N}, \mathbb{P}_{+1}^{\otimes N} \right) \right)$$
$$\geq \frac{\alpha \log 3}{\sqrt{N}} \left( 1 - \frac{\alpha \sqrt{2}}{\sqrt{1 - \frac{\alpha}{\sqrt{N}}}} \right).$$

When $\alpha = 1/4$, we get that

$$\max_{i \in \{-1,+1\}} \left( \mathbb{E}_{\mathbb{P}_i^{\otimes N}} \mathcal{K}\big(\mathbb{P}_i, Q_{\hat{m}(X_1^N)}\big) - \min_{j \in \{-1,+1\}} \mathcal{K}(\mathbb{P}_i, Q_j) \right) \geq \frac{\log 3}{8\sqrt{N}}.$$

This is to be compared with theorem 3.1.1 applied to a sample of size $N+1$ with $\pi(-1) = \pi(+1) = 1/2$ and $K = 0$, which gives

$$\sup_{\mathbb{P} \in \mathcal{M}_+^1(\{0,1\})} \left( \mathbb{E}_{\mathbb{P}^{\otimes N}(dX_1^N)} \mathcal{K}\big(\mathbb{P}, \mathbb{E}_{\rho(dm)} Q_m\big) - \inf_{j \in \{-1,+1\}} \mathcal{K}(\mathbb{P}, Q_j) \right) \leq \frac{\log 2}{N+1}.$$

The reason why this "counter example" works is of course that the map

$$B_\lambda \longmapsto \inf \left\{ \mathcal{K}(B_\lambda, B_{1/4}), \mathcal{K}(B_\lambda, B_{3/4}) \right\}$$

is not differentiable at its extremal point $\lambda = 1/2$, where it has two non zero directional first derivatives. In other words if a true selection rule gave an increase in the risk of order $1/N$, it would be possible to estimate $\lambda$ with a precision of order $1/N$.

This example may seem artificial, since we left a "gap" on purpose in the model we used. However, it resembles a lot of realistic non-convex situations. Indeed, when the data to analyse are high dimensional and complex (for example in applications to image or speech analysis), one cannot help using a large family of models with many gaps between them (because it is in practice impossible to test enough models to fill in all these gaps, even if it is theoretically easy to provide huge dense families of models). What we mean by a gap here is a situation where the true sample distribution is at a distance of order one from the model and where the minimum distance between the true distribution and the model is achieved (or almost achieved) at several model distributions that are themselves far away from each other.

## 3.6 Least square regression

In this section we will show how to apply the progressive mixture rule to least square regression. Let us consider a measurable space $(\mathcal{X}, \mathcal{B})$ and let $(X_n, Y_n)_{n=1}^N \overset{\text{def}}{=} (Z_n)_{n=1}^N$ be the canonical process on $(\mathcal{X} \times \mathbb{R})^N \overset{\text{def}}{=} \mathcal{Z}^N$. Let $\mathbb{P}_N$ be an exchangeable probability distribution on $\left( (\mathcal{X} \times \mathbb{R})^N, (\mathcal{B} \otimes \mathfrak{S})^{\otimes N} \right)$, where $\mathfrak{S}$ is the Borel sigma algebra on the real line.

Consider a family of regression estimators $\{f_m^K : \mathcal{X} \times \mathcal{Z}^K \to \mathbb{R}; m \in \mathfrak{M}, K \in \mathbb{N}\}$, (that are supposed to be measurable functions) and the least square aggregation problem of finding a posterior distribution $\rho(dm) \in \mathcal{M}_1^+(\mathfrak{M})$, such that

$$\mathbb{E}_{\mathbb{P}_N(dZ_1^N)} \left(Y_N - \mathbb{E}_{\rho(dm)} f_m^K(X_N \,|\, Z_1^K)\right)^2 \\ \leq \inf_{m \in \mathfrak{M}} \left( \mathbb{E}_{\mathbb{P}_N(dZ_1^N)} \left(Y_N - f_m^K(X_N \,|\, Z_1^K)\right)^2 + \frac{\gamma(m)}{N} \right).$$

We will solve this problem under the following assumptions :

- The conditional mean of $Y_N$ is bounded. Namely , we assume that there is a constant $B$ such that

$$\left|\mathbb{E}_{\mathbb{P}_N}(Y_N \,|\, X_N, Z_1^{N-1})\right| \leq B, \qquad \mathbb{P}_N \text{ a.s..}$$

When this is the case and $B$ is known, we may assume without loss of generality that the estimators $f_m^K$ are also taken within the same bounds (if it is not the case, we can threshold them without increasing their risk). Therefore we will also assume that for any $m \in \mathfrak{M}$, any $K < N$,

$$|f_m^K(X_N \,|\, Z_1^K)| \leq B, \qquad \mathbb{P}_N \text{ a.s..}$$

In the following we will use the short notation

$$\overline{Y}_N \stackrel{\text{def}}{=} \mathbb{E}_{\mathbb{P}_N}(Y_N \,|\, X_N, Z_1^{N-1}).$$

- The noise has a uniformly bounded exponential moment. More precisely, for any positive constant $\alpha$, we let $M_\alpha \in \mathbb{R} \cup \{+\infty\}$ and $V_\alpha \in \mathbb{R} \cup \{+\infty\}$ be such that for any values of $X_N$ and $Z_1^{N-1}$,

$$\mathbb{E}_{\mathbb{P}_N} \left(\exp\left(\alpha |Y_N - \overline{Y}_N|\right) \,|\, X_N, Z_1^{N-1}\right) \leq M_\alpha, \quad \mathbb{P}_N \text{ a.s.} \qquad (3.6.1)$$

and

$$\frac{\mathbb{E}_{\mathbb{P}_N}\left((Y_N - \overline{Y}_N)^2 e^{\alpha|Y_N - \overline{Y}_N|} \,|\, X_N, Z_1^{N-1}\right)}{\mathbb{E}_{\mathbb{P}_N}\left(e^{\alpha|Y_N - \overline{Y}_N|} \,|\, X_N, Z_1^{N-1}\right)} \leq V_\alpha, \quad \mathbb{P}_N \text{ a.s.} \qquad (3.6.2)$$

and we assume that for some small enough positive $\alpha > 0$ we have $M_\alpha < +\infty$ and $V_\alpha < +\infty$.

We will also use a prior distribution $\pi \in \mathcal{M}_1^+(\mathfrak{M})$, and consider for some positive parameter $\lambda$ to be chosen afterwards the conditional Gaussian densities

$$q_m^K(Y_N \,|\, X_N, Z_1^K) = \left(\frac{\lambda}{2\pi}\right)^{1/2} \exp\left(-\frac{\lambda}{2}(Y_N - f_m^K(X_N \,|\, Z_1^K))^2\right).$$

**Theorem 3.6.1.** *Under the previous hypotheses, for any parameter* $\lambda$ *such that*

$$\lambda^{-1} \geq M_{2\lambda B} \left(17\, B^2 + 3.4\, V_{2\lambda B}\right), \tag{3.6.3}$$

*and in particular for*

$$\lambda = \min \left\{ \frac{\alpha}{2B}, \left( M_\alpha \left(17\, B^2 + 3.4\, V_\alpha\right)\right)^{-1} \right\},$$

*the posterior distribution on* $\mathfrak{M}$

$$\rho(dm) = \frac{1}{N-K} \sum_{M=K}^{N-1} \frac{\displaystyle\prod_{k=K+1}^{M} q_m^K(Y_k \mid X_k, Z_1^K)}{\mathbb{E}_{\pi(dm')} \displaystyle\prod_{k=K+1}^{M} q_{m'}^K(Y_k \mid X_k, Z_1^K)} \pi(dm)$$

*is such that*

$$\begin{aligned}\mathbb{E}_{\mathbb{P}_N(dZ_1^N)} \left(\overline{Y}_N - \mathbb{E}_{\rho(dm)} f_m^K(X_N \mid Z_1^K)\right)^2 \\ \leq \inf_{m \in \mathfrak{M}} \left( \mathbb{E}_{\mathbb{P}_N(dZ_1^N)} \left(\overline{Y}_N - f_m^K(X_N \mid Z_1^K)\right)^2 + \frac{2}{\lambda(N-K)} \log \frac{1}{\pi(m)} \right).\end{aligned} \tag{3.6.4}$$

Before proving this theorem, let us establish a useful technical lemma.

**Lemma 3.6.1.** *Let us consider some probability distribution* $\nu$ *on some measurable space* $(\Theta, \mathcal{T})$, *and let* $h : \Theta \to \mathbb{R}$ *be some bounded measurable real valued function. For any positive parameter* $\gamma$, *let us consider the probability measure*

$$\nu_\gamma(d\theta) = \frac{\exp\left(\gamma h(\theta)\right)}{\mathbb{E}_{\nu(d\theta')} \exp\left(\gamma h(\theta')\right)} \nu(d\theta).$$

*Let us put also*

$$\begin{aligned}\mathbb{V}_{\nu_\gamma}(h) &= \mathbb{V}_{\nu_\gamma(d\theta)}(h(\theta)) \overset{\text{def}}{=} \mathbb{E}_{\nu_\gamma(d\theta)} \left(h(\theta) - \mathbb{E}_{\nu_\gamma} h\right)^2 \quad \textit{and} \\ \mathbb{M}^3_{\nu_\gamma}(h) &= \mathbb{M}^3_{\nu_\gamma(d\theta)}(h(\theta)) \overset{\text{def}}{=} \mathbb{E}_{\nu_\gamma(d\theta)} \left(h(\theta) - \mathbb{E}_{\nu_\gamma} h\right)^3.\end{aligned}$$

*We have the following bound for the "free energy" function:*

$$\log \mathbb{E}_{\nu(d\theta)} \exp(\lambda h(\theta)) \leq \lambda \mathbb{E}_\nu(h) + \frac{\lambda^2}{2} \mathbb{V}_\nu(h) \exp\left( \lambda \max \left\{ 0, \sup_{\gamma \in [0,\lambda]} \frac{\mathbb{M}^3_{\nu_\gamma}(h)}{\mathbb{V}_{\nu_\gamma}(h)} \right\} \right)$$

*Proof.* Taking derivatives with respect to $\lambda$, we have

$$\log \mathbb{E}_{\nu(d\theta)} \exp(\lambda h(\theta)) = \lambda \mathbb{E}_\nu(h) + \int_0^\lambda \int_0^\gamma \mathbb{V}_{\nu_\beta}(h) d\beta d\gamma$$

and

$$\frac{\partial}{\partial \beta} \log \mathbb{V}_{\nu_\beta}(h) = \frac{\mathbb{M}^3_{\nu_\beta}(h)}{\mathbb{V}_{\nu_\beta}(h)}.$$

Therefore when $\beta \leq \lambda$ we have

$$\mathbb{V}_{\nu_\beta}(h) \leq \mathbb{V}_\nu(h) \exp\left( \lambda \max \left\{ 0, \sup_{\xi \in [0,\lambda]} \frac{\mathbb{M}^3_{\nu_\xi}(h)}{\mathbb{V}_{\nu_\xi}(h)} \right\} \right),$$

and the lemma follows. □

*Proof of theorem 3.6.1.* The proof of theorem 3.1.1 does not require the density estimators $q_m^K$ to be normalised to 1, therefore it extends verbatim to the conditional densities defined above. We thus have that

$$\begin{aligned} - \mathbb{E}_{\mathbb{P}_N(dZ_1^N)} \log \mathbb{E}_{\rho(dm)} \exp\left( -\frac{\lambda}{2} (Y_N - f_m^K(X_N))^2 \right) \\ \leq \inf_{m \in \mathfrak{M}} \left( \mathbb{E}_{\mathbb{P}_N(dZ_1^N)} \frac{\lambda}{2} (Y_N - f_m^K(X_N))^2 + \frac{1}{N-K} \log \frac{1}{\pi(m)} \right), \quad (3.6.5) \end{aligned}$$

where $f_m^K(X_N)$ is a short notation for $f_m^K(X_N \mid Z_1^K)$. Using the previous lemma and putting

$$\rho_\gamma(dm) = \frac{\exp\left(-\frac{\gamma}{2}\left(Y_N - f_m^K(X_N)\right)^2\right)}{\mathbb{E}_{\rho(dm')} \exp\left(-\frac{\gamma}{2}\left(Y_N - f_{m'}^K(X_N)\right)^2\right)} \rho(dm)$$

$$\text{and} \quad \chi = 0 \vee \sup_{\gamma \in [0,\lambda]} \frac{\mathbb{M}^3_{\rho_\gamma(dm)}\left(-\frac{1}{2}(Y_N - f_m^K)^2\right)}{\mathbb{V}_{\rho_\gamma(dm)}\left(-\frac{1}{2}(Y_N - f_m^K)^2\right)},$$

we have

$$\begin{aligned} \log \mathbb{E}_{\rho(dm)} \exp\left( -\frac{\lambda}{2} \left(Y_N - f_m^K(X_N)\right)^2 \right) \\ \leq -\frac{\lambda}{2} \mathbb{E}_{\rho(dm)} (Y_N - f_m^K(X_N))^2 + \frac{\lambda^2}{2} e^{\lambda \chi} \mathbb{V}_{\rho(dm)} \left( \frac{1}{2} (Y_N - f_m^K(X_N))^2 \right). \end{aligned}$$

Expending

$$\begin{aligned} (Y_N - f_m^K(X_N))^2 = (Y_N - \overline{Y}_N)^2 \\ + (\overline{Y}_N - f_m^K(X_N))^2 + 2(Y_N - \overline{Y}_N)(\overline{Y}_N - f_m^K(X_N)), \end{aligned}$$

we see that

$$\chi \leq 2B^2 + 2B|Y_N - \overline{Y}_N|.$$

Moreover

$$\begin{aligned}
&\mathbb{V}_{\rho(dm)}\left(\frac{1}{2}(Y_N - f_m^K(X_N))^2\right)\\
&\quad\le \frac{1}{4}\mathbb{V}_{\rho(dm)}\Big(\big(f - \mathbb{E}_\rho f_m^K(X_N)\big)^2 + 2\big(Y_N - \bar{Y}_N\big)\big(f - \mathbb{E}_\rho f_m^K(X_N)\big)\\
&\qquad\qquad + 2\big(\bar{Y}_N - \mathbb{E}_\rho f_m^K(X_N)\big)\big(f - \mathbb{E}_\rho f_m^K(X_N)\big)\Big)\\
&\le \frac{3}{4}\Big(\mathbb{V}_\rho\big(f_m^K(X_N) - \mathbb{E}_\rho f_m^K(X_N)\big)^2 + 4\big(Y_N - \bar{Y}_N\big)^2\mathbb{V}_\rho\big(f_m^K(X_N) - \mathbb{E}_\rho f_m^K(X_N)\big)\\
&\qquad\qquad + 4\big(\bar{Y}_N - \mathbb{E}_\rho f_m^K(X_N)\big)^2\mathbb{V}_\rho\big(f_m^K(X_N) - \mathbb{E}_\rho f_m^K(X_N)\big)\Big)\\
&\qquad\qquad\qquad\le \Big(15B^2 + 3\big(Y_N - \bar{Y}_N\big)^2\Big)\mathbb{V}_{\rho(dm)} f_m^K(X_N)
\end{aligned}$$

On the other hand

$$\mathbb{E}_{\rho(dm)}(Y_N - f_m^K(X_N))^2 = (Y_N - \mathbb{E}_{\rho(dm)}(f_m^K(X_N)))^2 + \mathbb{V}_{\rho(dm)}(f_m^K(X_N)),$$

therefore

$$\begin{aligned}
&\log \mathbb{E}_{\rho(dm)} \exp\left(-\frac{\lambda}{2}\left(Y_N - f_m^K(X_N)\right)^2\right)\\
&\qquad\qquad \le -\frac{\lambda}{2}\left(Y_N - \mathbb{E}_{\rho(dm)}(f_m^K(X_N))\right)^2\\
&\qquad\qquad\quad + \frac{\lambda}{2}\mathbb{V}_\rho(f_m^K(X_N))\Big(\big(15B^2 + 3(Y_N - \overline{Y}_N)^2\big)\lambda e^{\lambda\chi} - 1\Big).
\end{aligned}$$

Integrating with respect to $\mathbb{P}_N(dY_N \mid X_N, Z_1^{N-1})$, we get that when $\lambda$ satisfies

$$\lambda^{-1} \ge \left(15\,B^2 + 3V_{2\lambda B}\right) M_{2\lambda B} \exp\left(2\lambda B^2\right), \tag{3.6.6}$$

then

$$\begin{aligned}
&\mathbb{E}_{\mathbb{P}_N} \log \mathbb{E}_{\rho(dm)} \exp\left(-\frac{\lambda}{2}(Y_N - f_m^K(X_N))^2\right)\\
&\qquad \le -\frac{\lambda}{2}\mathbb{V}_{\mathbb{P}_N}(Y_N \mid X_N, Z_1^{N-1}) - \frac{\lambda}{2}\left(\overline{Y}_N - \mathbb{E}_{\rho(dm)}(f_m^K(X_N))\right)^2,
\end{aligned}$$

this combined with equation (3.6.5) gives the estimate (3.6.4). The condition on $\lambda$ (3.6.6) implies that

$$2\lambda B^2 \exp\left(2\lambda B^2\right) \le 2/15,$$

and therefore that

$$\exp\left(\frac{\lambda}{2}B^2\right) \le 1.13\,.$$

It is easy to deduce from this that condition (3.6.6) can be strengthened to condition (3.6.3). □

*Remark* 3.6.1. It is interesting that, although we use a Gaussian model to represent the noise, theorem 3.6.1 is valid for noises with a much heavier tail (exponential tails are allowed).

*Remark* 3.6.2. Another interesting point is that we do not need the variance of the noise to be independent of $X$, and we do not need to estimate it precisely, we only need to have a uniform upper bound, and we can do as though the noise were uniform.

*Remark* 3.6.3. It should be stressed that the "bias" term in theorem 3.6.1 is exact (whereas the "variance" term is only of the right order). Having the right bias term may be important for applications to pattern recognition. Indeed in this kind of applications, the sample distribution is far too complex to be modelled with a low bias. With an estimate with a sharp bias term, we can get informations on the $\mathbb{L}_2$ projection of the true regression function on the model (or rather on the convex hull of the model), even in the case of a large bias. Therefore the model can be a simplified sketch of the true sample distribution, where unwanted details have been omitted.

*Remark* 3.6.4. All the previous remarks make it possible to "do the right thing with the wrong model". We think this is an important issue in "statistical learning theory" (where we give to this expression the same meaning as Vapnik [76]).

*Remark* 3.6.5. Let us see what we get for a Gaussian noise. Let us assume that

$$Y_k = f(X_k) + \epsilon_k$$

where the distribution of $(X_1, \dots, X_N)$ is exchangeable and where the conditional distribution of $(\epsilon_1, \dots, \epsilon_N)$ knowing $X_1^N$ is a product of Gaussian distributions :

$$\mathcal{L}((\epsilon_i)_{i=1}^N \mid X_1^N) = \bigotimes_{i=1}^N \mathcal{N}(0, \sigma(X_i)).$$

Let us put for short $\sigma = \sup_{x \in \mathcal{X}} \sigma(x)$. An easy computation shows that in this case we can take

$$M_\alpha = 2 \exp\left(\frac{\alpha^2\sigma^2}{2}\right)$$
$$V_\alpha = \sigma^2 \left(1 + \alpha^2\sigma^2 + \sqrt{\frac{2}{\pi}}\alpha\sigma \exp\left(-\frac{\alpha^2\sigma^2}{2}\right)\right).$$

Let us put $\alpha = 2\lambda B$. It is clear that for any $\lambda$ satisfying (3.6.3)

$$\lambda \leq \frac{1}{17\, B^2} \wedge \frac{1}{3.4\sigma^2},$$

therefore

$$\alpha \leq \frac{2}{17\, B} \wedge \frac{2B}{3.4\sigma^2}$$

and

$$\alpha^2 \leq \frac{4}{17 \times 3.4} \frac{1}{\sigma^2}.$$

This shows that condition (3.6.3) can be strengthened to

$$\lambda^{-1} \geq 2 \exp\left(\frac{2}{17 \times 3.4}\right) \left(17\, B^2 + 3.4 \left(1 + \frac{4}{17 \times 3.4} + \sqrt{\frac{2}{\pi}} \frac{2}{\sqrt{17 \times 3.4}}\right) \sigma^2\right),$$

and eventually to

$$\lambda^{-1} \geq (35.2\, B^2 + 9.005\, \sigma^2).$$

Setting the value of $\lambda$ to the righthand side of this inequality, we get

$$\mathbb{E}_{\mathbb{P}_N(dZ_1^{N-1})} \left(\overline{Y}_N - \mathbb{E}_{\rho(dm)} f_m^K(X_N \mid Z_1^K)\right)^2 \\ \leq \inf_{m \in \mathfrak{M}} \left( \mathbb{E}_{\mathbb{P}_N(dZ_1^N)} \left(\overline{Y}_N - f_m^K(X_N \mid Z_1^K)\right)^2 + \frac{70.4\, B^2 + 18.01\, \sigma^2}{N-K} \log \frac{1}{\pi(m)} \right).$$

*Remark* 3.6.6. Theorem 3.6.1 can be compared with Theorem 1 in Yang [87], where the estimator is different (it is obtained by minimising the Hellinger distance), the bias bound is not exact (it is multiplied by a constant larger than one) and there is a third term in the righthand side depending on the accuracy of estimation of the variance of the noise (whereas we do not need to estimate this variance here, but only to know a possibly crude uniform upper bound for it).

## 3.7 Adaptive regression estimation in Besov spaces

To give an illustration to the previous section, we deal here with the problem of least square regression in Besov spaces on the interval. For an introduction to regression estimation in Besov spaces and its link with wavelet expansions, we refer to [42]. We consider an i.i.d. experiment $(\mathcal{X}^N, \mathcal{Y}^N, \mathfrak{B}^{\otimes N}, \mathbb{P}^{\otimes N})$, where $\mathcal{X} = [0,1]$, $\mathcal{Y} = \mathbb{R}$ and $\mathfrak{B}$ is the Borel sigma algebra on $[0,1] \times \mathbb{R}$. Moreover we assume that $\mathbb{P}(dX)$, the marginal of $\mathbb{P}$ on the first coordinate $X$, is known to be the uniform probability distribution $U$ on $[0,1]$ (this is realistic when the design is sampled by the statistician, the estimation of an unknown design distribution is not what we mean to discuss here, we chose a random design rather than a deterministic grid because our method is more straightforward to implement in this case). We put

$$g(X) = \mathbb{E}_{\mathbb{P}}(Y \mid X),$$

and we assume that $g \in \mathbb{L}_\infty([0,1])$. More precisely we will assume in the following that $\|g\|_\infty \leq B$ for some constant $B$.

We consider an orthogonal basis $(\psi_k)_{k \in \mathbb{N}}$ of $\mathbb{L}_2([0,1], U)$.

For any finite subset $S \subset \mathbb{N}$ of the integers, we consider the linear space $\Phi_S = \operatorname{Vect}\{\psi_k \,;\, k \in S\}$. We let $\hat{f}_S : \mathcal{X} \mapsto \mathcal{Y}$ be the linear estimator in $\Phi_S$ defined by

$$\hat{f}_S(x) = \sum_{k \in S} \hat{\theta}_k \psi_k(x),$$

where

$$\hat{\theta}_k = \frac{1}{N-1} \sum_{n=1}^{N-1} Y_n \psi_k(X_n).$$

It is easy to bound the mean quadratic risk of $\hat{f}_S$ by

$$\begin{aligned}
&\mathbb{E}_{\mathbb{P}^{\otimes N}} \left(g(X_N) - \hat{f}_S(X_N)\right)^2 \\
&\quad = \sum_{k \notin S} \left( \int_0^1 g(x)\psi_k(x)dx \right)^2 \\
&\qquad + \sum_{k \in S} \mathbb{E}_{\mathbb{P}^{\otimes(N-1)}} \left( \int_0^1 g(x)\psi_k(x)dx - \hat{\theta}_k \right)^2 \\
&\quad = d(g, \Phi_S)^2 + \frac{1}{N-1} \sum_{k \in S} \operatorname{Var}_{\mathbb{P}}(Y\psi_k(X)) \\
&\quad = d(g, \Phi_S)^2 + \frac{1}{N-1} \sum_{k \in S} \mathbb{E}\left(\operatorname{Var}_{\mathbb{P}}(Y \,|\, X)\psi_k(X)^2\right) + \operatorname{Var}_{\mathbb{P}}(g(X)\psi_k(X)) \\
&\quad \leq d(g, \Phi_S)^2 + \left( \sup_{x \in \mathcal{X}} \operatorname{Var}_{\mathbb{P}}(Y \,|\, X = x) + \|g\|_\infty^2 \right) \frac{|S|}{N-1}
\end{aligned}$$

Let us consider an orthonormal wavelet basis $\{\psi_{j,k} \,|\, j \geq 0, k \in \Lambda_j\}$ of $\mathbb{L}_2([0,1], U)$, where $U$ is the Lebesgue measure on $[0,1]$. We assume that it is built from the scale function (also called father wavelet) $\phi = \psi_{0,0}$ and the wavelet function $\psi = \psi_{1,0}$, and that a suitable orthonormalization procedure (see [26]) has been used near the boundary of the interval. We assume that $\psi$ is compactly supported and have $R$ continuous derivatives. For some constant $L$ we have that $|\Lambda_j| \leq L2^j$.

We let for any $f \in \mathbb{L}_2([0,1])$

$$\theta_{j,k}(f) = \int_0^1 f(x)\psi_{j,k}(x)dx, \qquad j \geq 0, k \in \Lambda_j.$$

For any $p \in [1,2]$ and any $s \in ]1/p - 1/2, R[$, we consider the intersection of Besov and $\mathbb{L}_\infty$ balls

$$B^s_{p,\infty}(B, C)$$

$$= \left\{ f \in \mathbb{L}_2([0,1]) \, : \, \sup_{j \geq 0} 2^{j(s+\frac{1}{2}-\frac{1}{p})} \left( \sum_{k \in \Lambda_j} |\beta_{j,k}|^p \right)^{\frac{1}{p}} \leq C, \|f\|_\infty \leq B \right\}.$$

*Remark* 3.7.1. We consider sets of bounded functions because we want to apply the results of the two previous sections. The restrictions on the range of $(p, s)$ are made to ensure that $B^s_{p,\infty}(+\infty, C)$ is indeed a Besov ball in the usual sens (more precisely, this makes sure that the norm on wavelet coefficients used in the definition of $B^s_{p,\infty}(+\infty, C)$ is equivalent to the usual Besov norm, see for instance [62] or [42] for a proof). However, our computations will only use the fact that $0 < p \leq 2$ and $s > 1/p - 1/2$. Note that when $s > 1/p$ the Besov ball $B^s_{p,\infty}(+\infty, C)$ is bounded in $\mathbb{L}_\infty$ (see [9]), and therefore is equal to $B^s_{p,\infty}(B, C)$ when $B/C$ is large enough (however this is not the most interesting case here, the adaptation problem being more difficult when $s + 1/2 - 1/p$ is close to zero). We did not cover the case $p > 2$ because it is easier and less interesting (linear estimators are asymptotically minimax in each ball).

We let $\Lambda = \bigcup_{j \geq 0} \Lambda_j$ be the set of indices of the above defined wavelet expansion.

We are looking for a family $\mathcal{S}$ of finite subsets $S$ of $\Lambda$, such that the aggregation of the linear estimators $\{\hat{f}_S \,;\, S \in \mathcal{S}\}$ is simultaneously minimax in order in all the Besov balls

$$B^s_{p,\infty}(B, C), \quad p \in [1, 2] \,,\, s \in \left] \frac{1}{p} - \frac{1}{2}, R \right[ .$$

The hypotheses on the noise are the same as in section 3.6. Namely we assume that (3.6.1) and (3.6.2) hold for some value of $\alpha > 0$ such that $M_\alpha < \infty$ and $V_\alpha < \infty$. Then $V_0$ is also finite and (3.6.4) provides an "oracle inequality".

The following computations follow classical lines, we make them for the sake of being self contained and to see what constants we get. We have learnt the method in I. Johnstone's lectures at ENS [44] and in [11]. The acquainted reader will notice the link with the problem of optimal recovery in Besov bodies.

Let us first compute for a given $g \in B^s_{p,\infty}(B, C)$ the set $S$ for which the righthand side of

$$\mathbb{E}_{\mathbb{P}^{\otimes N}}(g(X_N) - \hat{f}_S(X_N))^2 \leq d(g, \Phi_S)^2 + (V_0 + B^2)\frac{|S|}{N-1},$$

is minimum. This set is obviously obtained by thresholding the coefficients $\theta_{j,k}(g)$ of the wavelet decomposition of $g$ at level $\epsilon = \sqrt{(V_0 + B^2)/(N-1)}$. Let us put

$$S_\epsilon(g) = \{(j,k) \,;\, |\theta_{j,k}(g)| > \epsilon\},$$

then

$$\inf_{S \subset \Lambda} d(g, \Phi_S)^2 + \epsilon^2 |S| = d(g, \Phi_{S_\epsilon(g)}) + \epsilon^2 |S_\epsilon(g)|.$$

Let us put $\alpha = s + 1/2 - 1/p$. We can bound $|S_\epsilon(g) \cap \Lambda_j|$ by

$$|S_\epsilon(g) \cap \Lambda_j| \leq C^p 2^{-jp\alpha} \epsilon^{-p}.$$

This suggests to put

$$\mathcal{S}_{m,J} = \{S \subset \Lambda;\ |S \cap \Lambda_j| = \left\lfloor L 2^{J+(J-j)/m} \right\rfloor \wedge |\Lambda_j|\},$$
$$\mathcal{S} = \bigcup_{m \in \mathbb{N}^*, J \in \mathbb{N}^*} \mathcal{S}_{m,J}.$$

For any $g \in B^s_{p,\infty}(B, C)$, there is $\bar{S}(g) \in \mathcal{S}_{m(g),J(g)}$ with

$$m(g) = \lceil (p\alpha)^{-1} \rceil,$$
$$J(g) = \left\lceil \frac{1}{p(s+1/2)} \log_2 \left( \frac{1}{L} \left( \frac{C}{\epsilon} \right)^p \right) \right\rceil,$$

such that $S_\epsilon(g) \subset \bar{S}(g)$. Moreover

$$\begin{aligned} |\bar{S}(g)| &\leq L 2^{J+1} + L 2^J \left(1 - 2^{-1/m}\right)^{-1} \\ &\leq 2 L 2^J (m+1) \\ &\leq 4(m+1) L^{1-p^{-1}(s+1/2)^{-1}} \left( \frac{C}{\epsilon} \right)^{(s+1/2)^{-1}}, \end{aligned}$$

and therefore

$$(V_0 + B^2) \frac{|\bar{S}(g)|}{N-1} \leq 4(m+1) L^{1-p^{-1}(s+1/2)^{-1}} C^{(s+1/2)^{-1}} \left( \frac{V_0 + B^2}{N-1} \right)^{2s/(2s+1)}.$$

To obtain an upper bound for $d(g, \Phi_{\bar{S}(g)})^2$, let us remark that $d(g, \Phi_{\bar{S}(g)}) \leq d(g, \Phi_{S_\epsilon(g)})$ and that generally speaking, as soon as $0 < p \leq 2$,

$$\begin{aligned} &\sup\{\sum_{k \in A} \gamma_k^2;\ \sup_{k \in A} |\gamma_k| \leq \epsilon,\ \sum_{k \in A} |\gamma_k|^p \leq G^p\} \\ &= \begin{cases} |A| \epsilon^2 & \text{if } |A| \epsilon^p \leq G^p \\ \left\lfloor \left( \frac{G}{\epsilon} \right)^p \right\rfloor \epsilon^2 + \left( \left( \frac{G}{\epsilon} \right)^p - \left\lfloor \left( \frac{G}{\epsilon} \right)^p \right\rfloor \epsilon^p \right)^{2/p} & \text{otherwise.} \end{cases} \\ &\leq \left( |A| \wedge \left( \frac{G}{\epsilon} \right)^p \right) \epsilon^2. \end{aligned}$$

Applying these remarks to each $\Lambda_j$, we get that

$$d(g, \Phi_{\bar{S}(g)})^2 \leq \epsilon^2 2^J + \sum_{j=J}^{\infty} C^p 2^{-jp\alpha} \epsilon^{2-p}$$

$$
\begin{aligned}
&\leq \epsilon^2 2^J + C^p \epsilon^{2-p} (1 - 2^{-p\alpha})^{-1} 2^{-J\alpha p} \\
&\leq 2 L^{p^{-1}(s+1/2)^{-1}} C^{(s+1/2)^{-1}} \epsilon^{2-(s+1/2)^{-1}} \\
&\qquad + 2mC^p L^{\alpha(s+1/2)^{-1}} C^{-p\alpha(s+1/2)^{-1}} \epsilon^{-p\alpha(s+1/2)^{-1}+2-p} \\
&\leq 2(Lm+1) L^{-p^{-1}(s+1/2)^{-1}} C^{2/(2s+1)} \left( \frac{V_0 + B^2}{N-1} \right)^{2s/(2s+1)}.
\end{aligned}
$$

We have now to choose a prior distribution $\pi$ on $\mathcal{S}$. Let us put for any $S \in \mathcal{S}_{m,J}$

$$
\pi(S) = \frac{6}{\pi^2} 2^{-m} J^{-2} |\mathcal{S}_{m,J}|^{-1},
$$

and let us find an upper bound for $|\mathcal{S}_{m,J}|$. We have

$$
\log |\mathcal{S}_{m,J}| \leq \sum_{j=J+1}^{+\infty} \log \binom{L2^j}{\lfloor L2^{J+(J-j)/m} \rfloor}.
$$

Using the inequality $\log \binom{n}{\lfloor \gamma n \rfloor} \leq \gamma n (1 - \log \gamma)$, we get that

$$
\begin{aligned}
\log |\mathcal{S}_{m,J}| &\leq \sum_{j=J+1}^{\infty} L 2^{J+(J-j)/m} \left( \log 2^{(j-J)(1+m^{-1})} + 1 \right) \\
&= L2^J \left(1 + \tfrac{1}{m}\right) \log 2 \frac{2^{-1/m}}{\left(1 - 2^{-1/m}\right)^2} + L2^J \frac{2^{-1/m}}{\left(1 - 2^{-1/m}\right)} \\
&\leq 2mL2^J \left(2(m+1)\log 2 + 1\right).
\end{aligned}
$$

Putting everything together, we obtain the following theorem:

**Theorem 3.7.1.** *Let us consider, as above, a regression experiment of the type*

$$
Y_i = g(X_i) + \epsilon_i, \quad i = 1, \dots, N
$$

*where $g \in B^s_{p,\infty}(B, C)$, with $p \in [1, 2]$, $s \in ]1/p - 1/2, R[$, and where $(X_i, \epsilon_i)_{i=1}^N$ are i.i.d., with*

$$
\begin{aligned}
X_i &\sim \text{Lebesgue measure on } [0,1], \\
\mathbb{E}(\epsilon_i \mid X_i) &= 0, \quad a.s., \\
\mathbb{E}(\exp(\alpha|\epsilon_i|) \mid X_i) &\leq M_\alpha, \quad a.s., \\
\frac{\mathbb{E}(\epsilon_i^2 \exp(\alpha|\epsilon_i|) \mid X_i)}{\mathbb{E}(\exp(\alpha|\epsilon_i|) \mid X_i)} &\leq V_\alpha, \quad a.s..
\end{aligned}
$$

*Let us assume that $B$ and $R$ are known, that for some positive value of $\alpha$, $M_\alpha$ and $V_\alpha$ are known and finite, and that $V_0$ is known as well. (On the other hand $s$, $p$ and $C$ are not assumed to be known.)*

*Let us choose some $K < N$, and let $\hat{f}_S^K$ be the above defined linear estimator based on the index set $S$ and on the observations $\{(X_i, Y_i)\,;\, 1 \leq i \leq K\}$. Let us threshold $\hat{f}_S^K$ and put*

$$\tilde{f}_S^K(x) = \max\{\min\{\hat{f}_S^K(x), B\}, -B\}, \quad x \in [0,1].$$

*Let us choose*

$$\lambda = \min\left\{\frac{\alpha}{2B}, \Big(M_\alpha\big(17B^2 + 3.4V_\alpha\big)\Big)^{-1}\right\}.$$

*Let us define the family of index sets $\mathcal{S} = \bigcup \mathcal{S}_{m,J}$ as above and let us put for any $S \in \mathcal{S}$*

$$q_S^K(Y_i \,|\, X_1^K, Y_1^K, X_i) = \exp\left(-\frac{\lambda}{2}\Big(Y_i - \tilde{f}_S^K(X_i)\Big)^2\right), \qquad K < i \leq N.$$

*Let us choose the prior distribution $\pi$ on $\mathcal{S}$ as above, and let us consider the posterior distribution on $\mathcal{S}$*

$$\rho(S) = \frac{1}{N+1-K}\sum_{M=K}^{N} \frac{\pi(S)\displaystyle\prod_{i=K+1}^{N} q_S^K(Y_i \,|\, X_1^K, Y_1^K, X_i)}{\displaystyle\sum_{S' \in \mathcal{S}} \pi(S') \prod_{i=K+1}^{N} q_{S'}^K(Y_i \,|\, X_1^K, Y_1^K, X_i)}$$

*When $L = 1$ and $K = N/2$, putting $m = \lceil (ps + p/2 - 1)^{-1} \rceil$, we have that*

$$\begin{aligned}
&\mathbb{E}_{\mathbb{P}^{\otimes N}} \|g - \mathbb{E}_{\rho(dS)} \tilde{f}_S\|^2 \\
&\qquad \leq \left(6(m+1) + \frac{8m\,(2(m+1)\log 2 + 1)}{\lambda(V_0 + B^2)}\right) C^{(s+1/2)^{-1}} \\
&\qquad\qquad \times \left(\frac{2(V_0 + B^2)}{N}\right)^{2s/(2s+1)} \\
&+ \frac{4(V_0 + B^2)}{\lambda(V_0 + B^2)N} \\
&\quad \times \left(\log\frac{\pi^2}{6} + m\log 2 + 2\log\left\lceil \log_2\left(L^{-1}\left(\frac{C^2 N}{2(V_0 + B^2)}\right)^{(2s+1)^{-1}}\right)\right\rceil\right).
\end{aligned}$$

*Remark* 3.7.2. The bound obtained for arbitrary values of $L$ and $K$ is

$$\begin{aligned}
&\mathbb{E}_{\mathbb{P}^{\otimes N}} \|g - \mathbb{E}_{\rho(dS)} \tilde{f}_S\|^2 \\
&\qquad \leq 6(m+1) L^{1 - p^{-1}(s+1/2)^{-1}} C^{(s+1/2)^{-1}} \left(\frac{V_0 + B^2}{K}\right)^{2s/(2s+1)}
\end{aligned}$$

$$
\begin{aligned}
&+ \frac{8m(2(m+1)\log 2 + 1)}{\lambda(V_0 + B^2)} L^{1-p^{-1}(s+1/2)^{-1}} C^{(s+1/2)^{-1}} \\
&\qquad \times \frac{V_0 + B^2}{N - K + 1} \left( \frac{V_0 + B^2}{K} \right)^{-(2s+1)^{-1}} \\
&+ \left( \log \frac{\pi^2}{6} + m \log 2 + 2 \log \left[ \log_2 \left( L^{-1} \left( \frac{C^2 K}{(V_0 + B^2)} \right)^{(2s+1)^{-1}} \right) \right] \right) \\
&\qquad \times \frac{2}{\lambda(V_0 + B^2)} \frac{V_0 + B^2}{N - K + 1}.
\end{aligned}
$$

*Remark* 3.7.3. We put $\lambda(V_0 + B^2)$ together because it will be bounded by a numerical constant in many applications (this is for instance true in the Gaussian case studied in the previous section).

*Remark* 3.7.4. Our aggregation rule is similar to what is known in the wavelet statistical literature as an "oracle inequality". This idea was introduced by D. Donoho and I. Johnstone [33].

*Remark* 3.7.5. We do not assume that the noise is Gaussian (however our result covers the Gaussian case, and also noises with heavier exponential tails).

*Remark* 3.7.6. We do not assume that the variance of the noise $\epsilon_i$ is independent of $X_i$. We do not assume that it is known either, but only that a uniform upper bound is known.

*Remark* 3.7.7. We get the right minimax order with respect to $N$ for any $(s, p)$ in the specified range. The rate in $C$ is the minimax rate for a deterministic equispaced design in the case of a homoscedastic Gaussian noise (and also for the white noise model) : see [35]. It is therefore reasonable to conjecture that it is also the minimax rate in our setting that cannot be expected to be easier.

*Remark* 3.7.8. Similar results can be found in Yang [87], where the noise is assumed to be Gaussian and a more involved estimator is used.

# 4

# Gibbs estimators

## 4.1 General framework

The Gibbs estimator is a method to aggregate estimators which we will first describe in a general and abstract case dealing with the mixture of "non normalized density functions". Expected benefits when compared with the progressive mixture estimator are twofold:

- building a posterior distribution which is faster to compute,
- building efficient posterior distributions in the case of a continuous family of fixed distributions, thus avoiding the use of sample splitting schemes.

Let $(\mathcal{X}, \mathcal{F})$ be some measurable space and $\mathbb{P} \in \mathcal{M}_+^1(\mathcal{X}^N, \mathcal{F}^{\otimes N})$ some exchangeable probability distribution on sequences of length $N$. Let $(\Theta, \mathcal{T})$ be a measurable space of parameters and $\pi \in \mathcal{M}_+(\Theta, \mathcal{T})$ a finite positive measure on $\Theta$. Let $q : \Theta \times \mathcal{X} \to [\epsilon, \epsilon^{-1}]$ be a positive measurable function bounded away from 0 and $+\infty$. The small positive parameter $\epsilon$ is introduced as in the case of the progressive mixture estimator to avoid discussing integrability issues, but will not play a significant role in forthcoming results, which will later be extended to weaker but more cumbersome integrability hypotheses.

As the prior measure $\pi$ will in many cases be a probability measure, we will use the probabilistic notation

$$\mathbb{E}_{\pi(d\theta)}(h) \stackrel{\text{def}}{=} \int_{\theta \in \Theta} h(\theta)\pi(d\theta), \qquad h \in \mathbb{L}^1(\pi).$$

**Definition 4.1.1.** The Gibbs estimator can be described as a posterior probability distribution $\rho \in \mathcal{M}_+^1(\Theta, \mathcal{T})$ built from the likelihood functions $\prod_{i=1}^{N-1} q(\theta, X_i)$, from the prior distribution $\pi$ and from an *inverse temperature* positive real parameter $\beta$ according to the formula

$$\rho(d\theta) \overset{\text{def}}{=} \frac{\left[\prod_{i=1}^{N-1} q(\theta, X_i)\right]^{\beta}}{\mathbb{E}_{\pi(d\theta')}\left\{\left[\prod_{i=1}^{N-1} q(\theta', X_i)\right]^{\beta}\right\}} \pi(d\theta).$$

As in the case of the progressive mixture estimator, we will compute an upper bound for

$$-\mathbb{E}_{\mathbb{P}}\Big\{\log\Big[\mathbb{E}_{\rho(d\theta)}\left[q(\theta, X_N)\right]\Big]\Big\}.$$

It will be described with the help of the two following quantities: the estimation error due to aggregation will be controled by

$$\gamma_\beta(\theta) \overset{\text{def}}{=} \mathbb{E}_{\mathbb{P}}\left\{ \beta \log\left[\prod_{i=1}^{N} q(\theta, X_i)\right] - \frac{\mathbb{E}_{\pi(d\theta')}\left\{\left[\prod_{i=1}^{N} q(\theta', X_i)\right]^{\beta} \beta\log\left[\prod_{i=1}^{N} q(\theta', X_i)\right]\right\}}{\mathbb{E}_{\pi(d\theta')}\left\{\left[\prod_{i=1}^{N} q(\theta', X_i)\right]^{\beta}\right\}} \right\}.$$

This definition may seem slightly artificial. We will nonetheless see that $\gamma_\beta(\theta)$ can itself be upper bounded by quantities having a natural interpretation. The expression of $\gamma_\beta(\theta)$ shows that it can be controled through the evaluation of a Laplace integral on the parameter space.

The second quantity needed to specify the valid choices of parameters for the Gibbs estimator is best defined in terms of the following generating function, (which would appear as some free energy function in a statistical mechanics context):

$$\eta \mapsto \mathcal{E}^{X_1^N}(\eta) \overset{\text{def}}{=} \log\left\{\mathbb{E}_{\pi(d\theta)}\left[\left(\prod_{i=1}^{N-1} q(\theta, X_i)\right)^{\beta} q(\theta, X_N)^{\eta}\right]\right\}.$$

The ratio between the third and the second derivatives of this generating function appears as a critical quantity when bounding the risk of the Gibbs estimator. This justifies the introduction of

$$\chi(\beta, \alpha) \overset{\text{def}}{=} 0 \vee \operatorname*{ess\,sup}_{\xi \in ]0,\alpha]} \left\{ \frac{-\mathbb{E}_{\mathbb{P}}\left[\frac{\partial^3}{\partial \xi^3}\mathcal{E}^{X_1^N}(\xi)\right]}{\mathbb{E}_{\mathbb{P}}\left[\frac{\partial^2}{\partial \xi^2}\mathcal{E}^{X_1^N}(\xi)\right]} \mathbb{1}\left[\mathbb{E}_{\mathbb{P}}\left(\frac{\partial^2}{\partial \xi^2}\mathcal{E}^{X_1^N}(\xi) > 0\right)\right]\right\}.$$

**Theorem 4.1.1.** *For any exchangeable probability distribution $\mathbb{P} \in \mathcal{M}_+^1(\mathcal{X}^N)$, for any real positive constant $\beta \in ]0, 1/2[$ satisfying*

$$\inf_{\alpha \in ]\beta,1]} \frac{\beta^2}{(\alpha-\beta)(2-\alpha-\beta)} \exp\bigl[\alpha\chi(\alpha,\beta)\bigr] \leq 1, \tag{4.1.1}$$

*the risk of the Gibbs estimator is such that*

$$\begin{aligned}\mathbb{E}_{\mathbb{P}}\Bigl\{\log\Bigl[\mathbb{E}_{\rho(d\theta)}\bigl[q(\theta, X_N)\bigr]\Bigr]\Bigr\} &= \mathbb{E}_{\mathbb{P}}\Bigl[\mathcal{E}^{X_1^N}(1) - \mathcal{E}^{X_1^N}(0)\Bigr] \\ &\geq \mathbb{E}_{\mathbb{P}}\left\{\frac{\partial}{\partial\eta}_{|\eta=\beta} \mathcal{E}^{X_1^N}(\eta)\right\} = \sup_{\theta\in\Theta} \mathbb{E}_{\mathbb{P}}\Bigl\{\log\bigl[q(\theta, X_N)\bigr]\Bigr\} - \frac{\gamma_\beta(\theta)}{\beta N}.\end{aligned}$$

*Remark* 4.1.1. Condition (4.1.1) can also be written as

$$\beta \leq \sup_{\alpha\in]\beta,1]} \frac{1}{e^{\alpha\chi(\beta,\alpha)}-1}\left(\sqrt{1+\alpha(e^{\alpha\chi(\beta,\alpha)}-1)(2-\alpha)}-1\right).$$

When $\chi(\beta, 1)$ is uniformly small, it is suitable to take $\alpha = 1$, leading to the sufficient condition:

$$\beta \leq \Bigl\{1 + \exp\bigl[\chi(\beta,1)/2\bigr]\Bigr\}^{-1}.$$

When $\chi(\beta, 1)$ is big, it is on the contrary suitable to take $\alpha = \chi(\beta,1)^{-1}\log\bigl[\chi(\beta,1)\bigr]$, leading to the sufficient condition (let us remind that $\alpha \mapsto \chi(\beta, \alpha)$ is non-decreasing):

$$\begin{aligned}\beta \leq\; & \\ & \frac{1}{\chi(\beta,1)-1}\left\{\sqrt{1+\bigl[\chi(\beta,1)-1\bigr]\left[2-\frac{\log\bigl[\chi(\beta,1)\bigr]}{\chi(\beta,1)}\right]\frac{\log\bigl[\chi(\beta,1)\bigr]}{\chi(\beta,1)}}-1\right\} \\ & \underset{\chi(\beta,1)\to+\infty}{\sim} \frac{\sqrt{2\log\bigl[\chi(\beta,1)\bigr]}}{\chi(\beta,1)}.\end{aligned}$$

*Remark* 4.1.2. There is an *independent of N* simple lower bound for the seemingly cumbersome quantity $\chi(\beta, \alpha)$:

$$\sup_{\beta,\alpha}\chi(\beta,\alpha) \leq \sup_{\theta,\theta'\in\Theta, x\in\mathcal{X}} \log\left[\frac{q(\theta,x)}{q(\theta',x)}\right].$$

Therefore it should be remembered that $\beta$ can be chosen to be independent of $N$ in the case when the log likelihoods are bounded.

*Remark* 4.1.3. Let us put

$$h(\theta,\theta',x_1^N) = \frac{\beta}{N}\log\left[\prod_{i=1}^N \frac{q(\theta,x_i)}{q(\theta',x_i)}\right].$$

The upper bound $\gamma_\beta(\theta)$ can be written as

$$\gamma_\beta(\theta) = \mathbb{E}_{\mathbb{P}} \left\{ \frac{\mathbb{E}_{\pi(d\theta')}\Bigl\{Nh(\theta,\theta',X_1^N)\exp\bigl[-Nh(\theta,\theta',X_1^N)\bigr]\Bigr\}}{\mathbb{E}_{\pi(d\theta')}\Bigl\{\exp\bigl[-Nh(\theta,\theta',X_1^N)\bigr]\Bigr\}} \right\},$$

and therefore can be upper-bounded by

$$\gamma_\beta(\theta) \leq \sup_{x_1^N \in \mathcal{X}^N} \frac{\mathbb{E}_{\pi(d\theta')}\Bigl\{Nh(\theta,\theta',x_1^N)\exp\bigl[-Nh(\theta,\theta',x_1^N)\bigr]\Bigr\}}{\mathbb{E}_{\pi(d\theta')}\Bigl\{\exp\bigl[-Nh(\theta,\theta',x_1^N)\bigr]\Bigr\}}.$$

In some cases the right-hand member of this inequality can be explicitely computed (e.g. when Gaussian density functions are used). In other cases it can be estimated as a Laplace integral when the sample size $N$ is large. The following lemma is also useful, especially when the parameter space is discrete:

**Lemma 4.1.1.** *For any lower-bounded measurable function $g : \Theta \to \mathbb{R}$, any prior probability distribution $\pi \in \mathcal{M}_+^1(\Theta, \mathcal{T})$,*

$$\frac{\mathbb{E}_{\pi(d\theta)}\Bigl\{g(\theta)\exp\bigl[-g(\theta)\bigr]\Bigr\}}{\mathbb{E}_{\pi(d\theta)}\Bigl\{\exp\bigl[-g(\theta)\bigr]\Bigr\}} \leq \inf_{\mu \in \mathbb{R}} \left\{\mu - \log\bigl[\pi(\Theta_\mu^g)\bigr]\right\},$$

*where* $\Theta_\mu^g \overset{\text{def}}{=} \{\theta \in \Theta : g(\theta) \leq \mu\}$.

*Remark* 4.1.4. In the discrete case, this lemma shows in particular that

$$\frac{\mathbb{E}_{\pi(d\theta)}\Bigl\{g(\theta)\exp\bigl[-g(\theta)\bigr]\Bigr\}}{\mathbb{E}_{\pi(d\theta)}\Bigl\{\exp\bigl[-g(\theta)\bigr]\Bigr\}} \leq \inf_{\theta \in \Theta} \Bigl\{g(\theta) - \log\bigl[\pi(\{\theta\})\bigr]\Bigr\}.$$

*Proof.* For a given value of $\mu \in \mathbb{R}$, let us define the threshold value $\epsilon = -\log\bigl[\pi\bigl(\Theta_\mu^g\bigr)\bigr] - 1$. Using the obvious upper bound

$$g(\theta) \leq \mu + \epsilon + \bigl(g(\theta) - \mu - \epsilon\bigr)_+,$$

we see that

$$\begin{aligned}
\frac{\mathbb{E}_{\pi(d\theta)}\Bigl\{g(\theta)\exp\bigl[-g(\theta)\bigr]\Bigr\}}{\mathbb{E}_{\pi(d\theta)}\Bigl\{\exp\bigl[-g(\theta)\bigr]\Bigr\}} &\leq \mu + \epsilon \\
&\quad + \frac{\mathbb{E}_{\pi(d\theta)}\Bigl\{\bigl[g(\theta) - \mu - \epsilon\bigr]_+ \exp\bigl[-g(\theta)\bigr]\Bigr\}}{\mathbb{E}_{\pi(d\theta)}\Bigl\{\exp\bigl[-g(\theta)\bigr]\Bigr\}} \\
&\leq \mu + \epsilon + \frac{\exp(-\epsilon)}{\pi(\Theta_\mu^g)} \sup_{\lambda \in \mathbb{R}} \bigl[\lambda \exp(-\lambda)\bigr]
\end{aligned}$$

$$= \mu + \epsilon + \frac{\exp(-\epsilon - 1)}{\pi(\Theta^g_\mu)}$$
$$= \mu - \log\big[\pi(\Theta^g_\mu)\big].$$

As the value of $\mu \in \mathbb{R}$ can be arbitrarily chosen, the lemma is proved. □

Applying this lemma to $h(\theta, \theta', X_1^N)$ gives

$$\gamma_\beta(\theta) \leq \inf_{\mu \in \mathbb{R}} \left\{ \mu - \log\left[\pi\left(\left\{\theta' : \sup_{x_1^N \in \mathcal{X}^N} \beta \log\left[\frac{\prod_{i=1}^N q(\theta, x_i)}{\prod_{i=1}^N q(\theta', x_i)}\right] \leq \mu\right\}\right)\right]\right\}.$$

As a special case, we can put $\mu = 0$, to show that

$$\gamma_\beta(\theta) \leq -\log\big[\pi(\{\theta\})\big].$$

*Proof of theorem 4.1.1.* Let us expand the function $\mathcal{E}^{X_1^N}(\eta)$ in the neighbourhood of $\beta$:

$$\begin{aligned}\mathcal{E}^{X_1^N}(1) - \mathcal{E}^{X_1^N}(0) &= \int_0^1 \frac{\partial}{\partial\eta} \mathcal{E}^{X_1^N}(\eta) d\eta \\ &= \frac{\partial}{\partial\eta}_{|\eta=\beta} \mathcal{E}^{X_1^N}(\eta) + \int_{\eta=\beta}^1 (1-\eta) \frac{\partial^2}{\partial\eta^2} \mathcal{E}^{X_1^N}(\eta) d\eta \qquad (4.1.2) \\ &\quad - \int_{\eta=0}^{\beta} \eta \frac{\partial^2}{\partial\eta^2} \mathcal{E}^{X_1^N}(\eta) d\eta.\end{aligned}$$

From Fubini's theorem, taking expectations and derivations can be swapped (the quantity whose expectation is taken being equal to the integral of its derivative), leading to:

$$\frac{\partial}{\partial\eta} \mathcal{E}^{X_1^N}(\eta) = \frac{\mathbb{E}_{\pi(d\theta)}\left[\left(\prod_{n=1}^{N-1} q(\theta, X_n)\right)^\beta (q(\theta, X_N))^\eta \log q(\theta, X_N)\right]}{\mathbb{E}_{\pi(d\theta)}\left[\left(\prod_{n=1}^{N-1} q(\theta, X_n)\right)^\beta (q(\theta, X_N))^\eta\right]},$$

$$\frac{\partial^2}{\partial\eta^2} \mathcal{E}^{X_1^N}(\eta) =$$
$$\frac{\mathbb{E}_{\pi(d\theta)}\left[\left(\prod_{n=1}^{N-1} q(\theta, X_n)\right)^\beta (q(\theta, X_N))^\eta \left(\log(q(\theta, X_N)) - \frac{\partial}{\partial\eta}\mathcal{E}^{X_1^N}(\eta)\right)^2\right]}{\mathbb{E}_{\pi(d\theta)}\left[\left(\prod_{n=1}^{N-1} q(\theta, X_n)\right)^\beta (q(\theta, X_N))^\eta\right]}.$$

Thus $\mathbb{E}_{\mathbb{P}}\left\{\frac{\partial^2}{\partial\eta^2}\mathcal{E}^{X_1^N}(\eta)\right\} = 0$ in the sole case when $\theta \mapsto q(\theta, X_N)$ is $\pi$ almost everywhere $\mathbb{P}$ almost surely constant. Therefore there are only two possible alternatives: either $\mathbb{E}_{\mathbb{P}}\left\{\frac{\partial^2}{\partial\eta^2}\mathcal{E}^{X_1^N}(\eta)\right\} > 0$ for any $\eta \in [0,1]$, either $\frac{\partial^2}{\partial\eta^2}\mathcal{E}^{X_1^N}(\eta) = 0$ $\mathbb{P}$ almost surely for any $\eta \in [0,1]$. In this last case

$$\left[\mathcal{E}^{X_1^N}(1) - \mathcal{E}^{X_1^N}(0)\right] = \frac{\partial}{\partial\eta}_{|\eta=\beta} \mathcal{E}^{X_1^N}(\eta), \qquad \mathbb{P} \text{ p.s.,}$$

proving theorem 4.1.1. In the other case, Fubini's theorem allowing to exchange expectations and derivatives shows that

$$\begin{aligned}
&\log\left\{\mathbb{E}_{\mathbb{P}}\left[\frac{\partial^2}{\partial\eta^2}_{|\eta=\xi}\mathcal{E}^{X_1^N}(\eta)\right]\right\} \\
&\qquad = \log\left\{\mathbb{E}_{\mathbb{P}}\left[\frac{\partial^2}{\partial\eta^2}_{|\eta=\zeta}\mathcal{E}^{X_1^N}(\eta)\right]\right\} + \int_{\eta=\xi}^{\zeta} \frac{\mathbb{E}_{\mathbb{P}}\left[\frac{\partial^3}{\partial\eta^3}\mathcal{E}^{X_1^N}(\eta)\right]}{\mathbb{E}_{\mathbb{P}}\left[\frac{\partial^2}{\partial\eta^2}\mathcal{E}^{X_1^N}(\eta)\right]} d\eta \\
&\qquad\qquad \geq \log\left\{\mathbb{E}_{\mathbb{P}}\left[\frac{\partial^2}{\partial\eta^2}_{|\eta=\zeta}\mathcal{E}^{X_1^N}(\eta)\right]\right\} - \alpha\chi(\beta,\alpha). \qquad (4.1.3)
\end{aligned}$$

Equation (4.1.2) can moreover be weakened to

$$\begin{aligned}
\mathcal{E}^{X_1^N}(1) - \mathcal{E}^{X_1^N}(0) &\geq \frac{\partial}{\partial\eta}_{|\eta=\beta}\mathcal{E}^{X_1^N}(\eta) \\
&\quad + \int_{\eta=\beta}^{\alpha}(1-\eta)\frac{\partial^2}{\partial\eta^2}\mathcal{E}^{X_1^N}(\eta)d\eta - \int_{\eta=0}^{\beta}\eta\frac{\partial^2}{\partial\eta^2}\mathcal{E}^{X_1^N}(\eta)d\eta.
\end{aligned}$$

Integrating with respect to $\mathbb{P}$ and using (4.1.3) shows that

$$\begin{aligned}
&\mathbb{E}_{\mathbb{P}}\left[\mathcal{E}^{X_1^N}(1) - \mathcal{E}^{X_1^N}(0)\right] \geq \mathbb{E}_{\mathbb{P}}\left[\frac{\partial}{\partial\eta}_{|\eta=\beta}\mathcal{E}^{X_1^N}(\eta)\right] \\
&\quad + \sup_{\eta\in[0,\beta]}\mathbb{E}_{\mathbb{P}}\left[\frac{\partial^2}{\partial\eta^2}\mathcal{E}^{X_1^N}(\eta)\right]\left\{\exp\left[-\alpha\chi(\beta,\alpha)\right]\int_{\eta=\beta}^{\alpha}(1-\eta)d\eta - \int_{\eta=0}^{\beta}\eta d\eta\right\} \\
&\qquad\qquad \geq \mathbb{E}_{\mathbb{P}}\left[\frac{\partial}{\partial\eta}_{|\eta=\beta}\mathcal{E}^{X_1^N}(\eta)\right] \\
&\quad + \sup_{\eta\in[0,\beta]}\mathbb{E}_{\mathbb{P}}\left[\frac{\partial^2}{\partial\eta^2}\mathcal{E}^{X_1^N}(\eta)\right]\left[\frac{(\alpha-\beta)(2-\alpha-\beta)}{2}\exp\left[-\alpha\chi(\beta,\alpha)\right] - \frac{\beta^2}{2}\right] \\
&\qquad\qquad\qquad \geq \mathbb{E}_{\mathbb{P}}\left[\frac{\partial}{\partial\eta}_{|\eta=\beta}\mathcal{E}^{X_1^N}(\eta)\right].
\end{aligned}$$

□

## 4.2 Dichotomic histograms

We will deal with dichotomic histograms using the Gibbs estimator, which leads to faster computations. (The progressive mixture rule could also be used.) We will focus on a two step scheme, with an adaptive choice of $\beta$. A single step scheme could also be derived, along the lines described in [77] where a more thorough study of this subject can be found. The best theoretical bound available is established for a two step progressive mixture scheme.

As in the general case of histograms described above, we consider some measurable space $(\mathcal{X}, \mathcal{F})$, equipped with a reference probability measure $\mu \in \mathcal{M}_+^1(\mathcal{X}, \mathcal{F})$. We assume moreover that $\mathcal{X}$ has been divided into a family of measurable cells $\{I_s \in \mathcal{F} : s \in \mathcal{S}\}$ indexed by a set of binary words $\mathcal{S}$ equal to the set of (non necessarily strict) prefix words of some prefix dictionary $\overline{\mathcal{D}} \subset \{0,1\}^* \cup \{\varnothing\}$. We chose the binary case for simplicity, the case of a finite alphabet being a straightforward extension. We assume that the cells $I_s$ define nested partitions, and more precisely that

$$\begin{cases} I_\varnothing = \mathcal{X}, \\ I_{(s,0)} \cap I_{(s,1)} = \varnothing, & s \in \mathring{\mathcal{D}}, \\ I_s = I_{(s,0)} \cup I_{(s,1)}, & s \in \mathring{\mathcal{D}}, \\ \mu(I_s) > 0, & s \in \mathcal{S}. \end{cases}$$

The following lemma is then the consequence of an easy induction argument:

**Lemma 4.2.1.** *For any complete prefix dictionary $\mathcal{D} \subset \mathcal{S}$, $\{I_s : s \in \mathcal{D}\}$ is a measurable partition of $\mathcal{X}$. Consequently, to any $x$ in $\mathcal{X}$ corresponds a word $s \in \bar{\mathcal{D}}$ defined by the relation $x \in I_s$, which we will call $\sigma(x)$.*

Let us define as above the probability densities

$$q_{\mathcal{D},\theta}(x) = \sum_{s \in \mathcal{D}} \frac{\theta(s)}{\mu(I_s)} \mathbb{1}(x \in I_s),$$

where $\theta \in \mathcal{M}_+^1(\mathcal{D})$. In order to estimate $\theta$, let us write it as

$$\theta(s) = \prod_{j=1}^{\ell(s)} \theta\big(s_j \,|\, s_1^{j-1}\big). \tag{4.2.1}$$

This decomposition comes from the fact that $\ell(s)$ is a stopping time. (The words of the dictionnary $\mathcal{D}$ can be made of constant length by adding repeats of an extra letter at the end of shorter ones, the decomposition formula is thus identified with the usual factorization formula into a product of conditional probabilities. It can be truncated to $\ell(s)$ because all the subsequent factors are equal to one, $\ell(s)$ being a stopping time.)

Conditional probabilities can then be estimated with the help of the Laplace estimator:

$$\hat{\theta}^{X_1^K}(s_j \mid s_1^{j-1}) = \frac{K\bar{P}_{\sigma(X_1^K)}(s_1^j) + 1}{K\bar{P}_{\sigma(X_1^K)}(s_1^{j-1}) + 2},$$

where $\bar{P}_{\sigma(X_1^K)}$ is the empirical distribution of the sequence $\sigma(X_1), \dots, \sigma(X_K)$, namely $\frac{1}{K}\sum_{j=1}^{K} \delta_{\sigma(X_j)} \in \mathcal{M}_+^1(\overline{\mathcal{D}})$.

Similarly to the case of the Laplace estimator, it can be established that

**Theorem 4.2.1.** *With the previous notations, for any complete prefix dictionary* $\mathcal{D} \subset \mathcal{S}$, *for any exchangeable probability distribution* $\mathbb{P} \in \mathcal{M}_+^1(\mathcal{X}^{K+1}, \mathcal{F}^{\otimes(K+1)})$,

$$\mathbb{E}_{\mathbb{P}}\Bigl\{-\log\bigl[q_{\mathcal{D},\hat{\theta}^{X_1^K}}(X_{K+1})\bigr]\Bigr\} \leq \inf_{\theta \in \mathcal{M}_+^1(\mathcal{D})} \mathbb{E}_{\mathbb{P}}\Bigl\{-\log\bigl[q_{\mathcal{D},\theta}(X_{K+1})\bigr]\Bigr\} + \frac{|\mathcal{D}| - 1}{K+1}.$$

*Proof.* For any $x \in \mathcal{X}$, let $\mathcal{D}(x)$ be the word $s$ of $\mathcal{D}$ defined by the relation $x \in I_{\mathcal{D}(x)}$. From the expression of $q_{\mathcal{D},\theta}(x)$, it is seen that we have only to prove that

$$\mathbb{E}_{\mathbb{P}}\Bigl\{-\log\Bigl[\hat{\theta}^{X_1^K}\bigl[\mathcal{D}(X_{K+1})\bigr]\Bigr]\Bigr\} \leq \mathbb{E}_{\mathbb{P}}\Bigl\{-\log\Bigl[\theta\bigl[\mathcal{D}(X_{K+1})\bigr]\Bigr]\Bigr\} + \frac{|\mathcal{D}| - 1}{K+1}.$$

Let us introduce the counters

$$b(s) = \sum_{j=1}^{K+1} \mathbb{1}\bigl[X_j \in I_s\bigr],$$

and let us use the exchangeability of $\mathbb{P}$ to write that

$$\begin{aligned}
&\mathbb{E}_{\mathbb{P}}\Bigl\{-\log\Bigl[\hat{\theta}^{X_1^K}\bigl[\mathcal{D}(X_{K+1})\bigr]\Bigr]\Bigr\} \\
&\quad = \mathbb{E}\left\{\frac{-1}{K+1}\sum_{j=1}^{K+1}\log\Bigl[\hat{\theta}^{(X_1^{j-1}, X_{j+1}^{K+1})}\bigl[\mathcal{D}(X_j)\bigr]\Bigr]\right\} \\
&\quad = -\mathbb{E}\left\{\sum_{s \in \mathcal{D}} \frac{b(s)}{K+1} \log\left[\prod_{j=1}^{\ell(s)} \frac{b(s_1^j)}{b(s_1^{j-1}) + 1}\right]\right\} \\
&\quad = \mathbb{E}\Biggl\{-\sum_{s \in \mathcal{D}} \frac{b(s)}{K+1} \log\left[\frac{b(s)}{K+1}\right] \\
&\qquad + \sum_{s \in \mathcal{D}} \frac{b(s)}{K+1} \sum_{j=1}^{\ell(s)} \log\left(1 + \frac{1}{b(s_1^{j-1})}\right)\Biggr\}
\end{aligned}$$

$$
\begin{aligned}
&= \mathbb{E}\Bigg\{ - \sum_{s \in \mathcal{D}} \frac{b(s)}{K+1} \log\left[\frac{b(s)}{K+1}\right] \\
&\qquad + \sum_{s \in \overset{\circ}{\mathcal{D}}} \frac{b(s)}{K+1} \log\left(1 + \frac{1}{b(s_1^{j-1})}\right)\Bigg\} \\
&\leq \mathbb{E}\Bigg\{ \inf_{\theta \in \mathcal{M}_+^1(\mathcal{D})} \left[ - \sum_{s \in \mathcal{D}} \frac{b(s)}{K+1} \log\big[\theta(s)\big]\right]\Bigg\} + \frac{|\overset{\circ}{\mathcal{D}}|}{K+1} \\
&= \mathbb{E}\left\{ \inf_{\theta \in \mathcal{M}_+^1(\mathcal{D})} \left[ - \frac{1}{K+1} \sum_{j=1}^{K+1} \log\Big[\theta\big[\mathcal{D}(X_j)\big]\Big]\right]\right\} + \frac{|\mathcal{D}|-1}{K+1} \\
&\leq \inf_{\theta \in \mathcal{M}_+^1(\mathcal{X})} \mathbb{E}\Big\{ - \log\Big[\theta\big[\mathcal{D}(X_{K+1})\big]\Big]\Big\} + \frac{|\mathcal{D}|-1}{K+1}.
\end{aligned}
$$

□

The next step is now to aggregate the estimators $q_{\mathcal{D},\hat{\theta}X_1^K}$, using a second sample $X_{K+1}^N$ and the Gibbs rule.

Let us define $\pi(\mathcal{D}) = 2^{-|\mathcal{D}|+1-|\mathcal{D}\setminus\overline{\mathcal{D}}|}$. (More generally we could consider $\pi(\mathcal{D}) = \alpha^{|\mathcal{D}|-1}(1-\alpha)^{|\mathcal{D}\setminus\overline{\mathcal{D}}|}$, however, the value 1/2, which is the critical branching rate, is asymptotically optimal when $|\overline{\mathcal{D}}|$ is growing large and the models of intermediate dimensions are the ones we would like to weight as most as possible.) Let $\mathfrak{D}$ be the set of all the complete prefix dictionaries included in $\mathcal{S}$. Let us put

$$
\chi = \max_{\mathcal{D},\mathcal{D}' \in \mathfrak{D}} \max_{x \in \mathcal{X}} \log\left[\frac{q_{\mathcal{D},\hat{\theta}X_1^K}(x)}{q_{\mathcal{D}',\hat{\theta}X_1^K}(x)}\right]
$$

and

$$
\beta = \sup_{\alpha \in ]0,1]} \frac{1}{e^{\alpha\chi}-1}\left[\sqrt{1+\alpha(2-\alpha)(e^{\alpha\chi}-1)} - 1\right].
$$

Let us mention that here $\beta$ is a random variable (being a function of $X_1^K$). Let us consider the Gibbs estimator

$$
\hat{q}(x) \overset{\text{def}}{=} \frac{\displaystyle\sum_{\mathcal{D}\in\mathfrak{D}} \pi(\mathcal{D}) \left(\prod_{i=K+1}^{N} q_{\mathcal{D},\hat{\theta}X_1^K}(X_i)\right)^{\beta} q_{\mathcal{D},\hat{\theta}X_1^K}(x)}{\displaystyle\sum_{\mathcal{D}\in\mathfrak{D}} \pi(\mathcal{D}) \left(\prod_{i=K+1}^{N} q_{\mathcal{D},\hat{\theta}X_1^K}(X_i)\right)^{\beta}}.
$$

From theorem 4.1.1 it follows that

**Theorem 4.2.2.** *Under the previous hypotheses, for any exchangeable probability distribution* $\mathbb{P} \in \mathcal{M}_+^1(\mathcal{X}^{N+1}, \mathcal{F}^{\otimes(N+1)})$,

$$\mathbb{E}\Bigl[-\log\bigl[\hat{q}(X_{N+1})\bigr]\Bigr] \leq \inf_{\mathcal{D}\in\mathfrak{D}} \inf_{\theta\in\mathcal{M}_+^1(\mathcal{D})} \mathbb{E}\Bigl[-\log\bigl[q_{\mathcal{D},\theta}(X_{N+1})\bigr]\Bigr] + \frac{|\mathcal{D}|-1}{K+1} - \frac{\log\pi(\mathcal{D})}{N-K+1}\mathbb{E}(\beta^{-1}).$$

*Remark* 4.2.1. Let us notice that $-\log\bigl[\pi(\mathcal{D})\bigr] \leq \bigl(2|\mathcal{D}|-1\bigr)\log(2)$. Let us notice also that in any case $\chi \leq \log(K+1)$. Considering $\alpha = \frac{\log(\chi)}{\chi}$, we deduce that

$$\mathbb{E}(\beta^{-1}) \leq \frac{\log(K+1)-1}{\sqrt{1+\frac{\log[\log(K+1)]}{\log(K+1)}\left(2-\frac{\log[\log(K+1)]}{\log(K+1)}\right)\Bigl(\log(K+1)-1\Bigr)}-1} \underset{K\to\infty}{\sim} \frac{\log(K+1)}{\sqrt{2\log[\log(K+1)]}}.$$

Let us now describe a fast factorized algorithm to compute $\hat{q}(x)$. Let us define the counters

$$c(s) = \sum_{i=K+1}^{N} \mathbb{1}\bigl[X_i \in I_s\bigr], \qquad s \in \mathcal{S}.$$

Let us attach to each word $s \in \mathcal{S}$ a weight $\Upsilon_s(x)$ defined by the following induction

$$\Upsilon_s(x) = \begin{cases} (1-\alpha)+\alpha \displaystyle\prod_{a\in\{0,1\}} \left[\Upsilon_{sa}(x)\left(\frac{\hat{\theta}^{X_1^K}(a|s)}{\mu(I_{sa}|I_s)}\right)^{\beta c(sa)+\mathbb{1}(x\in I_{sa})}\right], \\ \qquad\qquad \text{when } s \in \mathring{\overline{\mathcal{D}}}, \\ 1 \qquad\quad \text{when } s \in \overline{\mathcal{D}}. \end{cases}$$

The Gibbs estimator $\hat{q}$ is then computed from the formula

$$\hat{q}(x) = \frac{\Upsilon_\varnothing(x)}{\int_{\mathcal{X}} \Upsilon_\varnothing(y)\mu(dy)}.$$

It is constant on each cell $I_s$, $s \in \overline{\mathcal{D}}$ of the finest partition, which is defined by the maximal dictionary $\overline{\mathcal{D}}$. So there are in practice $|\overline{\mathcal{D}}|$ numbers to be computed. Computing $\hat{q}(x)$ for $x \in I_s$, and $s \in \overline{\mathcal{D}}$ can be done simply through updating $\Upsilon_{s_1^k}(x)$, for $k = \ell(s), \dots, 1$, starting from the weights $\Upsilon_s(\varnothing)$, where "$\mathbb{1}\left[x \in I_s\right]$ has been set to null everywhere". More precisely, the weights $\Upsilon_s(\varnothing)$ can be defined by the following induction :

$$\Upsilon_s(\varnothing) = \begin{cases} (1-\alpha)+\alpha \displaystyle\prod_{a \in \{0,1\}} \left[ \Upsilon_{sa}(\varnothing) \left( \frac{\hat{\theta}^{X_1^K}(a|s)}{\mu(I_{sa}|I_s)} \right)^{\beta c(sa)} \right], \\ \qquad\qquad \text{where } s \in \mathring{\overline{\mathcal{D}}}, \\ 1 \qquad\quad \text{where } s \in \overline{\mathcal{D}}. \end{cases}$$

The number of operations needed to compute $\hat{q}$ is of order $|\overline{\mathcal{D}}| \max_{s \in \overline{\mathcal{D}}} \ell(s)$. Let us mention by the way that

$$\int_{\mathcal{X}} \Upsilon_{\varnothing}(y)\mu(dy) = \Upsilon_{\varnothing}(\varnothing).$$

There we show some simulations made in the case when $\mathcal{X} = [0,1]$, where $\mu$ is the Lebesgue measure on the unit intervall and where $I_s = \sum_{k=1}^{\ell(s)} s_k 2^{-k} + [0, 2^{-\ell(s)}[$.

In the first example of [fig. 4.1], the distribution $\mathbb{P}$ to be estimated belongs to one of the parametric models used for estimation.

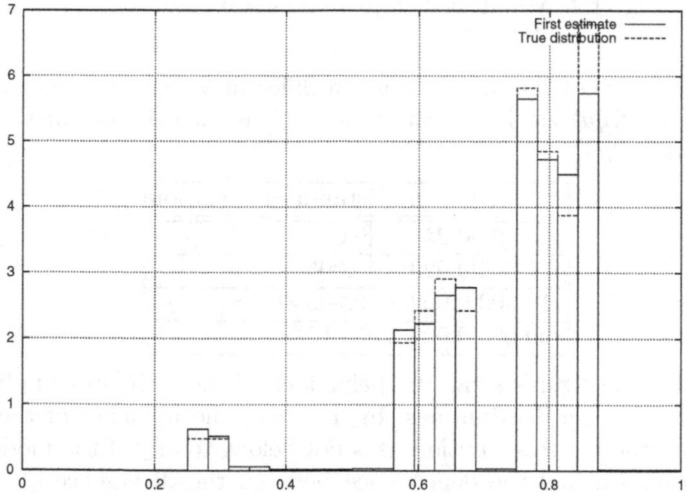


**Fig. 4.1.** $N = 1000$, $\overline{\mathcal{D}} = \{0,1\}^5$, $\mathcal{K} = 0,021$, $\beta = 0,159$

Upon 1000 independent trials of the same experiment, we have obtained a mean divergence of 0.027 with a standard deviation of 0.005. It can then be checked that overfitting is avoided by trying $\overline{\mathcal{D}} = \{0,1\}^{10}$. Again, out of 1000 trials, a mean divergence of 0.029 is obtained, with a standard deviation still of order 0.005. Next comes [fig. 4.2] an example of estimation with an overstated finer partition, one sees that estimation accuracy does not collapse significantly.

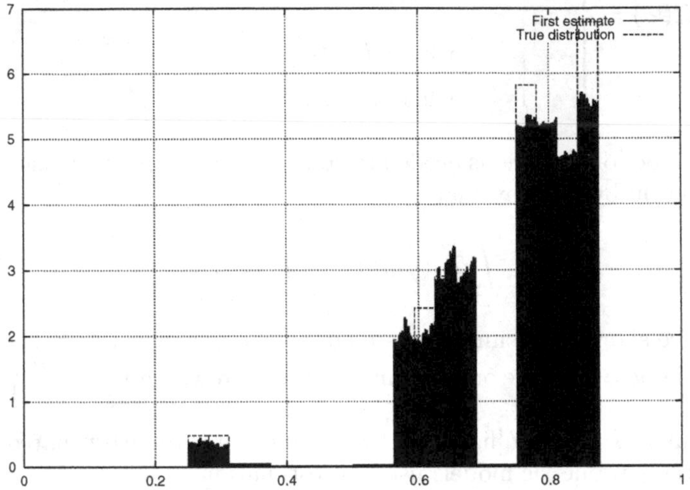


**Fig. 4.2.** $N = 1000$, $\overline{\mathcal{D}} = \{0,1\}^{10}$, $\mathcal{K} = 0,026$, $\beta = 0,155$

When the size of the sample is given different values, it is seen that the divergence $\mathcal{K}(\mathbb{P}, \hat{q}\mu)$ has linear variations, as it is the case for our theoretical upper bound.

| N | mean $\mathcal{K}$ | standard deviation |
|---|---|---|
| 100 | 0.24 | 0.02 |
| 1 000 | 0.029 | 0.005 |
| 10 000 | 0.002 4 | 0.000 46 |
| 100 000 | 0.000 23 | $5 \cdot 10^{-5}$ |

The following figures show the behaviour of the estimator applied to a mixture of Gaussian distributions. In this case, the influence of the bias is felt, because the true distribution does not belong to any of the models used by the estimator, and the dependence between the divergence of the true distribution with respect to the estimated one is more complex.

The software used to produce these examples can be downloaded from the author's web page :
`http://www.proba.jussieu.fr/users/catoni/homepage/newpage.`

The most CPU time consuming function is the one computing the weights $\Upsilon_s(x)$ (let alone the functions computing the divergence values, which serve to

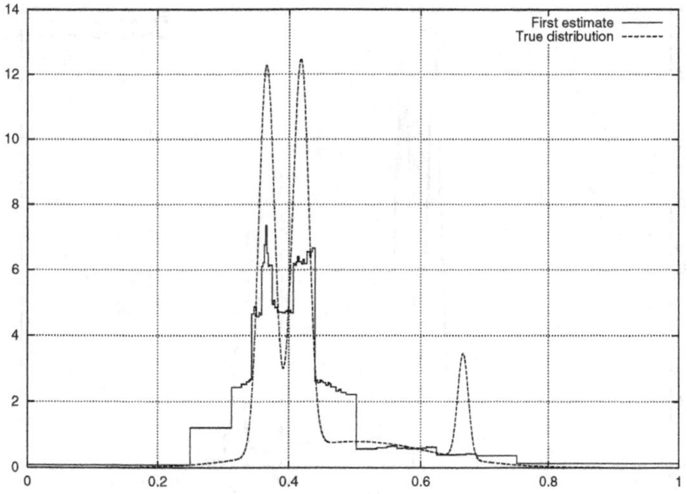


**Fig. 4.3.** $N = 100$, $\overline{\mathcal{D}} = \{0,1\}^9$, $\mathcal{K} = 0,321$, $\beta = 0,179$

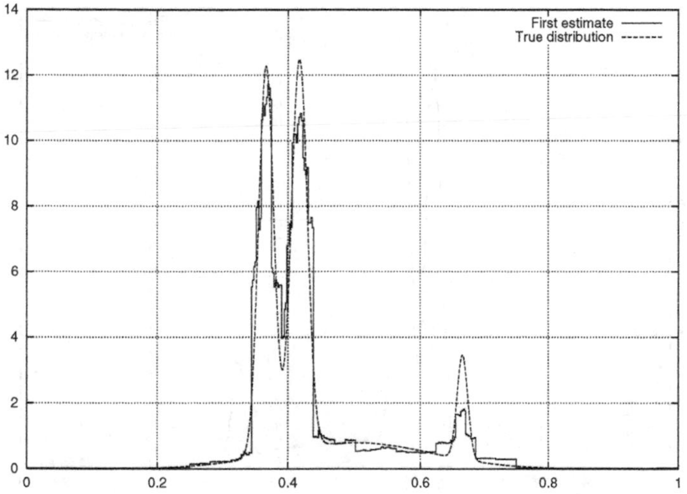


**Fig. 4.4.** $N = 1000$, $\overline{\mathcal{D}} = \{0,1\}^9$, $\mathcal{K} = 0,0533$, $\beta = 0,148$

monitor the performance of the algorithm in benchmark experiments where the true distribution of the sample is exactly known beforehand and can be compared with the estimated one).

To achieve a satisfactory numerical stability, it is necessary to carry on the computations on the logarithms of these weights, whose variations are exponential with the size of the sample. Representing in computer memory

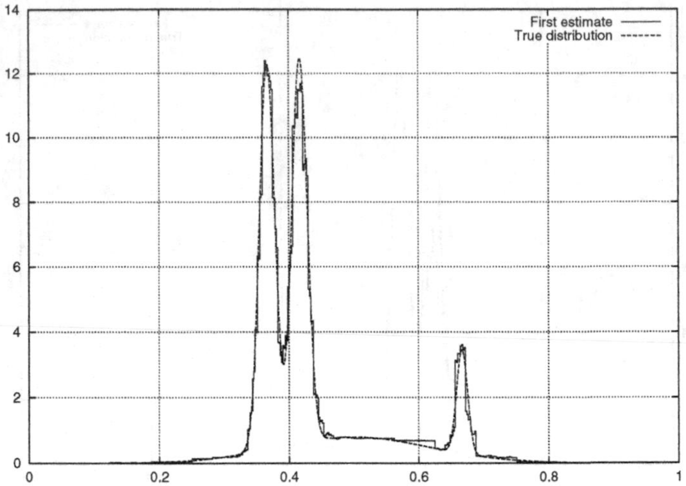


**Fig. 4.5.** $N = 10000$, $\overline{\mathcal{D}} = \{0,1\}^9$, $\mathcal{K} = 0,0111$, $\beta = 0,136$

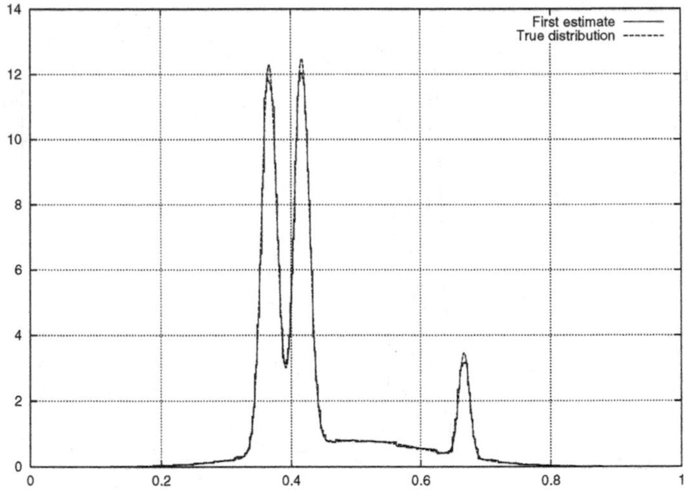


**Fig. 4.6.** $N = 100000$, $\overline{\mathcal{D}} = \{0,1\}^9$, $\mathcal{K} = 0,00232$, $\beta = 0,121$

the weights themselves would lead to impredictable results for large samples. The following code gave accurate results for all tested sample sizes (ranging from $10^2$ to $10^7$). In this implementation, trees are represented by static arrays, allowing fast indexation of the nodes through bit shifts and bit masks applied to array indices. This representation is well suited for histograms. For data compression (using context trees) a dynamic tree representation is usually prefered, as explained above, allowing for strongly unbalanced trees.

The function `WeightMix` takes as its arguments an array of weights `w` containing the conditional probabilities $\hat{\theta}^{X_1^K}(s_k \mid s_1^{k-1})$, an array of counters `c`, previously denoted as $c(s)$, $s \in \mathcal{S}$, a branching rate $\alpha$ (fixed to $1/2$ in the above discussion) and the value of the inverse temperature $\beta$.

```
#define LOGP(x,y) \
(((x)>(y))?(x)+log1p(exp((y)-(x))):(y)+log1p(exp((x-y))))
#define LASTB 1
#define OTHERB (~1)
#define BROTHER(i) (((i)&OTHERB)|((~(i))&LASTB))

typedef struct {
    int depth;
    int *first;
} Count;
/* a binary tree of counters.
 * tree nodes are indexed by integers, 1 for the root, 2 for
 * its left son, 3 for its right son, 4 for the left son of
 * its left son etc.
*/

typedef struct {
    int depth;
    double *first;
} Weight;
/* a binary tree of weights, could represent different
 * things : conditional probability densities, logarithms of
 *  conditional densities ...
*/

Weight *WeightMix(Weight *w, Count *c,
                    double alpha, double beta) {
    int depth, dd;
    int i,j,M,brother;
    Weight *mixW;
    double buff, *wp, *mixWp;
    int *cp;
    double sup;
    double ac, al, acl;
    double right;
    depth = w->depth;
    if (c->depth < depth) {
        depth = c->depth;
```

```
    }
    mixW = WeightNew(depth);
    M = 1 << depth;
    ac = 1 - alpha;
    al = log(alpha);
    acl = log(1-alpha);
    cp=c->first;
    wp=w->first;
    mixWp=mixW->first;
    for (i=(M>>1);i<M;i++) {
        right = al+(beta*cp[i<<1])*wp[i<<1]
            + (beta*cp[(i<<1)+1])*wp[(i<<1)+1];
        mixWp[i] = LOGP(acl,right);
    }
    for(i=(M>>1)-1;i;i--) {
        right = al+mixWp[i<<1]+mixWp[(i<<1)+1]
            + (beta*cp[i<<1])*wp[i<<1]
            + (beta*cp[(i<<1)+1])*wp[(i<<1)+1];
        mixWp[i] = LOGP(acl,right);
    }
    M = 1 << (depth+1);
    sup = 0;
    for(i=(1<<depth);i<M;i++) {
        brother = BROTHER(i);
        right = al+((beta*cp[i])+1)*wp[i]
            + (beta*cp[brother])*wp[brother];
        buff = LOGP(acl,right);
        for(j=(i>>1),dd=depth-1;j>1;j>>=1,dd--) {
            brother = BROTHER(j);
            right = al+buff+mixWp[brother]
                + ((beta*cp[j])+1)*wp[j]
                + (beta*cp[brother])*wp[brother];
            buff = LOGP(acl,right);
        }
        mixWp[i] = buff;
        if (buff > sup) {
            sup = buff;
    }
    /* normalizing the weights in two steps to
        make things numerically more stable */
    for (i=(1<<depth);i<M;i++) {
        mixWp[i] -= sup;
    }
    for (i=(1<<depth)-1;i;i--) {
        mixWp[i] =
```

```
            LOGP(mixW->first[i<<1],mixW->first[(i<<1)+1]);
    }
    for (i=M-1;i;i--) { /* back from the log
                                    representation of weights */
        mixWp[i] = exp(mixWp[i]-mixWp[1]);
    }
    return mixW;
}
```

## 4.3 Mathematical framework for density estimation

We set here the mathematical framework for density estimation. We use the term density estimation, because it is well established, but in fact, we do not require the true sample distribution to have a density with respect to the reference measure we consider. It is only required that the estimator has a density with respect to a given reference measure on the sample space, and weak convergence may occur to a true distribution which does not satisfy this property.

Let us start with a convenient and classical

**Definition 4.3.1.** A regular conditional probability distribution $\nu(dX \mid Y)$ on the product $\mathcal{X} \times \mathcal{Y}$ of two measurable spaces $(\mathcal{X}, \mathcal{B})$ and $(\mathcal{Y}, \mathcal{F})$ will be a map $\nu : \mathcal{B} \times Y \to \mathbb{R}$ such that

1. for any $y \in Y$, the map $A \mapsto \nu(A, y)$ is a probability measure on $(\mathcal{X}, \mathcal{B})$,
2. for any $A \in \mathcal{B}$, the map $y \mapsto \nu(A, y)$ is measurable with respect to $\mathcal{F}$.

Let $(\mathcal{X}, \mathcal{B})$ be a measurable sample space, and let $(X_1, \dots, X_N)$ be an i.i.d. sample of observations drawn with respect to some distribution $P_N$. In practice $P_N = P^{\otimes N} \in \mathcal{M}_+^1(\mathcal{X}^N, \mathcal{B}^{\otimes N})$ will be a product measure, but from the technical point of view, we will only use the fact that $P_N$ is the marginal of rank $N$ of some exchangeable distribution defined on some larger power of $\mathcal{X}$, $\mathcal{X}^M$, with $M > N$. The question under investigation is to estimate $P$ (or $P_M(dX_{N+1} \mid X_1^N)$, depending on the setting). This means we want to build a regular conditional probability distribution $\hat{P}(dX \mid X_1, \dots, X_N)$ on $(\mathcal{X}^{N+1}, \mathcal{B}^{\otimes(N+1)})$ which is on the average close to $P$ (in some sense to be made more precise in the sequel).

It is well known that there is in general no answer to this question, unless we give it a more restricted meaning, because the space of all probability measures on $\mathcal{X}$ is usually too big (except when $\mathcal{X}$ is a small finite set). One classical way of making the question well posed would be to impose some restrictive hypotheses on the unknown distribution $P$. The alternative way is to impose some restrictions on the set of distributions among which the estimator is to be chosen. This is somehow more realistic, because this set is left to our choice, whereas the properties of $P$ are in most practical cases unknown. Dealing with an unknown sample distribution within a restricted class of estimators is an active research topic in more than one scientific community, among which we can cite researches in nonparametric adaptive statistics (too many prominent contributors to cite them all), in statistical learning theory (Vapnik's school), in information theory (Rissanen's MDL school) and in PAC learning (Valliant, Schapire, Mansour, McAllester, ...).

Accordingly to what we just explained, we will focus on some restricted family of distributions $\{Q_\theta \in \mathcal{M}_+^1(\mathcal{X}, \mathcal{B})\,;\, \theta \in \Theta\}$. This notation is somewhat redundent, we could have chosen to consider a subset $\mathcal{Q} \subset \mathcal{M}_+^1(\mathcal{X}, \mathcal{B})$. Anyhow, as we will have to define a prior probability measure on the set $\mathcal{Q}$, and therefore a sigma-algebra on this set, we prefer to use the setting of conditional probability measures on a product space $\Theta \times \mathcal{X}$. More precisely, we will assume that $(\Theta, \mathcal{B}')$ is a measurable space, and that the family $\{Q_\theta : \theta \in \Theta\}$ is a regular conditional probability measure on $\bigl(\mathcal{X} \times \Theta, \mathcal{B} \otimes \mathcal{B}'\bigr)$. We will also assume that we can find a reference measure $\mu \in \mathcal{M}_+^1(\mathcal{X}, \mathcal{B})$ such that for any $\theta \in \Theta$, the measure $Q_\theta$ is absolutely continuous with respect to $\mu$ and such that there exists for each $\theta$ a version of the Radon-Nikodym derivative $\frac{Q_\theta}{\mu}$ such that the map $(\theta, x) \mapsto \frac{Q_\theta}{\mu}(x)$ is measurable with respect to the product sigma algebra $\mathcal{B}' \otimes \mathcal{B}$ on $\Theta \times \mathcal{X}$.

Another important specificity of the estimation scheme we will introduce is that the estimator $\hat{P}(dX \,|\, X_1, \dots, X_N)$ will not be chosen from the family $\{Q_\theta : \theta \in \Theta\}$ itself, but from the mixtures $\{\int_{\theta \in \Theta} Q_\theta \rho(d\theta \,|\, X_1, \dots, X_N)\}$, where $\rho(d\theta \,|\, X_1, \dots, X_N)$ ranges among all the possible regular conditional probability distributions on $\Theta \times \mathcal{X}^N$. This framework is familiar in the setting of Bayesian estimation. However, we will not use a Bayesian estimator, as the reader will see, but some pseudo-Bayesian posterior $\rho$ which is more suited to obtain a non asymptotic oracle inequality under weak hypotheses. Understanding why it is often more efficient to estimate a posterior $\rho$ on the parameter space rather than a given value of the parameter $\hat{\theta}(X_1, \dots, X_N)$ is in our opinion a deep and still partially open question. Anyhow, we produce here some toy counter example showing that there is no parameter estimator $\hat{\theta}(X_1, \dots, X_N)$ achieving in all circumstances the same performance as the posterior $\rho(d\theta \,|\, X_1, \dots, X_N)$ we will use.

Our approach is "pseudo-Bayesian" in the sense that we use of some prior probability measure $\pi \in \mathcal{M}_+^1(\Theta, \mathcal{B}')$ on the parameter space. However, it is to be understood that this measure has nothing to do with what our expectations might be about the true sample distribution $P$ we are confronted with: we will obtain worst case bounds for the average risk, meaning bounds which hold for any specific value of $P$. The choice of $\pi$ will on the other hand influence the value of the bound we will get for any particular $P$: changing the prior $\pi$ will make the bound tighter for some values of $P$, while it will make it looser for other values of $P$.

All these preliminary explanations being given, we are ready for the definition of the Gibbs estimator:

**Definition 4.3.2.** Let $\pi \in \mathcal{M}_+(\Theta)$ be a positive sigma-finite measure on the parameter space $\Theta$. Let $\beta \in ]0,1]$, $\alpha \in ]\beta, 1]$ and $N \in \mathbb{N}$ be given. Let us assume that the map

$$(\theta, x) \mapsto q(\theta, x) \stackrel{\text{def}}{=} \frac{Q_\theta}{\mu}(x)$$

satisfies the set of hypotheses $\mathcal{H}(N+1, \beta, \alpha)$ given in appendix 4.9. (Let us remind the purists that we chose once for all to work with a given realisation of the derivatives $\frac{Q_\theta}{\mu}$. The hypotheses of appendix 4.9 are for instance fulfilled when this realisation is such that the function $(\theta, x) \mapsto \log\left(\frac{Q_\theta}{\mu}(x)\right)$ is bounded and $\pi$ is a finite measure.) We define the Gibbs estimator at inverse temperature $\beta \in \mathbb{R}_+$ to be the regular conditional probability distribution

$$G_\beta^N(dX \mid X_1, \dots, X_N) = \frac{\mathbb{E}_{\pi(d\theta)}\left(\left(\prod_{n=1}^N \frac{Q_\theta}{\mu}(X_n)\right)^\beta Q_\theta(dX)\right)}{\mathbb{E}_{\pi(d\theta)}\left(\left(\prod_{n=1}^N \frac{Q_\theta}{\mu}(X_n)\right)^\beta\right)}.$$

The fact that it is for any $(X_1, \dots, X_N)$ a probability measure is a consequence of the monotone convergence theorem. The fact that it is regular is part of Fubini's theorem (applied separately to the numerator and denominator).

In the study of the Gibbs estimator, an important role will be played by the log-Laplace transform introduced in appendix 4.9:

$$\mathcal{E}^{X_1^{N+1}}(\alpha) = \log \mathbb{E}_{\pi(d\theta)}\left(\left(\prod_{n=1}^N \frac{Q_\theta}{\mu}(X_n)\right)^\beta \left(\frac{Q_\theta}{\mu}(X_{N+1})\right)^\alpha\right),$$

where a dummy (i.e. non observed) variable $X_{N+1}$ has been introduced for convenience. Note also that $X_1^{N+1}$ is a shorthand for $(X_1, \dots, X_{N+1})$. Depending on his background, the reader may also like to see the function $\mathcal{E}^{X_1^N}$ as a free-energy function, or as the logarithm of the moment generating function of an exponential family of distributions. As it can be seen from the computations made in appendix 4.9, computing derivatives of $\mathcal{E}^{X_1^N}$ involves the family of posterior regular conditional distributions

$$\rho_{\beta,\alpha}^{X_1^{N+1}}(d\theta) = \frac{\left(\prod_{n=1}^N \frac{Q_\theta}{\mu}(X_n)\right)^\beta \left(\frac{Q_\theta}{\mu}(X_{N+1})\right)^\alpha \pi(d\theta)}{\mathbb{E}_{\pi(d\theta')}\left(\left(\prod_{n=1}^N \frac{Q_{\theta'}}{\mu}(X_n)\right)^\beta \left(\frac{Q_{\theta'}}{\mu}(X_{N+1})\right)^\alpha\right)}$$

on the parameter space $\Theta$. The notation $\rho_{\beta,\alpha}^{X_1^{N+1}}(d\theta)$ is a short hand for

$$\rho_{\beta,\alpha}(d\theta \mid X_1, \dots, X_{N+1}),$$

which will be useful to make equations fit in the page width.

Note that $\rho_{\beta,0}^{X_1^{N+1}}$ does not depend on $X_{N+1}$: we will also write it as $\rho_\beta^{X_1^N}$. The Gibbs estimator can then be written as

$$G_\beta^N(dX_{N+1} \mid X_1, \dots, X_N) = \mathbb{E}_{\rho_\beta^{X_1^N}(d\theta)}\big(Q_\theta(dX_{N+1})\big).$$

This justifies the name "Gibbs estimator": indeed the measure $\rho_\beta^{X_1^N}$ can be seen as a Gibbs measure, related to the hamiltonian:

$$h(\theta\,;\,X_1^N) = -\frac{1}{N}\sum_{n=1}^{N} \log\left(\frac{Q_\theta}{\mu}(X_n)\right),$$

since it can be written as

$$\rho_\beta^{X_1^N}(d\theta) = \frac{\exp\Big(-\beta N h(\theta\,;\,X_1^N)\Big)\pi(d\theta)}{\mathbb{E}_{\pi(d\theta')}\Big(\exp\Big(-\beta N h(\theta'\,;\,X_1^N)\Big)\Big)}.$$

(A Gibbs measure is simply a measure obtained by an exponential change of measure involving an energy function — here $h(\theta\,;\,X_1, \dots, X_N)$ — and an exponent called the inverse temperature — here $N\beta$, — from a reference measure — here $\pi$.) The normalisation given to $h$ in this attempt to justify our terminology is quite natural since when the sample distribution is i.i.d., $h$ will converge $P^{\otimes \mathbb{N}}$ a.s. to a finite limit from the strong law of large numbers (at least when $\log \frac{Q_\theta}{\mu}(X)$ has a first moment).

The theorems in this section are concerned with the high temperature region $\beta \in ]0, 1/2]$. They provide oracle inequalities for the mean divergence of the true sample distribution with respect to the estimator. Namely, assuming that there exists a regular version of $P(dX_{N+1} \mid X_1, \dots, X_N)$, they provide a bound for

$$\mathbb{E}_{P_{N+1}}\Big(\mathcal{K}\big(P_{N+1}(dX_{N+1} \mid X_1, \dots, X_N), G_\beta^N(dX_{N+1} \mid X_1, \dots, X_N)\big)\Big),$$

where $\mathcal{K}$ is the Kullback Leibler divergence: let us recall that the divergence $\mathcal{K}(m_1, m_2)$ of two probability measures $m_1$ and $m_2$ is defined as

$$\mathcal{K}(m_1, m_2) \stackrel{\text{def}}{=} \begin{cases} \mathbb{E}_{m_1}\left(\frac{m_1}{m_2}\right) & \text{if } m_1 \ll m_2 \\ +\infty & \text{otherwise.} \end{cases}$$

To put things into perspective, let us remark that when $\beta = 1$ the Gibbs estimator $G_1^N$ is a Bayesian estimator: it minimizes the Bayesian risk

$$\mathbb{E}_{\pi(d\theta)}\mathbb{E}_{Q_\theta^{\otimes N}(dX_1^N)}\Big(\mathcal{K}\big(Q_\theta(dX_{N+1}), \hat{Q}(dX_{N+1} \mid X_1, \dots, X_N)\big)\Big)$$

over all possible choices of regular conditional probabilities $\hat{Q}(dX_{N+1} \mid X_1, \ldots, X_N)$. Moreover, when the maximum likelihood is achieved for a single value of the parameter $\theta$, the Gibbs estimator converges to the maximum likelihood estimator when $\beta$ tends to $+\infty$. Therefore we can see the Gibbs estimator as a "thermalized" version of both the Bayesian and the maximum likelihood estimators. Working in the high temperature region $\beta < 1$ can also be interpreted as a deliberate underestimation of the sample size: To compute the Gibbs estimator, we plug the empirical distribution of $N$ observations into the Bayesian estimator for a sample of size $\beta N$.

The main reason for introducing the Gibbs estimator is that it satisfies a rather sharp non asymptotic oracle inequality which is the statistical pendent of the universal compression properties of universal codes in information theory.

## 4.4 Main oracle inequality

**Theorem 4.4.1.** *In the setting and under the hypotheses described in definition 4.3.2,*

$$\mathbb{E}_{P_{N+1}}\left(\log \frac{\mu}{G_\beta^N}(X_{N+1} \mid X_1^N)\right) \leq \inf_{\theta \in \Theta_1} \left\{ \mathbb{E}_{P_{N+1}}\left(\log \frac{\mu}{Q_\theta}(X_{N+1})\right) + \frac{\gamma_\beta(\theta)}{\beta(N+1)} \right\}, \quad (4.4.1)$$

*where $\Theta_1$ and $\gamma_\beta(\theta)$ are defined by equations* (4.9.6) *and* (4.9.7) *in appendix 4.9. Moreover, according to what is assumed in defintion 4.3.2, the "inverse temperature" parameter $\beta$ is assumed to satisfy condition* (4.9.9)*, which itself depends on condition* (4.9.8)*.*

*As a consequence, in the case when there exists a regular version of $P_{N+1}(dX_{N+1} \mid X_1^N)$,*

$$\mathbb{E}_{P_{N+1}(dX_1^N)} \mathcal{K}\big(P_{N+1}(dX_{N+1} \mid X_1^N), G_\beta^N(dX_{N+1} \mid X_1^N)\big) \leq \inf_{\theta \in \Theta_1} \left\{ \mathbb{E}_{P_{N+1}(dX_1^N)} \mathcal{K}\big(P_{N+1}(dX_{N+1} \mid X_1^N), Q_\theta(dX_{N+1})\big) + \frac{\gamma_\beta(\theta)}{\beta(N+1)} \right\}, \quad (4.4.2)$$

*where $\mathcal{K}(.,.)$ is the Kullback Leibler divergence.*

*Proof.* The first part (4.4.1) is just a rewriting of theorem 4.9.1. To prove the second part, let us remark that (4.4.2) is trivial when

$$\inf_{\theta \in \Theta_1} \mathbb{E}_{P_{N+1}(dX_1^N)} \mathcal{K}\big(P_{N+1}(dX_{N+1} \mid X_1^N), Q_\theta(dX_{N+1})\big) = +\infty.$$

Otherwise, there is $\theta_1 \in \Theta_1$ such that

$$\mathbb{E}_{P_{N+1}(dX_1^N)} \mathcal{K}(P_{N+1}(dX_{N+1} \,|\, X_1^N), Q_{\theta_1}(dX_{N+1})) < +\infty.$$

This means that

$$X_1^{N+1} \mapsto \log \left( \frac{P(dX_{N+1} \,|\, X_1^N)}{Q_{\theta_1}(dX_{N+1})} \right) \in \mathbb{L}^1(\mathcal{X}^{N+1}, P_{N+1}).$$

Moreover by definition of $\Theta_1$,

$$X_1^{N+1} \mapsto \log \left( \frac{Q_{\theta_1}(dX_{N+1})}{\mu(dX_{N+1})} \right) \in \mathbb{L}^1(\mathcal{X}^{N+1}, P_{N+1}).$$

Thus the sum of the two previous functions is also integrable :

$$X_1^{N+1} \mapsto \log \left( \frac{P_{N+1}(dX_{N+1} \,|\, X_1^N)}{\mu(dX_{N+1})} \right) \in \mathbb{L}^1(\mathcal{X}^{N+1}, P_{N+1}),$$

and consequently we go from (4.4.1) to (4.4.2) by substracting on both side the finite constant

$$\mathbb{E}_{P_{N+1}(dX_1^{N+1})} \left[ \log \left( \frac{P_{N+1}(dX_{N+1} \,|\, X_1^N)}{\mu(dX_{N+1})} \right) \right].$$

□

*Remark* 4.4.1. Note that the definition of $G_\beta^N$ depends on the choice of the reference measure $\mu$ only through the assumption that for each $\theta$, $Q_\theta$ is absolutely continuous with respect to $\mu$. Replacing $\mu$ by an equivalent probability measure $\mu'$ does not change the definition of $G_\beta^N$ (by equivalent, we mean that $\mu \ll \mu'$ and $\mu' \ll \mu$). However it may change the definition of $\Theta_1$ as well as the integrability assumption (4.9.4).

*Remark* 4.4.2. The hypotheses of the theorem are for instance fulfilled for any sample distribution $P_{N+1}$ when $\pi$ is a finite measure and the function $(\theta, x) \mapsto \log \frac{Q_\theta}{\mu}(x)$ is bounded. In this case $\Theta_1 = \Theta$ and the value of $\beta$ can be set independently from $P_{N+1}$ using remark 4.9.3.

*Remark* 4.4.3. In the case of an exponential model, we can give for $\chi_{\beta,\alpha}$ (see (4.9.8)), $\beta$ (see (4.9.9)) and $\gamma_\beta(\theta)$ (see (4.9.7)) worst case upper bounds with respect to $P_{N+1}$ which depend only on the entropy structure of the model. Let us restrict for simplicity to the case when $\mathcal{X}$ is finite, and assume that $\{Q_\theta \,;\, \theta \in \Theta \subset \mathbb{R}^d\}$ is defined by

$$Q_\theta(dx) = \frac{\exp\langle \theta, f(x) \rangle}{Z(\theta)} \mu(dx),$$

where $f : \mathcal{X} \to \mathbb{R}^d$, $\mu$ is everywhere positive, and where $\Theta$ is the support of the prior distribution $\pi$. For any values of $(x_1, \dots, x_{N+1}) \in \mathcal{X}^{N+1}$, $(\beta, \xi) \in [0,1]^2$, we can define $\widehat{Q}_{\beta,\xi}^{x_1^{N+1}}$ in the closure $\overline{\{Q_\theta \,;\, \theta \in \mathbb{R}^d\}}$ of $\{Q_\theta \,;\, \theta \in \mathbb{R}^d\}$ such that

$$\mathbb{E}_{\widehat{Q}_{\beta,\xi}^{x_1^{N+1}}(dX)}\bigl(f(X)\bigr) = \frac{1}{\beta N + \xi}\left(\sum_{i=1}^{N} \beta f(x_i) + \xi f(x_{N+1})\right).$$

It is not necessarily unique and can be obtained by minimization of the non-negative convex function

$$\theta \longmapsto \mathcal{K}(\widehat{P}_{\beta,\xi}^{x_1^{N+1}}, Q_\theta),$$

where

$$\widehat{P}_{\beta,\xi}^{x_1^N} = \frac{1}{\beta N + \xi}\left(\sum_{i=1}^{N} \beta \delta_{x_i} + \xi \delta_{x_{N+1}}\right).$$

Then we can write $\rho_{\beta,\xi}^{x_1^{N+1}}(d\theta)$ as

$$d\rho_{\beta,\xi}^{x_1^{N+1}}(\theta) \propto \exp\Bigl(-(\beta N + \xi)\mathcal{K}(\widehat{Q}_{\beta,\xi}^{x_1^{N+1}}, Q_\theta)\Bigr)\, d\pi(\theta),$$

where the symbol $\propto$ means that we have omitted to write the normalizing constant that turns $\rho_{\beta,\xi}^{x_1^{N+1}}$ into a probability distribution. Accordingly, we can bound $\gamma_\beta(\theta)$ by

$$\sup_{\theta \in \Theta} \gamma_\beta(\theta) \leq$$
$$\sup_{\bar{Q} \in \overline{\{Q_\theta\,;\,\theta \in \mathbb{R}^d\}}} \frac{\mathbb{E}_{\pi(d\theta)}\Bigl[\beta(N+1)\mathcal{K}(\bar{Q}, Q_\theta)\exp\Bigl(-\beta(N+1)\mathcal{K}(\bar{Q}, Q_\theta)\Bigr)\Bigr]}{\mathbb{E}_{\pi(d\theta)}\Bigl[\beta(N+1)\mathcal{K}(\bar{Q}, Q_\theta)\Bigr]}.$$

In the same way, putting

$$\rho_{\beta,\xi}^{\bar{Q}}(d\theta) \propto \exp\Bigl(-(\beta N + \xi)K(\bar{Q}, Q_\theta)\Bigr)\pi(d\theta),$$

and

$$M^3_{\rho(d\theta)}\bigl(g(\theta)\bigr) = E_{\rho(d\theta)}\bigl(g(\theta) - E_{\rho(d\theta')}g(\theta')\bigr)^3,$$

we can bound $\chi_{\beta,\alpha}$ by

$$\chi_{\beta,\alpha} \leq 0 \vee \sup_{\xi \in [0,\alpha]} \sup_{\bar{Q} \in \overline{\{Q_\theta\,;\,\theta \in \mathbb{R}^d\}}} \sup_{x \in \mathcal{X}} \frac{-M^3_{\rho_{\beta,\xi}^{\bar{Q}}(d\theta)}\left(\log \frac{Q_\theta}{\mu}(x)\right)}{\mathrm{Var}_{\rho_{\beta,\xi}^{\bar{Q}}(d\theta)}\left(\log \frac{Q_\theta}{\mu}(x)\right)}.$$

In conclusion, when theorem 4.4.1 is applied to an exponential model $\{Q_\theta : \theta \in \Theta\}$, the coefficients $\sup_{\theta \in \Theta} \gamma_\beta(\theta)$ and $\chi_{\beta,\alpha}$ corresponding to an arbitrary exchangeable sample distribution $P_{N+1}$ can be bounded by expressions where the empirical distribution has been replaced by the worst case distribution $\bar{Q}$ in the closure $\overline{\{Q_\theta\,;\,\theta \in \mathbb{R}^d\}}$ of the larger model obtained by letting the parameter $\theta$ range in the whole $\mathbb{R}^d$.

Theorem 4.4.1 is concerned with values of $\beta$ in the region $]0, 1/2[$. Another interesting value of the temperature is $\beta = 1/2$. To cover this case, we can adopt the set of hypotheses $\mathcal{H}'(N+1, \beta)$ described in appendix 4.10 and define the third centered moment of the function $h(\theta)$ with respect to the distribution $\rho \in \mathcal{M}_+^1(\Theta)$ to be

$$M^3_{\rho(d\theta)}\bigl[h(\theta)\bigr] = E_{\rho(d\theta)}\Bigl[\bigl[h(\theta) - E_{\rho(d\theta')}h(\theta')\bigr]^3\Bigr].$$

Theorem 4.10.1 of appendix 4.10 reads in this framework as

**Theorem 4.4.2.** *In the setting described in definition 4.3.2, when hypothesis $\mathcal{H}(N+1, \beta, \alpha)$ is replaced by $\mathcal{H}'(N+1, 1/2)$ described in appendix 4.10,*

$$\begin{aligned}\mathbb{E}_{P_{N+1}}&\left[\log \frac{\mu}{G^N_{1/2}}\bigl(X_{N+1} \,|\, X_1^N\bigr)\right] \\ &\leq \inf_{\theta \in \Theta_1} \left\{ \mathbb{E}_{P_{N+1}} \left( \log \frac{\mu}{Q_\theta}(X_{N+1}) \right) + \frac{2\gamma(\theta)}{(N+1)} \right\} \\ &\quad + \frac{1}{24} \sup_{\eta \in [0,1]} -\mathbb{E}_{P_{N+1}} M^3_{\rho^{X_1^{N+1}}_{1/2,\eta}(d\theta)} \left[ \log \left( \frac{Q_\theta}{\mu}(X_{N+1}) \right) \right]. \end{aligned} \quad (4.4.3)$$

*As a consequence, in the case when there exists a regular version of $P(dX_{N+1} \,|\, X_1^N)$,*

$$\begin{aligned}&\mathbb{E}_{P_{N+1}(X_1^N)} \mathcal{K}\bigl(P_{N+1}(dX_{N+1} \,|\, X_1^N), G^N_{1/2}(dX_{N+1} \,|\, X_1^N)\bigr) \\ &\leq \inf_{\theta \in \Theta_1} \left\{ \mathbb{E}_{P_{N+1}(dX_1^N)} \mathcal{K}\bigl(P_{N+1}(dX_{N+1} \,|\, X_1^N), Q_\theta(dX_{N+1})\bigr) + \frac{2\gamma_{1/2}(\theta)}{N+1} \right\} \\ &\quad + \frac{1}{24} \sup_{\eta \in [0,1]} -\mathbb{E}_{P_{N+1}} M^3_{\rho^{X_1^{N+1}}_{1/2,\eta}(d\theta)} \left[ \log \left( \frac{Q_\theta}{\mu}(X_{N+1}) \right) \right]. \end{aligned} \quad (4.4.4)$$

## 4.5 Checking the accuracy of the bounds on the Gaussian shift model

It is instructive to illustrate theorems 4.4.1 and 4.4.2 with the Gaussian shift model, where we can compute everything explicitly. In this exponential model $\mathcal{X} = \mathbb{R}^d$,

$$\mu(dx) = (2\pi)^{-d/2} (\det H)^{-1/2} \exp\left( -\frac{1}{2} \langle x, H^{-1} x \rangle \right) dx$$

where "$dx$" is the Lebesgue measure on $\mathbb{R}^d$ and $H$ is a symmetric positive definite matrix, $f(x) = x$, so that

$$Q_\theta(dx) = (2\pi)^{-d/2} (\det H)^{-1/2} \exp(-\frac{1}{2} \|H^{-1/2}(x - H\theta)\|^2) dx,$$

and $\pi(d\theta) = d\theta$ (the Lebesgue measure).

Here the relative entropy takes the form

$$\mathcal{K}(Q_\theta, Q_{\theta'}) = \frac{1}{2}\|H^{1/2}(\theta - \theta')\|^2.$$

Moreover

$$\gamma_\beta(\theta) \leq \sup_{\hat{\theta}} \frac{E_{d\theta'} \frac{\beta(N+1)}{2}\|H^{1/2}(\theta' - \theta)\|^2 \exp(-\frac{\beta(N+1)}{2}\|H^{1/2}(\theta' - \hat{\theta})\|^2)}{E_{d\theta'} \exp(-\frac{\beta(N+1)}{2}\|H^{1/2}(\theta' - \hat{\theta})\|^2)} - \frac{\beta(N+1)}{2}\|H^{1/2}(\hat{\theta} - \theta)\|^2 = \frac{d}{2}.$$

We have also

$$\rho_{\beta,\xi}^{x_1^(N+1)}(d\theta) = \left(\frac{\beta N + \xi}{2\pi}\right)^{d/2} (\det H)^{1/2} \exp\left(-\frac{\beta N + \xi}{2}\|H^{1/2}(\theta - \bar{\theta}_{\beta,\xi})\|^2\right) d\theta,$$

where

$$\bar{\theta}_{\beta,\xi} = H^{-1}\left(\frac{1}{\beta N + \xi}\left(\sum_{i=1}^{N} \beta x_i + \xi x_{N+1}\right)\right).$$

Making the computation in a orthonormal basis of eigenvectors of $H$, we obtain easily that

$$\begin{aligned}
&\mathrm{Var}_{\rho_{\beta,\xi}^{x_1^{N+1}}(d\theta)}\left[\log\left(\frac{Q_\theta}{\mu}(x_{N+1})\right)\right] \\
&\quad = \mathrm{Var}_{\rho_{\beta,\xi}^{x_1^{N+1}}(d\theta)}\left[-\frac{1}{2}\|H^{-1/2}(x_{N+1} - H\theta)\|^2\right] \\
&\quad = (\beta N + \xi)^{-2}\left(\frac{d}{2} + (\beta N + \xi)\|H^{-1/2}(x_{N+1} - H\bar{\theta}_{\beta,\xi})\|^2\right),
\end{aligned}$$

$$\begin{aligned}
&M^3_{\rho_{\beta,\xi}^{x_1^{N+1}}(d\theta)}\left[\log\left(\frac{Q_\theta}{\mu}(x_{N+1})\right)\right] \\
&\quad = M^3_{\rho_{\beta,\xi}^{x_1^{N+1}}(d\theta)}\left[-\frac{1}{2}\|H^{-1/2}(x_{N+1} - H\theta)\|^2\right] \\
&\quad = -(\beta N + \xi)^{-3}\left(d + 3(\beta N + \xi)\|H^{-1/2}(x_{N+1} - H\bar{\theta}_{\beta,\xi})\|^2\right).
\end{aligned}$$

Thus $\chi_{\beta,1} \leq \frac{3}{\beta N}$ and for $N > 9$ we can take in theorem 4.4.1

$$\beta = \frac{1}{2} - \frac{1}{N+1}.$$

We obtain that

$$E_{P_{N+1}}\mathcal{K}(P_{N+1}(dX_{N+1}\,|\,X_1^N), G_{1/2-1/(N+1)}(dX_{N+1}\,|\,X_1^N))$$
$$\leq \inf_{\theta\in\Theta} E_{P_{N+1}}\mathcal{K}(P(dX_{N+1}\,|\,X_1^N), Q_\theta(dX_{N+1})) + \frac{d}{N+1}\left(1-\frac{2}{N+1}\right)^{-1}.$$

In the case when $P_{N+1} = P^{\otimes(N+1)}$ is a product measure having a second moment $E_P(\|X\|^2) < +\infty$,

$$E_{P^{\otimes(N+1)}}\|H^{-1/2}(X_{N+1} - H\bar{\theta}_{\beta,\xi})\|^2$$
$$= E_P\|H^{-1/2}(X - E_P(X))\|^2\left(\left(1-\frac{\xi}{\beta N+\xi}\right)^2 + \frac{\beta^2 N}{(\beta N+\xi)^2}\right)$$
$$\leq \frac{N+1}{N}E_P\|H^{-1/2}(X-E_P(X))\|^2.$$

Therefore applying theorem 4.4.2 we obtain that

$$E_{P^{\otimes N}(dX_1^N)}\mathcal{K}(P, G_{1/2}^N(dX_{N+1}\,|\,X_1^N))$$
$$\leq \inf_{\theta\in\Theta}\mathcal{K}(P,Q_\theta) + \frac{d}{N+1}$$
$$+ \frac{N+1}{2N^3}E_P\|H^{-1/2}(X-E_P(X))\|^2 + \frac{d}{3N^3}$$

As in this case $G_\beta^N(dX_{N+1}\,|\,X_1^N)$ can be explicitly computed, we can check to which extent theorems 4.4.1 and 4.4.2 are sharp for the Gaussian shift model: putting

$$\bar{\theta} = H^{-1}\left(\frac{1}{N}\sum_{i=1}^N x_i\right)$$

we have

$$G_\beta^N(dx_{N+1}\,|\,x_1^N) = E_{\rho_{\beta,0}^{x_1^{N+1}}(d\theta)} Q_\theta(dx_{N+1})$$
$$= \int_{\theta\in\mathbb{R}^d}\left(\frac{\beta N}{2\pi}\right)^{d/2}(\det H)^{1/2}\exp\left(-\frac{\beta N}{2}\|H^{1/2}(\theta-\bar{\theta})\|^2\right)$$
$$\times(2\pi)^{-d/2}(\det H)^{-1/2}\exp\left(-\frac{1}{2}\|H^{-1/2}(x_{N+1}-H\theta)\|^2\right)d\theta dx_{N+1}$$
$$= \int_\theta\left(\frac{\beta N}{(2\pi)^2}\right)^{d/2}\exp\left(-\frac{\beta N+1}{2}\|H^{1/2}(\theta-\bar{\theta})\|^2\right.$$
$$\left.+\langle(x_{N+1}-H\bar{\theta}),(\theta-\bar{\theta})\rangle - \frac{1}{2}\|H^{-1/2}(x_{N+1}-H\bar{\theta})\|^2\right)d\theta dx_{N+1}$$
$$= (2\pi)^{-d/2}(\det H)^{-1/2}\left(\frac{\beta N}{\beta N+1}\right)^{d/2}$$

$$\times \exp\left(-\frac{1}{2}\frac{\beta N}{\beta N+1}\|H^{-1/2}\left(x_{N+1}-\frac{1}{N}\sum_{i=1}^{N}x_i\right)\|^2\right)dx_{N+1}.$$

We can then explicitly compute

$$\begin{aligned}
&\sup_\theta E_{P^{\otimes(N+1)}(dX_1^{N+1})}\left[\log\left(\frac{Q_\theta(dX_{N+1})}{G_\beta^N(dX_{N+1}\,|\,X_1^N)}\right)\right]\\
&= \frac{d}{2}\log\frac{\beta N+1}{\beta N}\\
&\quad+\frac{1}{2}\frac{\beta N}{\beta N+1}E_{P^{\otimes(N+1)}(dX_1^{N+1})}\|H^{-1/2}\left(X_{N+1}-\frac{1}{N}\sum_{i=1}^{N}X_i\right)\|^2\\
&\quad-\inf_\theta\frac{1}{2}E_{P(dX_{N+1})}\|H^{-1/2}(X_{N+1}-H\theta)\|^2\\
&=\frac{d}{2}\log\left(1+\frac{1}{\beta N}\right)\\
&\quad-\frac{1-\beta}{2(\beta N+1)}E_{P(dX_{N+1})}\|H^{-1/2}(X_{N+1}-E_P(X))\|^2.
\end{aligned}$$

This shows that in case $\beta = 1/2 - 1/(N+1)$ the evaluation given by theorem 4.4.1 is sharp for the worst $P$ (the case when its variance is going to zero). In theorem 4.4.2 the leading term is also sharp in the worst case. It is also interesting to notice that the average risk of the Bayesian estimator $Q_1^{N+1}$ is independent of $P$ and that $G_\beta^{N+1}$ is asymptotically minimax when the sample distribution $P$ is supposed to be drawn from the model $\{Q_\theta\,;\,\theta\in\Theta\}$ for any $\beta \in ]0,1]$. Indeed, the choice of $\beta$ does not influence the prediction made for the mean of $P$. This is a good thing, because in some situations it is necessary to make a "pessimistic" choice of $\beta$ to meet the hypotheses of theorem 4.4.1, and it is hopeful that such a pessimistic choice would not systematically prevent the estimator from having an optimal behaviour. In the case when $P \notin \{Q_\theta\,;\,\theta\in\Theta\}$, smaller values of $\beta$ do a better job when the variance of $P$ is larger than in the model and larger values of $\beta$ do a better job when the variance of $P$ is smaller than in the model.

These easy computations made on the Gaussian shift model can be expected to be typical of more general exponential models in which, under broad conditions, $\theta$ will be asymptotically normal under $\rho_{\beta,\gamma}^{X_1^{N+1}}(d\theta)$, given $X_1,\dots,X_{N+1}$.

We would like now to illustrate the use of theorem 4.4.1 in a totally different completely discrete setting.

## 4.6 Application to adaptive classification

In this application to classification, we consider a product space $(\mathcal{X}\times\mathcal{Y},\mathcal{B}_1\otimes\mathcal{B}_2)$, where $\mathcal{Y}$ is a finite set with $C$ elements and $\mathcal{B}_2$ is the discrete sigma algebra. We let $\{f_\theta : \mathcal{X} \to \mathcal{Y}; \theta \in \Theta\}$ be a family of classification rules. Our goal is to find a classification rule $f : \mathcal{X} \to \mathcal{Y}$ which minimizes the error of classification $P(f(X) \neq Y)$, for some probability measure $P \in \mathcal{M}_+^1(\mathcal{X} \times \mathcal{Y})$, which we know through the observation of an i.i.d. sample of classified patterns $(X_i, Y_i)_{i=1}^N$ distributed according to $P^{\otimes N}$.

We introduce the family of conditional "non-normalised" likelihood functions

$$q_\theta(y \,|\, x) = \exp\Bigl(-\lambda \mathbf{1}\bigl(f_\theta(x) \neq y\bigr)\Bigr),$$

where $\lambda \in ]0,1[$ is a fixed constant (to be optimized later on). We have

$$-E_P \log q_\theta(Y \,|\, X) = \lambda P(f_\theta(X) \neq Y).$$

Applying theorem 4.9.1 of appendix 4.9 to $q_\theta(y \,|\, x)$, noticing that in this situation $\chi_{\beta,\alpha} \leq \lambda$ (for any values of $\beta$ and $\alpha$), we obtain that

$$\begin{aligned} &- E_{P^{\otimes(N+1)}} \log g_\beta^N(Y_{N+1} \,|\, X_{N+1}) \\ &\leq \inf_{\theta \in \Theta} \left( -E_{P^{\otimes(N+1)}} \log q_\theta(Y_{N+1} \,|\, X_{N+1}) - \frac{1 + \exp \lambda/2}{N} \log \pi(\{\theta\}) \right), \end{aligned} \tag{4.6.1}$$

where

$$g_\beta^N(Y_{N+1} \,|\, X_{N+1}) = \frac{\mathbb{E}_{\pi(d\theta)} \left[ \left( \prod_{i=1}^N q_\theta(Y_i \,|\, X_i) \right)^\beta q_\theta(Y_{N+1} \,|\, X_{N+1}) \right]}{\mathbb{E}_{\pi(d\theta)} \left[ \left( \prod_{i=1}^N q_\theta(Y_i \,|\, X_i) \right)^\beta \right]},$$

with $\beta = \dfrac{1}{1 + \exp(\lambda/2)}$.

From this main inequality, we are going to derive four different results. One is concerned with a randomised classification rule in the general case, the second one is concerned with the performance of the same classification rule in the case when it is known that the conditional distribution of the classes is sufficiently far from the uniform distribution, the last two results are concerned with the derivation of a non randomised classification rule.

### 4.6.1 Randomized classification rule, general case

Let us define a notation for the posterior probability distribution associated with the Gibbs estimator:

$$\rho_\beta(d\theta) \overset{\text{def}}{\propto} \left( \prod_{i=1}^N q_\theta(Y_i \,|\, X_i) \right)^\beta \pi(d\theta).$$

We can see that

$$\begin{aligned}
-\log g_\beta^N(Y_{N+1}\,|\,X_{N+1}) &= -\log \mathbb{E}_{\rho_\beta(d\theta)} q_\theta(Y_{N+1}\,|\,X_{N+1}) \\
&= -\log\Bigl[\mathbb{E}_{\rho_\beta(d\theta)} \exp\bigl(-\lambda \mathbb{1}(f_\theta(X_{N+1}) \neq Y_{N+1})\bigr)\Bigr] \\
&= -\log\Bigl[1-(1-e^{-\lambda})\mathbb{E}_{\rho_\beta(d\theta)}\bigl[\mathbb{1}(f_\theta(X_{N+1}) \neq Y_{N+1})\bigr]\Bigr] \\
&\geq (1-e^{-\lambda})\mathbb{E}_{\rho_\beta(d\theta)}\Bigl[\mathbb{1}(f_\theta(X_{N+1}) \neq Y_{N+1})\Bigr].
\end{aligned}$$

This proves

**Theorem 4.6.1.**

$$\begin{aligned}
&\mathbb{E}_{P^{\otimes N}(dZ_1^N)}\mathbb{E}_{\rho_\beta(d\theta)} P\bigl(f_\theta(X_{N+1}) \neq Y_{N+1}\bigr) \\
&\qquad \leq \inf_{\theta\in\Theta}\left(\frac{\lambda}{1-e^{-\lambda}} P\bigl(f_\theta(X_{N+1}) \neq Y_{N+1}\bigr) - \frac{1+e^{\lambda/2}}{1-e^{-\lambda}}\frac{\log \pi(\{\theta\})}{N+1}\right).
\end{aligned}$$

### 4.6.2 Randomized classification rule for "non-ambiguous" classification problems

We will see further below that the order of the bound obtained in theorem 4.6.1 is optimal in the general case. However, another type of result can be obtained when the classification problem is not too ambiguous in the following sense : Assuming that we have chosen some regular version of $P(dY\,|\,X)$, let us consider an "ideal" classification rule $\tilde{f}: \mathcal{X} \to \mathcal{Y}$, such that

$$P(Y=\tilde{f}(x)\,|\,X=x) = \max_{y\in\mathcal{Y}} P(Y=y\,|\,X=x), \qquad x \in \mathcal{X}.$$

Let us introduce for any $x \in \mathcal{X}$ the notation

$$\mu_x = \min_{y\in\mathcal{Y}\setminus\{\tilde{f}(x)\}} \Bigl[P\bigl(Y=\tilde{f}(x)\,|\,X=x\bigr) - P\bigl(Y=y\,|\,X=x\bigr)\Bigr].$$

Assume for the moment that the value of the observation $(X_1, Y_1), \ldots, (X_N, Y_N)$ is fixed, so that $\rho(d\theta)$ is fixed, and remark that for any $x \in \mathcal{X}$

$$\begin{aligned}
&-\mathbb{E}_{P(dY\,|\,X=x)}\biggl\{\log\biggl[\mathbb{E}_{\rho(d\theta)}\Bigl\{\exp\Bigl[-\lambda\Bigl(\mathbb{1}\bigl(Y\neq f_\theta(X)\bigr) - \mathbb{1}\bigl(Y\neq\tilde{f}(X)\bigr)\Bigr)\Bigr]\Bigr\}\biggr]\biggr\} \\
&\qquad \geq \mathbb{E}_{P(dY\,|\,X=x)}\mathbb{E}_{\rho(d\theta)}\Bigl[-(e^\lambda-1)\mathbb{1}\bigl(Y=f_\theta(X) \text{ and } Y\neq\tilde{f}(X)\bigr) \\
&\qquad\qquad + (1-e^{-\lambda})\mathbb{1}\bigl(Y\neq f_\theta(X) \text{ and } Y=\tilde{f}(X)\bigr)\Bigr] \\
&\geq \mathbb{E}_{\rho(d\theta)}\Bigl[(1-e^{-\lambda})\Bigl(P\bigl(Y\neq f_\theta(X)\,|\,X=x\bigr) - P\bigl((Y\neq\tilde{f}(X)\,|\,X=x\bigr)\Bigr) \\
&\qquad - (e^\lambda-1)(1-e^{-\lambda})P\bigl(Y=f_\theta(X) \text{ and } Y\neq\tilde{f}(X)\,|\,X=x\bigr)\Bigr] \qquad (4.6.2)
\end{aligned}$$

As moreover

$$P\bigl(Y = f_\theta(X) \text{ and } Y \neq \tilde{f}(X) \,|\, X = x\bigr) \leq \frac{1-\mu_x}{2} \mathbb{1}\bigl(f_\theta(X) \neq \tilde{f}(X)\bigr)$$

and

$$P\bigl(Y \neq f_\theta(X) \,|\, X = x\bigr) - P\bigl(Y \neq \tilde{f}(X) \,|\, X = x\bigr) \geq \mu_x \mathbb{1}\bigl(f_\theta(X) \neq \tilde{f}(X)\bigr),$$

it is necessarily true that

$$\begin{aligned} &P\bigl(Y = f_\theta(X) \text{ and } Y \neq \tilde{f}(X) \,|\, X = x\bigr) \\ &\qquad \leq \frac{(1-\mu_x)}{2\mu_x} \Bigl[P\bigl(Y \neq f_\theta(X) \,|\, X = x\bigr) - P\bigl(Y \neq \tilde{f}(X) \,|\, X = x\bigr)\Bigr]. \end{aligned}$$

Coming back to (4.6.2) we obtain

$$\begin{aligned} &- \mathbb{E}_{P(dY \,|\, X=x)} \biggl\{ \log \biggl[ \mathbb{E}_{\rho(d\theta)} \Bigl\{ \exp \Bigl[ -\lambda \Bigl( \mathbb{1}\bigl(Y \neq f_\theta(X)\bigr) - \mathbb{1}\bigl(Y \neq \tilde{f}(X)\bigr) \Bigr) \Bigr] \Bigr\} \biggr] \biggr\} \\ &\quad \geq \mathbb{E}_{\rho(d\theta)} \biggl[ \bigl(1 - e^{-\lambda}\bigr) \left( 1 - \frac{1-\mu_x}{2\mu_x} (e^{\lambda} - 1) \right) \Bigl( P\bigl(Y \neq f_\theta(X) \,|\, X = x\bigr) \\ &\qquad\qquad - P\bigl(Y \neq \tilde{f}(X) \,|\, X = x\bigr) \Bigr) \biggr]. \qquad (4.6.3) \end{aligned}$$

This will be useful when $\mu_x$ is not too small. For small values of $\mu_x$, we would better use the bound established in the "general" case, which can be written as

$$\begin{aligned} &- \mathbb{E}_{P(dY \,|\, X=x)} \biggl\{ \log \biggl[ \mathbb{E}_{\rho(d\theta)} \Bigl\{ \exp \Bigl[ -\lambda \Bigl( \mathbb{1}\bigl(Y \neq f_\theta(X)\bigr) - \mathbb{1}\bigl(Y \neq \tilde{f}(X)\bigr) \Bigr) \Bigr] \Bigr\} \biggr] \biggr\} \\ &\quad \geq (1 - e^{-\lambda}) \mathbb{E}_{\rho(d\theta)} \Bigl[ P\bigl(Y \neq f_\theta(X) \,|\, X = x\bigr) - P\bigl(Y \neq \tilde{f}(X) \,|\, X = x\bigr) \Bigr] \\ &\qquad\qquad - (\lambda - 1 + e^{-\lambda}) P\bigl(Y \neq \tilde{f}(X) \,|\, X = x\bigr) \\ &\quad \geq (1 - e^{-\lambda}) \mathbb{E}_{\rho(d\theta)} \Bigl[ P\bigl(Y \neq f_\theta(X) \,|\, X = x\bigr) - P\bigl(Y \neq \tilde{f}(X) \,|\, X = x\bigr) \Bigr] \\ &\qquad\qquad - \frac{1}{2} (\lambda - 1 + e^{-\lambda}). \qquad (4.6.4) \end{aligned}$$

Let us define for any real positive constant $\mu$ the set

$$\Omega_\mu \overset{\text{def}}{=} \Bigl\{ x \in \mathcal{X} \,:\, P\bigl(Y = \tilde{f}(x) \,|\, X = x\bigr) < \max_{z \in \mathcal{Y} \setminus \{\tilde{f}(x)\}} P\bigl(Y = z \,|\, X = x\bigr) + \mu \Bigr\}.$$

Combining equations (4.6.3) and (4.6.4), we get

$$- \mathbb{E}_{P(dX, dY)} \biggl\{ \log \biggl[ \mathbb{E}_{\rho(d\theta)} \Bigl\{ \exp \Bigl[ -\lambda \Bigl( \mathbb{1}\bigl(Y \neq f_\theta(X)\bigr) - \mathbb{1}\bigl(Y \neq \tilde{f}(X)\bigr) \Bigr) \Bigr] \Bigr\} \biggr] \biggr\}$$

$$\geq \mathbb{E}_{\rho(d\theta)}\Big[(1-e^{-\lambda})\Big(1-\frac{1-\mu}{2\mu}(e^{\lambda}-1)\Big)\Big(P(Y\neq f_\theta(X)) \\ - P((Y\neq \tilde{f}(X))\Big)\Big] - (\lambda-1+e^{-\lambda})\frac{P(\Omega_\mu)}{2}. \quad (4.6.5)$$

This proves the following theorem :

**Theorem 4.6.2.** *In the setting described above,*

$$\mathbb{E}_{P^{\otimes N}(dZ_1^N)}\Big\{\mathbb{E}_{\rho_\beta(d\theta)}\big[P(f_\theta(X_{N+1})\neq Y_{N+1})\big]\Big\} \leq P(\tilde{f}(X_{N+1})\neq Y_{N+1}) \\ + \inf_{\theta\in\Theta}\Big\{C_1(\lambda,\mu)\big(P(f_\theta(X_{N+1})\neq Y_{N+1}) \\ - P(\tilde{f}(X_{N+1})\neq Y_{N+1})\big) - C_2(\lambda,\mu)\frac{\log(\pi(\theta))}{(N+1)}\Big\} + C_3(\lambda,\mu)$$

*where*

$$C_1(\lambda) = \frac{\lambda}{1-e^{-\lambda}}\Big(1-\frac{1-\mu}{2\mu}(e^{\lambda}-1)\Big)^{-1},$$
$$C_2(\lambda) = \frac{(1+e^{\lambda/2})}{(1-e^{-\lambda})}\Big(1-\frac{1-\mu}{2\mu}(e^{\lambda}-1)\Big)^{-1},$$
$$C_3(\lambda) = \frac{\lambda-1+e^{-\lambda}}{1-e^{-\lambda}}\Big(1-\frac{1-\mu}{2\mu}(e^{\lambda}-1)\Big)^{-1}\frac{P(\Omega_\mu)}{2}.$$

*Remark* 4.6.1. Note that for any fixed value of $\mu \in ]0,1[$,

$$C_1(\lambda,\mu) \underset{\lambda\downarrow 0}{\sim} 1,$$
$$C_2(\lambda,\mu) \underset{\lambda\downarrow 0}{\sim} \frac{2}{\lambda},$$
$$C_3(\lambda,\mu) \underset{\lambda\downarrow 0}{\sim} \frac{\lambda P(\Omega_\mu)}{4}.$$

*Remark* 4.6.2. In the case when

- the best classification rule is achieved : $\tilde{f} = f_{\theta_0}$ for some value $\theta_0 \in \Theta$,
- the classification problem is not ambiguous in the sense that $P(\Omega_\mu) = 0$ for some positive value of $\mu$,

then the Gibbs estimator achieves the best possible classification rate up to an additive loss factor of order $1/N$.

### 4.6.3 Deterministic classification rule

If we want a non random classification rule, we can take

$$\hat{f}_\beta(x) = \arg\max_y \mathbb{E}_{\rho_\beta(d\theta)}\big[\mathbf{1}(f_\theta(x) = y)\big].$$

As $\mathbb{E}_{\rho_\beta(d\theta)}\delta_{f_\theta(X_{N+1})}(dY_{N+1}) \leq 1/2$ when $\hat{f}_\beta(X_{N+1}) \neq Y_{N+1}$, we get

$$\begin{aligned}\mathbb{E}_{\rho_\beta(d\theta)}\big[\mathbf{1}(f_\theta(X_{N+1}) \neq Y_{N+1})\big] &= \frac{1}{2}E_{\rho_\beta(d\theta)}\|\delta_{f_\theta(X_{N+1})} - \delta_{Y_{N+1}}\|_{\mathrm{Var}} \\ &\geq \frac{1}{2}\|E_{\rho_\beta(d\theta)}\delta_{f_\theta(X_{N+1})} - \delta_{Y_{N+1}}\|_{\mathrm{Var}} \\ &\geq \frac{1}{2}\mathbf{1}(\hat{f}_\beta(X_{N+1}) \neq Y_{N+1}),\end{aligned}$$

and therefore

$$\begin{aligned}-\log\Big(g_\beta^N(Y_{N+1}\,|\,X_{N+1})\Big) &\geq -\log\left(1 - \frac{1}{2}(1-e^{-\lambda})\mathbf{1}\big(\hat{f}_\beta(X_{N+1}) \neq Y_{N+1}\big)\right) \\ &= -\mathbf{1}\big(\hat{f}_\beta(X_{N+1}) \neq Y_{N+1}\big)\log\left(1 - \frac{1}{2}(1-e^{-\lambda})\right).\end{aligned}$$

Thus

**Proposition 4.6.1.** *The deterministic classification rule $\hat{f}_\beta$ defined above satisfies*

$$\begin{aligned}&P^{\otimes(N+1)}\big(\hat{f}_\beta(X_{N+1}) \neq Y_{N+1}\big) \\ &\leq \inf_\theta\left(\log\frac{2}{1+e^{-\lambda}}\right)^{-1}\left(\lambda P\big(f_\theta(X_{N+1}) \neq Y_{N+1}\big) - \frac{(1+e^{\lambda/2})\log\pi(\{\theta\})}{N+1}\right).\end{aligned}$$

In the case of a "non-ambiguous" classification problem, it is also possible to establish an oracle inequality for the deterministic classification rule $\hat{f}$. we can remark that for any $x \in \mathcal{X}$,

$$\begin{aligned}\mathbb{E}_{\rho(d\theta)}&\Big[P\big(\tilde{f}(x) = Y\,|\,X = x\big) - P\big(f_\theta(x) = Y\,|\,X = x\big)\Big] \\ &\geq \mu_x\mathbb{E}_{\rho(d\theta)}\Big[\mathbb{1}\big(\tilde{f}(x) \neq f_\theta(x)\big)\Big] \\ &\geq \frac{\mu_x}{2}\mathbb{1}\big(\hat{f}(x) \neq \tilde{f}(x)\big) \\ &\geq \frac{\mu_x}{1+\mu_x}\Big[P\big(\tilde{f}(x) = Y\,|\,X = x\big) \\ &\qquad - P\big(\hat{f}(x) = Y\,|\,X = x\big)\Big].\end{aligned}$$

The last inequality is due to the fact that

$$2P\big(\tilde{f}(x) = Y\,|\,X = x\big) - \mu_x \leq 1.$$

Thus

$$
\begin{aligned}
&P(Y \neq \hat{f}(X) \,|\, X = x) - P(Y \neq \tilde{f}(X) \,|\, X = x) \\
&\quad \leq \frac{1+\mu_x}{\mu_x} \mathbb{E}_{\rho(d\theta)} \Big[ P(Y \neq f_\theta(X) \,|\, X = x) - P(Y \neq \tilde{f}(X) \,|\, X = x) \Big] \\
&\quad \leq -\frac{1+\mu_x}{\mu_x} \big(1 - e^{-\lambda}\big)^{-1} \left( 1 - \frac{1-\mu_x}{2\mu_x} (e^{\lambda} - 1) \right)^{-1} \\
&\quad \times \mathbb{E}_{P(dY \,|\, X=x)} \log \mathbb{E}_{\rho(d\theta)} \Big\{ \exp\Big( -\lambda \big[ \mathbb{1}(Y \neq f_\theta(X)) - \mathbb{1}(Y \neq \tilde{f}(X)) \big] \Big) \Big\}.
\end{aligned}
$$

We will use this chain of inequalities when $x \notin \Omega_\mu$. When $x \in \Omega_\mu$, we will use the general case inequality

$$
\begin{aligned}
&P(Y \neq \hat{f}(X) \,|\, X = x) - P(Y \neq \tilde{f}(X) \,|\, X = x) \\
&\qquad \leq 2\mathbb{E}_{\rho(d\theta)} \Big[ P(Y \neq f_\theta \,|\, X = x) \Big] - P(Y \neq \tilde{f}(X) \,|\, X = x) \\
&\leq -\frac{2}{1-e^{-\lambda}} \log \Big\{ \mathbb{E}_{\rho(d\theta)} \exp \Big[ -\lambda \mathbb{1}(Y \neq f_\theta(X)) \Big] \Big\} - P(Y \neq \tilde{f}(X) \,|\, X = x) \\
&\leq -\frac{2}{1-e^{-\lambda}} \log \Big\{ \mathbb{E}_{\rho(d\theta)} \exp \Big[ -\lambda \mathbb{1}(Y \neq f_\theta(X) - \mathbb{1}(Y \neq \tilde{f}(X)) \Big] \Big\} \\
&\qquad\qquad + \left( \frac{2\lambda}{1-e^{-\lambda}} - 1 \right) P(Y \neq \tilde{f}(X) \,|\, X = x).
\end{aligned}
$$

Combining the two last equations proves a new theorem about the deterministic aggregation rule $\hat{f}$.

**Theorem 4.6.3.** *In the setting described above,*

$$
\begin{aligned}
&P(\hat{f}(X_{N+1}) \neq Y_{N+1}) \leq P(\tilde{f}(X_{N+1}) \neq Y_{N+1}) \\
&\quad + \inf_{\theta \in \Theta} \Big\{ C_1(\lambda, \mu) \Big[ P(f_\theta(X_{N+1}) \neq Y_{N+1}) - P(\tilde{f}(X_{N+1}) \neq Y_{N+1}) \Big] \\
&\qquad\qquad - C_2(\lambda, \mu) \frac{\log(\pi(\theta))}{(N+1)} \Big\} + C_3(\lambda, \mu)
\end{aligned}
$$

*where*

$$
\begin{aligned}
C_1(\lambda) &= \frac{1+\mu}{\mu} \frac{\lambda}{1-e^{-\lambda}} \left( 1 - \frac{1-\mu}{2\mu} (e^{\lambda} - 1) \right)^{-1}, \\
C_2(\lambda) &= \frac{1+\mu}{\mu} \frac{(1+e^{\lambda/2})}{(1-e^{-\lambda})} \left( 1 - \frac{1-\mu}{2\mu} (e^{\lambda} - 1) \right)^{-1}, \\
C_3(\lambda) &= \left( \frac{2\lambda}{1-e^{-\lambda}} - 1 \right) P(\Omega_\mu).
\end{aligned}
$$

*Remark* 4.6.3. Note that for any fixed value of $\mu \in ]0,1[$,

$$
C_1(\lambda, \mu) \underset{\lambda \downarrow 0}{\sim} \frac{1+\mu}{\mu},
$$

$$C_2(\lambda,\mu) \underset{\lambda\downarrow 0}{\sim} \frac{2(1+\mu)}{\mu\,\lambda},$$
$$C_3(\lambda,\mu) \underset{\lambda\downarrow 0}{\sim} P(\Omega_\mu).$$

*Remark* 4.6.4. Here again, when $P(\Omega_\mu) = 0$ for some value of $\mu$ and $\tilde{f} = f_{\theta_0}$ for some $\theta_0 \in \Theta$, then the optimal probability of error $P(Y \neq \tilde{f}(X))$ can be reached at spead $1/N$.

### 4.6.4 Counter example

It is easy to build a counter example in which, for some positive constant $A$ and for any (randomized or not) estimator $\hat{f}$,

$$\sup_P \left\{ P^{\otimes(N+1)}\big(\hat{f}(X_{N+1}) \neq Y_{N+1}\big) - \inf_{\theta\in\Theta} P\big(f_\theta(X_{N+1}) \neq Y_{N+1}\big) \right\} \geq \frac{A}{\sqrt{N}}.$$

Take for example $\mathcal{X} = \{x\}$ to be a one point set, $\mathcal{Y} = \{0,1\}$ and consider the two sample distributions $\{P_i^{\otimes N}\,;\, i \in \{-1,+1\}\}$, where $P_i(x,0) = \frac{1}{2} + \frac{i\alpha}{\sqrt{N}} = 1 - P_i(x,1)$, $\Theta = \{0,1\}$ and $f_\theta(x) = \theta$. Then

$$\begin{aligned}
\max_{i\in\{-1,+1\}} &\left\{ P_i^{\otimes(N+1)}(\hat{f} \neq Y_{N+1}) - \min_{\theta\in\Theta} P_i(f_\theta \neq Y_{N+1}) \right\} \\
&= \max_{i\in\{-1,+1\}} \frac{2\alpha}{\sqrt{N}} P_i^{\otimes N}(\hat{f} \neq \tfrac{i+1}{2}) \\
&\geq \frac{\alpha}{\sqrt{N}} \sum_{i\in\{-1,+1\}} P_i^{\otimes N}(\hat{f} \neq \tfrac{i+1}{2}) \\
&\geq \frac{\alpha}{\sqrt{N}} \sum_{y_1^N\in\{+1,-1\}^N} \min_{i\in\{-1,+1\}} P_i^{\otimes N}(y_1^N) \\
&= \frac{\alpha}{\sqrt{N}}\Big(1 - \frac{1}{2}\,\|P_{+1}^{\otimes N} - P_{-1}^{\otimes N}\|\Big)
\end{aligned}$$

Moreover, according to Pinsker's inequality

$$\begin{aligned}
\frac{1}{2}\|P_{+1}^{\otimes N} - P_{-1}^{\otimes N}\|^2 &\leq \mathcal{K}(P_{+1}^{\otimes N}, P_{-1}^{\otimes N}) \\
&= N\mathcal{K}(P_{+1}, P_{-1}) \\
&= 2\alpha\sqrt{N}\log\left(1 + \frac{4\alpha}{\sqrt{N}\Big(1 - \frac{2\alpha}{\sqrt{N}}\Big)}\right) \\
&\leq \frac{8\alpha^2}{1 - \frac{2\alpha}{\sqrt{N}}}.
\end{aligned}$$

Therefore for any $\alpha < 1/2$, there is a constant $A$, independent of $N$, such that

$$\inf_{\hat{f}} \sup_i \left( P_i^{\otimes(N+1)}(\hat{f} \neq Y_{N+1}) - \inf_{\theta \in \Theta} P_i(f_\theta \neq Y_{N+1}) \right) \geq \frac{A}{\sqrt{N}}.$$

On the other hand if we take $\lambda = 1/\sqrt{N}$ in theorem 4.6.1, we get that

$$\mathbb{E}_{P^{\otimes N}} \left[ \mathbb{E}_{\rho_\beta(d\theta)} \mathbf{1}(f_\theta \neq Y_{N+1}) \right] \\ \leq \inf_{\theta \in \Theta} \left( 1 + \frac{1}{2\sqrt{N}} \right) P_N(f_\theta \neq Y_{N+1}) - \frac{2 \log \pi(\{\theta\})}{\sqrt{N}} + \mathcal{O}\left( \frac{1}{N} \right).$$

This shows that, when $\inf_\theta P(f_\theta(X_{N+1}) \neq Y_{N+1})$ is of order one a randomized classification with $\lambda$ of order $1/\sqrt{N}$ is almost as good as the best $f_\theta$ in the sense that the supplementary error rate is of the optimal order $1/\sqrt{N}$.

When $\inf_\theta P(f_\theta(X_{N+1}) \neq Y_{N+1})$ is of order $1/N$, we should take $\lambda$ of order 1, and we can build a deterministic adaptive classification rule with a probability of error of order $1/N$.

## 4.7 Two stage adaptive least square regression

Let us consider a product space $(\mathcal{X} \times \mathcal{Y}, \mathcal{B}_1 \otimes \mathcal{B}_2)$ where $\mathcal{Y} = \mathbb{R}^D$ and $\mathcal{B}_2$ is the Borel sigma algebra. Let $\{ f_m^K : \mathcal{X} \times (\mathcal{X} \times \mathcal{Y})^K \to \mathcal{Y} \,;\, m \in M \,,\, K \in \mathbb{N} \}$ be a countable family of regression estimators. We would like to select the "best" regression estimator according to the quadratic criterion

$$\mathbb{E}_{P_N} \| Y_{K+1} - f_m^K(X_{K+1} \,;\, Z_1^K) \|^2$$

where $P_N \in \mathcal{M}_+^1((\mathcal{X} \times \mathcal{Y})^N, (\mathcal{B}_1 \otimes \mathcal{B}_2)^{\otimes N})$ is some exchangeable distribution and where $Z_1^K = \{(X_i, Y_i) \,;\, 1 \leq i \leq K\}$ is the observation.

We will assume that $Y_{K+1} - f_m^K(X_{K+1} \,;\, Z_1^K)$ is $P_N$ almost surely bounded by some constant $B$:

$$\sup_{m \in M} \| Y_{K+1} - f_m^K(X_{K+1} \,;\, Z_1^K) \| \leq B, \quad P_N \text{ a.s.}$$

We will use a two stage scheme based on theorem 4.4.1. We consider an observation vector $(Z_1^N)$ which we split into $(Z_1^K)$ and $(Z_{K+1}^N)$. We use the Gaussian shift model

$$Q_m^K(dY_{K+1} \,|\, X_{K+1} \,;\, Z_1^K) = \\ \left( \frac{\lambda}{2\pi} \right)^{D/2} \exp\left( -\frac{\lambda}{2} \| Y_{K+1} - f_m^K(X_{K+1} \,;\, Z_1^K) \|^2 \right) dY_{K+1},$$

where $dY_{K+1}$ is the Lebesgue measure on $\mathbb{R}^D$. We extract from this model the non-normalized likelihood function :

$$q_m(Y_i \,|\, X_i; Z_1^K) = \exp\left( -\frac{\lambda}{2} \| Y_i - f_m^K(X_i) \|^2 \right).$$

(We could have kept the normalization, but it will be simpler to drop it, since we will not need it in the following application of theorem 4.9.1.) We form the Gibbs estimator

$$g_\beta^N(Y_{N+1} \,|\, X_{N+1}\,;\, Z_1^N) \stackrel{\text{def}}{=} \frac{\mathbb{E}_{\pi(dm)}\left[\left(\prod_{i=K+1}^N q_m^K(Y_i \,|\, X_i; Z_1^K)\right)^\beta q_m^K(Y_{N+1} \,|\, X_{N+1}\,;\, Z_1^K)\right]}{\mathbb{E}_{\pi(dm)}\left[\left(\prod_{i=K+1}^N q_m^K(Y_i \,|\, X_i\,;\, Z_1^K)\right)^\beta\right]},$$

where $\pi \in \mathcal{M}_+^1(M)$ is some prior probability distribution. Applying theorem 4.4.1 we get

$$\mathbb{E}_{P_{N+1}} \log g_\beta^N(Y_{N+1} \,|\, X_{N+1}\,;\, Z_1^N) \geq \sup_{m\in M}\left(\mathbb{E}_{P_{N+1}(dZ_1^{N+1})}\left[\log q_m^N(Y_{N+1} \,|\, X_{N+1}\,;\, Z_1^K)\right] - \frac{\log \pi(\{m\})}{\beta(N-K+1)}\right),$$

when $\beta = (1+\exp(\frac{\lambda B^2}{4}))^{-1}$. On the other hand

$$\log g_\beta^N(Y_{N+1} \,|\, X_{N+1}\,;\, Z_1^N) = \frac{D}{2}\log\left(\frac{\lambda}{2\pi}\right) + \log\left\{\mathbb{E}_{\rho(dm)}\left[\exp\left(-\frac{\lambda}{2}\|Y_{N+1} - f_m(X_{N+1}\,;\, Z_1^K)\|^2\right)\right]\right\},$$

where

$$\rho(dm) \propto \left(\prod_{i=K+1}^N q_m^K(Y_i \,|\, X_i\,;\, Z_1^K)\right)^\beta \pi(dm).$$

Choosing $\lambda = B^{-2}$, and using the fact that the Gaussian function

$$y \longmapsto \exp\left(-\frac{1}{2}\|y\|^2\right)$$

is concave in the unit ball of $\mathbb{R}^D$, we get that

$$-\log\left\{\mathbb{E}_{\rho(dm)}\left[\exp\left(-\frac{\lambda}{2}\|Y_{N+1} - f_m^K(X_{N+1}\,;\, Z_1^K)\|^2\right)\right]\right\} \geq \frac{\lambda}{2}\|Y_{N+1} - \mathbb{E}_{\rho(dm)} f_m^K(X_{N+1}\,;\, Z_1^N)\|^2.$$

This proves the following

**Theorem 4.7.1.** *With the previous notations and the choice of parameters*

$$\lambda = B^{-2},$$
$$\beta = (1 + e^{1/4})^{-1},$$

*we get*

$$\mathbb{E}_{P_{N+1}} \| Y_{N+1} - \mathbb{E}_{\rho(dm)} f_m^K(X_{N+1}\,;\,Z_1^K) \|^2$$
$$\leq \inf_{m \in M} \left( \mathbb{E}_{P_{N+1}} \| Y_{N+1} - f_m^K(X_{N+1}\,;\,Z_1^K) \|^2 - \frac{2B^2(1+e^{1/4})}{(N-K+1)} \log \pi(\{m\}) \right).$$

*Remark* 4.7.1. If we know only that

$$\| Y_{N+1} \| \leq B, \quad P_{N+1} \text{ a.s.}$$

then we should force $f_m^K(X_{N+1}\,;\,Z_1^K)$ to be bounded, using

$$\tilde{f}_m^K(X_{N+1}\,;\,Z_1^K) = T_B\big(f_m^K(X_{N+1}\,;\,Z_1^K)\big),$$

where

$$T_B(y) = \begin{cases} y & \text{if } \|y\| \leq B, \\ \frac{B}{\|y\|} y & \text{otherwise.} \end{cases}$$

As we have in this situation

$$\mathbb{E}_{P_{N+1}} \| Y_{N+1} - \tilde{f}_m^K(X_{N+1}\,;\,Z_1^K) \|^2 \leq \mathbb{E}_{P_{N+1}} \| Y_{N+1} - f_m^K(X_{N+1}\,;\,Z_1^K) \|^2$$

this truncation operation will always improve the estimator.

*Remark* 4.7.2. Using the progressive mixture estimator posterior distribution (see [21])

$$\tilde{\rho}(dm) = \frac{1}{N-K+1} \sum_{M=K+1}^{N+1} \frac{\prod_{i=M}^{N} q_m^K(Y_i \,|\, X_i\,;\,Z_1^K) \pi(dm)}{\mathbb{E}_{\pi(dm')} \prod_{i=M}^{N} q_{m'}^K(Y_i \,|\, X_i\,;\,Z_1^K)}, \tag{4.7.1}$$

we would have obtained the sharper result

$$\mathbb{E}_{P_{N+1}} \| Y_{N+1} - \mathbb{E}_{\tilde{\rho}(dm)} f_m^K(X_{N+1}\,;\,Z_1^K) \|^2$$
$$\leq \inf_{m \in M} \left( \mathbb{E}_{P_{N+1}} \| Y_{N+1} - f_m^K(X_{N+1}\,;\,Z_1^K) \|^2 - \frac{2B^2}{N-K+1} \log \pi(\{m\}) \right).$$

With the Gibbs estimator we loose a factor $1 + e^{1/4}$. Nevertheless, we think that the Gibbs estimator can be preferred to the progressive mixture estimator in many circumstances for at least three reasons:

- It can be computed faster than the progressive mixture estimator. Indeed, as shown in equation (4.7.1), the progressive mixture estimator is a Cesaro mean of $N-K+1$ terms, the last one having the same form as the Gibbs estimator, except for the introduction of the $\beta$ exponent.
- As we explained in the first section, the Gibbs posterior distribution $\rho(dm)$ is a Gibbs distribution at temperature $\beta N$. Therefore we can expect that it will be as a rule sharply peaked around only a few values of $m \in M$. Therefore, although the Gibbs estimator appears at first sight as a combination of all the regression models, it will in practice perform a selection, because the majority of the weights $\rho(dm)$ will be small with respect to $1/N$, which is the required precision to preserve the quality of approximation stated in the theorem. This is not the case for the progressive mixture estimator : indeed the first term in equation (4.7.1) is the "flat" distribution $\pi(dm)/(N-K+1)$, which gives comparable weights of orfer $1/N$ to all the regression models.
- The Gibbs estimator is "almost" the Bayesian estimator corresponding to the prior distribution $\pi(dm)$, therefore in the case when we expect the conditional mean $\mathbb{E}_{P_{N+1}}(Y_{N+1} \,|\, X_{N+1}\,;\, Z_1^N)$ to be close to $f_m^K(X_{N+1}\,;\, Z_1^K)$ with probability $\pi(m)$, but we are not completely sure of this guess and want a robust estimator, this is a good incitation to use the Gibbs estimator, which appears as a "cautious" variant of the Bayes estimator.

*Remark* 4.7.3. It would be impossible to obtain the same kind of result with a true selection rule (a rule which would select only one $m$ as opposed to a convex combination of several $f_m^K$). The counter example given in the previous section about classification applies to quadratic regression. We can also modify this example to make it look more like a regression problem.

Indeed, let $\mathcal{X}$ be a trivial one point set. Let $\mathcal{Y} = \mathbb{R}^2$. Consider the parameter set $\Theta = \Theta_1 \cup \Theta_2$, where $\Theta_1 = \mathbb{R} \times \{0\}$ and $\Theta_2 = \{0\} \times \mathbb{R}$. Let us consider the (constant !) regression functions $\{f_\theta = \theta\,;\, \theta \in \Theta\}$ and let us drop in the notations the trivial dependence on $X$. Splitting the sample in two halves and using

$$f_{+1}^K(Y_1^K) = \left(\frac{1}{K}\sum_{i=1}^K Y_i(1), 0\right),$$

$$f_{-1}^K(Y_1^K) = \left(0, \frac{1}{K}\sum_{i=1}^K Y_i(2)\right),$$

where $Y_i = (Y_i(1), Y_i(2)) \in \mathbb{R}^2$, we get that for some constant $A > 0$ and any exchangeable probability distribution $P_{N+1}$ such that $P_{N+1}$ almost surely $\|Y_i\| \le B$

$$\mathbb{E}_{P_{N+1}(dY_1^{N+1})} \|Y_{N+1} - \mathbb{E}_{\rho(dm)} f_m^K(Y_1^K)\|^2 - \inf_{\theta \in \Theta} \mathbb{E}_{P_{N+1}(dY_1^{N+1})} \|Y_{N+1} - f_\theta\|^2 \le \frac{AB^2}{N}.$$

To see that no estimator $\hat{\theta} : \mathcal{Y}^N \to \Theta$ could have the same rate of convergence in the worst case, let us consider in $\mathcal{Y}$ the two points

$$p_m = (1 + \frac{m}{2}, 1 - \frac{m}{2}), \quad m \in \{-1, +1\},$$

and the two distributions

$$P_m(dY) = \left(\frac{1}{2} + \frac{\alpha m}{\sqrt{N}}\right) \delta_{p_{+1}}(Y) + \left(\frac{1}{2} - \frac{\alpha m}{\sqrt{N}}\right) \delta_{p_{-1}}(Y), \quad m \in \{-1, +1\}.$$

We are going to prove that for some small enough positive constant $\alpha$ and some positive constant $C$

$$\inf_{\hat{\theta}} \sup_{m \in \{-1,+1\}} \Bigl( \mathbb{E}_{P_m^{\otimes(N+1)}(dY_1^{N+1})} \|Y_{N+1} - f_{\hat{\theta}}\|^2 \\ - \inf_{\theta \in \Theta} \mathbb{E}_{P_m(dY_{N+1})} \|Y_{N+1} - f_\theta\|^2 \Bigr) \geq \frac{C}{\sqrt{N}}.$$

Let

$$\bar{Y}_m = \mathbb{E}_{P_m(dY)}(Y) = \left(1 + \frac{\alpha\, m}{\sqrt{N}}, 1 - \frac{\alpha\, m}{\sqrt{N}}\right),$$

$$\hat{m}(Y_1^N) = \begin{cases} +1 & \text{if } \hat{\theta}(Y_1^N) \in \Theta_1 \\ -1 & \text{otherwise.} \end{cases}$$

Remarking that

$$\|\hat{\theta} - \bar{Y}_m\|^2 \geq \left(1 - \frac{\alpha\, m\, \hat{m}}{\sqrt{N}}\right)^2,$$

we get that

$$\begin{aligned} &\inf_{\hat{\theta}} \sup_m \Bigl( \mathbb{E}_{P_m^{\otimes N+1}} \|Y_{N+1} - f_{\hat{\theta}}\|^2 - \inf_\theta \mathbb{E}_{P_m(dY_{N+1})} \|Y_{N+1} - f_\theta\|^2 \Bigr) \\ &= \inf_{\hat{\theta}} \sup_m \mathbb{E}_{P_m^{\otimes N}} \|\bar{Y}_m - \hat{\theta}\|^2 - \left(1 - \frac{1}{\alpha\sqrt{N}}\right)^2 \\ &\geq \inf_{\hat{m}} \sup_m \frac{4\,\alpha}{\sqrt{N}} P_m^{\otimes N}(m \neq \hat{m}) \\ &\geq \frac{C}{\sqrt{N}}, \end{aligned}$$

the justification of the last inequality being the same as in the counter example given about adaptive pattern classification in section 4.6.4.

## 4.8 One stage piecewise constant regression

In this section, we consider a measurable product space $(\mathcal{X} \times \mathcal{Y}, \mathcal{B}_1 \otimes \mathcal{B}_2)$, where $\mathcal{Y} = \mathbb{R}^D$ and $\mathcal{B}_2$ is the Borel sigma algebra. We consider an exchangeable probability distribution $P_N \in \mathcal{M}_+^1((\mathcal{X}\times\mathcal{Y})^N, (\mathcal{B}_1\otimes\mathcal{B}_2)^{\otimes N})$. We let $\{(X_i, Y_i) = Z_i\,;\, i = 1, \dots, N\}$ be the canonical process. The observation is $Z_1^{N-1}$ and we want to estimate $P_N(Y_N \mid X_N\,;\, Z_1^{N-1})$.

We assume that $Y$ is a.s. bounded: there exists a positive constant $B$ such that

$$\|Y_i\| \leq B, \quad P_N \text{ a.s.} \quad i = 1, \dots, N.$$

To define a model, we consider a countable family $\mathcal{S}$ of measurable partitions $S \subset \mathcal{B}_1$ of $\mathcal{X}$. In connection with each partition $S \in \mathcal{S}$, we consider the parameter space $\Theta_S = \mathcal{Y}^S$ of the maps $\theta_S : S \to \mathcal{Y}$. We consider also the canonical projection $I_S : \mathcal{X} \to S$ defined by $X \in I_S(X)$, and we call $f_\theta : \mathcal{X} \to \mathcal{Y}$ the regression function $f_\theta(X) = f_{\theta_S}^S(X) = \theta_S(I_S(X))$, where $\theta = (S, \theta_S)$. On the global "structured" parameter set $\Theta = \{(S, \theta_S)\,;\, S \in \mathcal{S}, \theta_S \in \Theta_S\}$, we consider a prior probability distribution $\pi(S, d\theta_S)$ depending in a permutation invariant way on $X_1^N$. We take

$$\pi(d\theta_S \mid S) = \prod_{I \in S} \left(\frac{\epsilon(I)}{2\pi}\right)^{D/2} \exp\left(-\frac{\epsilon(I)}{2}\|\theta_S(I)\|^2\right) d\theta_S(I),$$

where the choice of the regularization parameter $\epsilon(I)$ will depend on $\sum_{i=1}^N \mathbf{1}(X_i \in I)$. The marginal distribution $\pi(dS)$ can be arbitrary at this stage.

We then consider the model $\{Q_\theta\,;\, \theta \in \Theta\}$ defined by

$$Q_\theta(dY \mid X) = \left(\frac{\lambda(I_S(X))}{2\pi}\right)^{D/2} \exp\left(-\frac{\lambda(I_S(X))}{2}\|Y - f_\theta(X)\|^2\right) dY,$$

where the choice of $\lambda(I)$ will depend on $\sum_{i=1}^N \mathbf{1}(X_i \in I)$. We will study the Gibbs estimator at inverse temperature $\beta$

$$\begin{aligned} G_\beta(dY_N \mid X_N\,;\, Z_1^{N-1}) &= g_\beta(Y_N \mid X_N\,;\, Z_1^{N-1}) dY_N \\ &= \mathbb{E}_{\rho_{\beta,0}(d\theta)} Q_\theta(dY_N \mid X_N), \end{aligned}$$

where

$$\rho_{\beta,\xi}(d\theta) \sim \left(\prod_{i=1}^{N-1} q_\theta(Y_i \mid X_i)\right)^\beta q_\theta(Y_N \mid X_N)^\xi\, \pi(d\theta).$$

Theorem 4.4.1 extends verbatim to the case when $\pi(d\theta)$ and $Q_\theta(dY \mid X)$ depend on the design $X_1^N$ in a permutation invariant way. Indeed we can apply the theorem to

$$\frac{1}{|\mathfrak{S}_N|}\sum_{\sigma\in\mathfrak{S}_N}\mathbf{1}(X_i=\bar{X}_{\sigma(i)}\,;1\leq i\leq N)P_N(Y_1^N\,|\,X_1^N)$$

and write $P_N$ as a combination of these more elementary distributions which have a fixed deterministic design.

We have to compute the posterior distributions $\rho_{\beta,\xi}(d\theta)$ and the constants $\gamma_\beta(\theta)$ and $\chi_{\beta,\alpha}$. Let

$$\begin{aligned}
a(I)&=\sum_{i=1}^{N-1}\mathbf{1}(X_i\in I),\\
b(I)&=\mathbf{1}(X_N\in I),\\
\eta(I)&=\lambda(I)\left(\beta a(I)+\xi b(I)\right)+\epsilon(I),\\
\bar{\theta}(I)&=\left(a(I)+\frac{\xi}{\beta}b(I)+\frac{\epsilon(I)}{\lambda(I)\beta}\right)^{-1}\left(\sum_{i=1}^{N-1}Y_i\,\mathbf{1}(X_i\in I)+\frac{\xi}{\beta}Y_N\,\mathbf{1}(X_N\in I)\right).
\end{aligned}$$

Let us choose $\epsilon(I)=\bar{\epsilon}\lambda(I)$ where $\bar{\epsilon}$ is a constant. With these notations

$$\rho_{\beta,\xi}^{Z_1^N}(d\theta_S\,|\,S)=\prod_{I\in S}\left(\frac{\eta(I)}{2\pi}\right)^{D/2}\exp\left(-\frac{\eta(I)}{2}\|\theta_S(I)-\bar{\theta}_S(I)\|^2\right)d\theta_S(I),$$

$$\begin{aligned}
\rho_{\beta,\xi}^{Z_1^N}(S)\sim\pi(S)\prod_{I\in S}&\left(\frac{\eta(I)}{2\pi}\right)^{-D/2}\left(\frac{\lambda(I)}{2\pi}\right)^{(\beta a(I)+\xi b(I))D/2}\\
\times\exp&\left(-\frac{\beta}{2}\sum_{i=1}^{N-1}\lambda(I_S(X_i))\|Y_i-f_{\bar{\theta}_S}^S(X_i)\|^2-\frac{\xi}{2}\lambda(I_S(X_N))\|Y_N-f_{\bar{\theta}_S}^S(X_N)\|^2\right.\\
&\left.-\sum_{I\in S}\frac{\epsilon(I)}{2}\|\bar{\theta}_I\|^2\right).
\end{aligned}$$

Let us put for short

$$h(\theta)=\frac{\lambda(I_S(X_N))}{2}\|Y_N-f_\theta(X_N)\|^2-\frac{D}{2}\log\lambda(I_S(X_N)).$$

We have

$$\chi_{\beta,1}\leq 0\vee\sup_{\xi\in[0,1]}\frac{\mathbb{E}_{P_N}M^3_{\rho_{\beta,\xi}(d\theta)}h(\theta)}{\mathbb{E}_{P_N}\operatorname{Var}_{\rho_{\beta,\xi}(d\theta)}h(\theta)}.$$

In the following computations, we put $\rho(d\theta)=\rho_{\beta,\xi}^{Z_1^N}(d\theta)$ for short. We can decompose the variance into

$$\operatorname{Var}_\rho h(\theta)=\mathbb{E}_\rho(h-\mathbb{E}_\rho(h\,|\,S))^2+\mathbb{E}_\rho(\mathbb{E}_\rho(h\,|\,S)-\mathbb{E}_\rho(h))^2.$$

In the same way, we can decompose the third moment into

$$\begin{aligned}M_\rho^3(h) &= \mathbb{E}(h - \mathbb{E}(h\,|\,S) + \mathbb{E}(h\,|\,S) - \mathbb{E}(h))^3 \\ &= \mathbb{E}(h - \mathbb{E}(h\,|\,S))^3 + 3\mathbb{E}((h - \mathbb{E}(h\,|\,S))^2(\mathbb{E}(h\,|\,S) - \mathbb{E}(h))) \\ &\quad + \mathbb{E}(\mathbb{E}(h\,|\,S) - \mathbb{E}(h))^3\end{aligned}$$

The conditional expectation of $h$ is equal to

$$\mathbb{E}_\rho(h\,|\,S) = \frac{\lambda(I_N)}{2}\left(\|Y_N - f^S_{\bar\theta_S}(X_N)\|^2 + \frac{D}{\eta(I_N)}\right) - \frac{D}{2}\log\lambda(I_N),$$

where we have put $I_N = I_S(X_N)$.

Let us choose

$$\lambda(I) = \bar\lambda\left(1 + \frac{1}{\beta(a(I) + b(I) - 1) + \bar\epsilon}\right),$$

where $\bar\lambda$ is a constant. We obtain that

$$\|\mathbb{E}(h\,|\,S) - \mathbb{E}(h)\| \le 2B^2\bar\lambda\left(1 + \frac{1}{\bar\epsilon}\right) + \frac{D}{2}\left(\frac{1}{\bar\epsilon} + \log\left(1 + \frac{1}{\bar\epsilon}\right)\right).$$

The terms $\mathbb{E}((h - \mathbb{E}(h\,|\,S))^2\,|\,S)$ and $\mathbb{E}((h - \mathbb{E}(h\,|\,S))^3\,|\,S)$ can be computed explicitly from the lemma:

**Lemma 4.8.1.** *Let $\theta \sim \mathcal{N}(0, \alpha^{-1}I)$ be a centered Gaussian random variable in $\mathbb{R}^D$ and let $Y \in \mathbb{R}^D$ be a fixed vector, then*

$$\begin{aligned}\mathbb{E}\left(\|Y - \theta\|^2 - \mathbb{E}(\|Y - \theta\|^2)\right)^2 &= \frac{1}{\alpha^2}\left(4\alpha\|Y\|^2 + 2D\right), \\ \mathbb{E}\left(\|Y - \theta\|^2 - \mathbb{E}(\|Y - \theta\|^2)\right)^3 &= \frac{1}{\alpha^3}\left(24\alpha\|Y\|^2 + 8D\right) \\ &\le \frac{6}{\alpha}\mathbb{E}\left(\|y - \theta\|^2 - \mathbb{E}(\|Y - \theta\|^2)\right)^2.\end{aligned}$$

Taking $\alpha = \eta(I_N)$, we get that

$$\begin{aligned}\mathbb{E}((h - \mathbb{E}(h\,|\,S))^3\,|\,S) &= \left(\frac{\lambda(I_N)}{2\eta(I_N)}\right)^3\left(24\eta(I_N)\|Y_N - \bar\theta_S(I_N)\|^2 + 8D\right) \\ &\le \frac{3\lambda(I_N)}{\eta(I_N)}\mathbb{E}((h - \mathbb{E}(h\,|\,S))^2\,|\,S) \\ &\le \frac{3}{\bar\epsilon}\mathbb{E}((h - \mathbb{E}(h\,|\,S))^2\,|\,S).\end{aligned}$$

Therefore we have

$$\begin{aligned}M_\rho^3(h) &\le \frac{3}{\bar\epsilon}\mathbb{E}((h - \mathbb{E}(h\,|\,S))^2) \\ &\quad + \left(2B^2\bar\lambda\left(1 + \frac{1}{\bar\epsilon}\right) + \frac{D}{2}\left(\frac{1}{\bar\epsilon} + \log\left(1 + \frac{1}{\bar\epsilon}\right)\right)\right)\end{aligned}$$

$$\times \left(3\mathbb{E}((h - \mathbb{E}(h \mid S))^2) + \mathbb{E}((\mathbb{E}(h \mid S) - \mathbb{E}(h))^2)\right)$$
$$\leq \left(3\left(2B^2\bar{\lambda}\left(1+\frac{1}{\bar{\epsilon}}\right) + \frac{D}{2}\left(\frac{1}{\bar{\epsilon}} + \log\left(1+\frac{1}{\bar{\epsilon}}\right)\right)\right) + \frac{3}{\bar{\epsilon}}\right) \mathrm{Var}_\rho h.$$

In conclusion, we have proved that

**Lemma 4.8.2.**

$$\chi_{\beta,1} \leq 6B^2\bar{\lambda}\left(1+\frac{1}{\bar{\epsilon}}\right) + \frac{3D}{2}\left(\frac{1}{\bar{\epsilon}} + \log\left(1+\frac{1}{\bar{\epsilon}}\right)\right) + \frac{3}{\bar{\epsilon}}.$$

Let us compute now $\gamma_\beta(\theta')$. We have

$$\mathbb{E}_{\rho_{\beta,\beta}(d\theta)} \log \left(\prod_{i=1}^N q_\theta(Y_i \mid X_i)\right)^{-\beta}$$
$$= \mathbb{E}_{\rho_{\beta,\beta}(d\theta)} \sum_{i=1}^N \beta \left(\frac{\lambda(I_S(X_i))}{2} \|Y_i - f_\theta(X_i)\|^2 - \frac{D}{2} \log \frac{\lambda(I_S(X_i))}{2\pi}\right).$$

Moreover, putting $c(I) = \sum_{i=1}^N \mathbf{1}(X_i \in I)$, we get that

$$\mathbb{E}_{\rho_{\beta,\beta}} \left(\sum_{i=1}^N \frac{\beta\lambda(I_S(X_i))}{2} \|Y_i - f_\theta(X_i)\|^2 \mid S\right)$$
$$= \sum_{i=1}^N \frac{\beta\lambda(I_S(X_i))}{2} \|Y_i - f^S_{\bar{\theta}_S}(X_i)\|^2 + \sum_{I \in S} \frac{Dc(I)}{2\left(c(I) + \frac{\bar{\epsilon}}{\beta}\right)}$$
$$\leq \sum_{i=1}^N \frac{\beta\lambda(I_S(X_i))}{2} \|Y_i - f^S_{\bar{\theta}_S}(X_i)\|^2 + \frac{D|S|}{2}.$$

We can now use the fact (proved in proposition 4.9.1) that

$$\frac{\displaystyle\sum_{S \in \mathcal{S}} \tilde{\pi}(S)\tilde{h}(S)e^{-\tilde{h}(S)}}{\displaystyle\sum_{S \in \mathcal{S}} \tilde{\pi}(S)e^{-\tilde{h}(S)}} \leq \tilde{h}(S') - \log \tilde{\pi}(S') + \log \sum_{S \in \mathcal{S}} \tilde{\pi}(S).$$

Putting

$$\tilde{\pi}(S) = \pi(S) \prod_{I \in S} \left(\frac{\eta(I)}{2\pi}\right)^{-D/2}$$
$$\times \exp\left(\frac{D|S|}{2} - \sum_{I \in S} \frac{\bar{\epsilon}\lambda(I)}{2} \|\bar{\theta}_S(I)\|^2\right),$$

$$\tilde{h}(S) = \sum_{i=1}^{N} \beta \left( \frac{\lambda(I_S(X_i))}{2} \|Y_i - f^S_{\bar{\theta}_S}(X_i)\|^2 - \frac{D}{2} \log \frac{\lambda(I_S(X_i))}{2\pi} \right) + \frac{D|S|}{2},$$

we get that

**Lemma 4.8.3.**

$$\mathbb{E}_{\rho_{\beta,\beta}(d\theta)} \log \left( \prod_{i=1}^{N} q_\theta(Y_i \,|\, X_i) \right)^{-\beta}$$
$$\leq \sum_{i=1}^{N} \frac{\beta\lambda(I_{S'}(X_i))}{2} \|Y_i - f^{S'}_{\bar{\theta}_{S'}}(X_i)\|^2 - \sum_{I \in S'} \frac{\beta D c(I)}{2} \log \frac{\lambda(I)}{2\pi}$$
$$+ \frac{D|S'|}{2} + |S'| \frac{\bar{\lambda}\bar{\epsilon}}{2} \left(1 + \frac{1}{\bar{\epsilon}}\right) B^2 - \log \bar{\pi}(S') + \log \sum_{S \in \mathcal{S}} \bar{\pi}(S),$$

*where*

$$\bar{\pi}(S) = \pi(S) \prod_{I \in S} \left( \frac{\eta(I)}{2\pi} \right)^{-D/2} \exp \frac{D|S|}{2}.$$

Now remembering that

$$\bar{\theta}_S(I) = \left( c(I) + \frac{\bar{\epsilon}}{\beta} \right)^{-1} \sum_{j=1}^{N} \mathbf{1}(X_j \in I) Y_j$$

it is easy to see that

$$\sum_{i=1}^{N} \frac{\beta\lambda(I_{S'}(X_i))}{2} \|Y_i - f^{S'}_{\bar{\theta}_{S'}}(X_i)\|^2$$
$$\leq \inf_{\theta_{S'}} \sum_{i=1}^{N} \frac{\beta\lambda(I_{S'}(X_i))}{2} \|Y_i - f^{S'}_{\theta_{S'}}(X_i)\|^2$$
$$+ \sum_{I \in S'} \frac{\beta\lambda(I)}{2} \left( \frac{\bar{\epsilon}}{\beta c(I) + \bar{\epsilon}} \right)^2 c(I) \,\| \frac{1}{c(I)} \sum_{j=1}^{N} Y_j \mathbf{1}(X_j \in I) \|^2$$
$$\leq \inf_{\theta_{S'}} \sum_{i=1}^{N} \frac{\beta\lambda(I_{S'}(X_i))}{2} \|Y_i - f^{S'}_{\theta_{S'}}(X_i)\|^2$$
$$+ \frac{1}{2} |S'| \bar{\lambda} (1 + \bar{\epsilon}) B^2.$$

This proves that

**Lemma 4.8.4.**

$$\sup_{\theta_{S'}\in\Theta_{S'}} \gamma_\beta(S',\theta_{S'}) \leq \frac{D|S'|}{2} + |S'|B^2\bar{\lambda}(1+\bar{\epsilon}) - \log\bar{\pi}(S') + \log\sum_{S\in\mathcal{S}}\bar{\pi}(S).$$

We can now compute $g_\beta$ more explicitly.

$$\begin{aligned} g_\beta(Y_N\,|\,X_N\,;\,Z_1^{N-1}) &= \mathbb{E}_{\rho_{\beta,0}(dS)}\left(\frac{\lambda(I_N)}{2\pi}\right)^{D/2}\left(\frac{\eta(I_N)}{2\pi}\right)^{D/2} \\ &\quad\times\exp\left(-\frac{\lambda(I_N)}{2}\|Y_N-\theta_S(I_N)\|^2\right. \\ &\quad\left.-\frac{\eta(I_N)}{2}\|\theta_S(I_N)-\bar{\theta}_S(I_N)\|^2\right)d\theta_S(I_N) \\ &= \mathbb{E}_{\rho_{\beta,0}(dS)}\left(2\pi\left(\frac{1}{\lambda(I_N)}+\frac{1}{\eta(I_N)}\right)\right)^{-D/2} \\ &\quad\times\exp\left(-\frac{1}{2}\left(\frac{1}{\lambda(I_N)}+\frac{1}{\eta(I_N)}\right)^{-1}\|Y_N-\bar{\theta}_S(I_N)\|^2\right), \end{aligned}$$

where

$$\begin{aligned} \lambda(I_N) &= \bar{\lambda}\left(1+\frac{1}{\beta a(I_N)+\bar{\epsilon}}\right), \\ \eta(I_N) &= \lambda(I_N)(\beta a(I_N)+\bar{\epsilon}), \\ \bar{\theta}_S(I) &= \frac{1}{a(I)+\bar{\epsilon}/\beta}\sum_{i=1}^{N-1}Y_i\,\mathbf{1}(X_i\in I). \end{aligned}$$

However

$$\frac{1}{\lambda(I_N)}+\frac{1}{\eta(I_N)}=\frac{1}{\bar{\lambda}},$$

(the value of $\lambda(I_N)$ was chosen for that!) and therefore:

$$g_\beta(Y_N\,|\,X_N;Z_1^{N-1}) = \mathbb{E}_{\rho_{\beta,0}(dS)}\left(\frac{\bar{\lambda}}{2\pi}\right)^{D/2}\exp\left(-\frac{\bar{\lambda}}{2}\|Y_N-f^S_{\bar{\theta}_S}(X_N)\|^2\right).$$

Let us assume from now that $\bar{\lambda}\leq\frac{1}{4B^2}$ and that $\bar{\epsilon}\geq\beta$. Let $\bar{Z}_1^N$ be fixed for a while and let $\bar{P}$ be the exchangeable distribution obtained by integrating over the permutations of the indices:

$$\bar{P}_{\bar{Z}_1^N}(dZ_1^N) = \frac{1}{|\mathfrak{S}_N|}\sum_{\sigma\in\mathfrak{S}_N}\delta_{(\bar{Z}_{\sigma(i)})}(dZ_1^N).$$

We can apply theorem 4.4.1 to $\bar{P}_{\bar{Z}_1^N}$ first and integrate over $\bar{Z}_1^N$ afterwards. Using moreover the fact that the Gaussian function is concave in the unit ball as in the previous section, we get that

$$\begin{aligned}
\mathbb{E}_{\bar{P}} \frac{\bar{\lambda}}{2} \| Y_N - \mathbb{E}_{\rho_{\beta,0}(dS)} f^S_{\bar{\theta}_S}(X_N) \|^2 - \frac{D}{2} \log \frac{\bar{\lambda}}{2\pi} \\
&\leq \mathbb{E}_{\bar{P}} \log g_\beta(Y_N \mid X_N \,;\, Z_1^{N-1})^{-1} \\
&\leq \inf_{S \in \mathcal{S}} \Biggl\{ \inf_{\theta_S \in \Theta_S} \mathbb{E}_{\bar{P}} \biggl[ \frac{\lambda(I_S(X_N))}{2} \| Y_N - f_\theta(X_N) \|^2 \\
&\qquad - \frac{D}{2} \log \frac{\lambda(I_S(X_N))}{2} \biggr] + \frac{\gamma_\beta(S)}{\beta N} \Biggr\},
\end{aligned}$$

as soon as $\beta \leq (1 + \exp \chi_{\beta,1}/2)^{-1}$. But

$$\lambda(I_S(X_N)) = \bar{\lambda} \left( 1 + \frac{1}{\beta a(I_S(X_N)) + \bar{\epsilon}} \right),$$

and therefore

$$\begin{aligned}
\mathbb{E}_{\bar{P}} \left( \frac{\lambda(I_S(X_N))}{\bar{\lambda}} - 1 \right) \| Y_N - f_\theta(X_N) \|^2 \\
&\leq 4B^2 \mathbb{E}_{\bar{P}} \frac{1}{\beta(c(I_N) - 1) + \bar{\epsilon}} \\
&= 4B^2 \frac{1}{N} \sum_{I \in S} \frac{c(I)}{\beta(c(I) - 1) + \bar{\epsilon}} \\
&\leq \frac{4B^2 |S|}{\beta N}.
\end{aligned}$$

Thus

$$\begin{aligned}
\mathbb{E}_{P_N} \| Y_N - \mathbb{E}_{\rho_{\beta,0}(dS)} f^S_{\bar{\theta}_S}(X_N) \|^2 \\
&\leq \inf_{S \in \mathcal{S}} \Biggl( \inf_{\theta_S \in \Theta_S} \mathbb{E}_{P_N} \| Y_N - f^S_{\theta_S}(X_N) \|^2 \\
&\qquad + \frac{4B^2 |S|}{\beta N} + \frac{2\gamma_\beta(S)}{\bar{\lambda} \beta N} \Biggr).
\end{aligned}$$

We can summarize what we have obtained by a theorem:

**Theorem 4.8.1.** *For any exchangeable probability distribution $P_N$ such that*

$$\|Y_i\| \leq B \quad P_N \text{ almost surely,}$$

*for any choice of parameters $\bar{\lambda}$, $\bar{\epsilon}$ and $\beta$ such that*

$$\begin{aligned}
&\bar{\lambda} \leq \frac{1}{4B^2}, \\
&\beta \leq \bar{\epsilon}, \\
&\inf_{\alpha > \beta} \frac{\beta^2}{(\alpha - \beta)(2 - \alpha - \beta)} \exp \alpha\chi \leq 1,
\end{aligned}$$

*where*

$$\chi = 6B^2\bar{\lambda}\left(1+\frac{1}{\bar{\epsilon}}\right) + \frac{3D}{2}\left(\frac{1}{\bar{\epsilon}} + \log\left(1+\frac{1}{\bar{\epsilon}}\right)\right) + \frac{3}{\bar{\epsilon}},$$

*for any choice of prior probability distribution $\bar{\pi}$ on $\mathcal{S}$, the posterior Gibbs distribution*

$$\rho(S) \sim \bar{\pi}(S) \prod_{I\in S}\left(1+\frac{1}{\beta(c(I)-1)+\bar{\epsilon}}\right)^{\beta a(I)D/2}\left(\frac{\beta c(I)+\bar{\epsilon}}{\beta a(I)+\bar{\epsilon}}\right)^{D/2}$$
$$\times \exp\left\{-\frac{D|S|}{2} - \frac{\bar{\lambda}}{2}\left[\sum_{i=1}^{N-1}\beta\left(1+\frac{1}{\beta(c(I_S(X_i))-1)+\bar{\epsilon}}\right)\|Y_i - f^S_{\bar{\theta}_S}(X_i)\|^2\right.\right.$$
$$\left.\left. +\bar{\epsilon}\sum_{I\in S}\left(1+\frac{1}{\beta(c(I)-1)+\bar{\epsilon}}\right)\|\bar{\theta}_S(I)\|^2\right]\right\}$$

*satisfies*

$$\mathbb{E}_{P_N}\|Y_N - \mathbb{E}_{\rho(dS)} f^S_{\bar{\theta}_S}(X_N)\|^2 \leq \inf_{S\in\mathcal{S}}\left(\inf_{\theta_S\in\Theta_S}\mathbb{E}_{P_N}\|Y_N - f^S_{\theta_S}(X_N)\|^2 + \frac{\bar{\gamma}(S)}{N}\right),$$

*with*

$$\bar{\gamma}(S) = \frac{D|S|}{\bar{\lambda}\beta} + \frac{B^2|S|}{\beta}(6+2\bar{\epsilon}) - \frac{2\log\bar{\pi}(S)}{\bar{\lambda}\beta}.$$

*Remark* 4.8.1. We can take for example

$$\bar{\epsilon} = 3(D+1),$$
$$\bar{\lambda} = \frac{1}{7B^2},$$
$$\beta = \frac{1}{1+e},$$

which gives

$$\bar{\gamma}(S) = (1+e)B^2\left((13D+12)|S| - 14\log\bar{\pi}(S)\right).$$

*Remark* 4.8.2. It is clear from the proof that the constants in this theorem are pessimistic, at least when $N$ is large with respect to $|S|$, since we have in many places used 0 as a lower bound for $a(I)$.

*Remark* 4.8.3. The expression of $\rho(S)$ is interesting, since it shows what an approximate "least favourable" prior distribution may look like. Of particular significance is the factor $\exp(-D|S|/2)$ which appears as a penalty term for the dimension of $\Theta_S$. We can find here an analogy with the penalized maximum likelihood estimator (see [6]). Note also that the use of a "double mixture" over a countable union of continuous parameter spaces finds its origin in information theory (see for example [67], [38], [82]).

## 4.9 Some abstract inference problem

In this appendix, we give a non-normalized form of the density estimation theorem, which covers all the cases studied in this chapter.

Let us consider a measurable *state space* $(\mathcal{X}, \mathcal{B})$ and an exchangeable probability measure $P_N$ on $\left(\mathcal{X}^N, \mathcal{B}^{\otimes N}\right)$. Let us consider also a measurable *parameter space* $(\Theta, \mathcal{B}')$, and a positive $\sigma$-finite measure $\pi \in \mathcal{M}_+(\Theta, \mathcal{B}')$ (more explicitely we assume that $\Theta$ is the union of a countable family of $\mathcal{B}'$ measurable sets of finite $\pi$-measure, in order to be able to apply Fubini's theorem to the product of $\pi$ with some other measures). Let $q : \Theta \times \mathcal{X} \to \mathbb{R}_+$ be a non negative measurable function (the reader may like to view it as a "non normalized" likelihood function). Let $\lambda$ be the Lebesgue measure on the real line. As the positive measure $\pi$ will be in many applications a probability measure, we will use for any function $h \in \mathbb{L}^1(\Theta, \pi)$ the notation

$$\mathbb{E}_{\pi(d\theta)}(h) \overset{\text{def}}{=} \int_{\theta \in \Theta} h(\theta)\pi(d\theta).$$

In the same way, for any $f \in \mathbb{L}^1(\mathcal{X}^N, P_N)$, $\mathbb{E}_{P_N}\left(f(X_1, \dots, X_N)\right)$ – or more explicitely $\mathbb{E}_{P_N(dX_1^N)}\left(f(X_1, \dots, X_N)\right)$ – will denote the expectation with respect to $P_N$ .

In order to impose some *integrability conditions* on the functions $q(\theta, x)$ related to our purpose, we will make use of the following simple lemma concerning the composition of absolutely continuous functions :

**Lemma 4.9.1.** *Let $I$ be an interval of the real line and let $F : I^2 \to \mathbb{R} \in \mathcal{C}^1(I^2)$ be a function with a continuous and* bounded *differential $dF \in \mathbb{L}^\infty$. Let $g : J \to I$ and $h : J \to I$ be two absolutely continuous functions defined on some other real interval $J$. Under this set of hypotheses, $x \mapsto F\big(g(x), h(x)\big)$ is also absolutely continuous and its Radon-Nikodym derivative is*

$$F(g,h)' = \frac{\partial F}{\partial x}(g,h)g' + \frac{\partial F}{\partial y}(g,h)h',$$

*where $g'$ and $h'$ are the Radon-Nikodym derivatives of $g$ and $h$.*

*Proof.* Let $\mu_g$ and $\mu_h$ be the signed measures defined by $g$ and $h$. Then as $F$ is Lipschitz of order 1, $F(g,h)$ is of bounded variations, and $\mu_{F(g,h)}$ is dominated by $|\mu_g| + |\mu_h|$, the sum of the total variation measures of $\mu_g$ and $\mu_h$. Therefore $\mu_{F(g,h)}$ is dominated by the Lebesgue measure, which means by definition that $F(g,h)$ is absolutely continuous. The composition formula comes from the fact that the Radon-Nikodym derivatives $g'$ and $h'$ are also almost everywhere the usual derivatives of $g$ and $h$. For more background on absolutely continuous functions, we refer the reader to [68, Chap 8]. □

Assume that the following *integrability conditions* are fulfilled for some positive real parameter $\beta \in ]0, 1/2]$ :

$$\theta \mapsto \left( \prod_{n=1}^{N-1} q(\theta, X_n) \right)^{\beta} \left( q(\theta, X_N) \right)^k \in \mathbb{L}^1(\Theta, \pi), \; k \in \{0,1\}, \; (X_1, \dots, X_N) \in \mathcal{X}^N, \quad (4.9.1)$$

$$(\theta, \eta) \mapsto \left( \prod_{n=1}^{N-1} q(\theta, X_n) \right)^{\beta} \left( q(\theta, X_N) \right)^{\eta} \left( \log q(\theta, X_N) \right)^3 \in \mathbb{L}^1(\Theta \times [\epsilon, 1], \pi \otimes \lambda), \quad (X_1, \dots, X_N) \in \mathcal{X}^N, \epsilon > 0. \quad (4.9.2)$$

Note that from the interpolation of norms (which is nothing but Hölder's inequality) $\|fg^{\eta}\| \leq \|fg^{\beta}\|^{\eta/\beta} \|f\|^{(1-\eta/\beta)}$, $\eta \in [0, \beta]$, hypothesis (4.9.1) (along with Fubini's theorem) implies

$$(\theta, \eta) \mapsto \left( \prod_{n=1}^{N-1} q(\theta, X_n) \right)^{\beta} \left( q(\theta, X_N) \right)^{\eta} \in \mathbb{L}^1(\Theta \times [0, 1], \pi \otimes \lambda), \quad (X_1, \dots, X_N) \in \mathcal{X}^N.$$

In view of this, hypothesis (4.9.2) implies that

$$(\theta, \eta) \mapsto \left( \prod_{n=1}^{N-1} q(\theta, X_n) \right)^{\beta} \left( q(\theta, X_N) \right)^{\eta} \left( \log q(\theta, X_N) \right)^k \in \mathbb{L}^1(\Theta \times [\epsilon, 1], \pi \otimes \lambda), \quad (X_1, \dots, X_N) \in \mathcal{X}^N, 0 \leq k \leq 3, \epsilon > 0.$$

Hypotheses (4.9.1), (4.9.2) and Fubini's theorem imply that

$$\begin{aligned} \mathbb{E}_{\pi(d\theta)} & \left[ \left( \prod_{n=1}^{N-1} q(\theta, X_n) \right)^{\beta} \Bigl( q(\theta, X_N) \Bigr)^{\eta} \Bigl( \log\bigl( q(\theta, X_N) \bigr) \Bigr)^k \right] \\ & = \mathbb{E}_{\pi(d\theta)} \left[ \left( \prod_{n=1}^{N} q(\theta, X_n) \right)^{\beta} \Bigl( \log\bigl( q(\theta, X_N) \bigr) \Bigr)^k \right] \\ & + \int_{\tau=\beta}^{\eta} \mathbb{E}_{\pi(d\theta)} \left[ \left( \prod_{n=1}^{N-1} q(\theta, X_n) \right)^{\beta} \Bigl( q(\theta, X_N) \Bigr)^{\tau} \Bigl( \log\bigl( q(\theta, X_N) \bigr) \Bigr)^{k+1} \right] d\tau \\ & \qquad (X_1, \dots, X_N) \in \mathcal{X}^N, \quad k \in \{0, 1, 2\}, \quad \eta \in ]0, 1]. \end{aligned}$$

Therefore, using Fubini's theorem again, we see that

$$\eta \mapsto \mathbb{E}_{\pi(d\theta)}\left[\left(\prod_{n=1}^{N-1} q(\theta, X_n)\right)^{\beta} (q(\theta, X_N))^{\eta}\right] \in \mathcal{C}^2(]0,1]),$$

$$(X_1, \dots, X_N) \in \mathcal{X}^N.$$

In order to consider its logarithm, we will introduce the following positivity condition :

$$\pi\left(\prod_{n=1}^{N} q(\theta, X_n) > 0\right) > 0, \quad (X_1, \dots, X_N) \in \mathcal{X}^N. \tag{4.9.3}$$

Note that this condition implies that

$$\mathbb{E}_{\pi(d\theta)}\left[\left(\prod_{n=1}^{N-1} q(\theta, X_n)\right)^{\beta} (q(\theta, X_N))^{\eta}\right] > 0, \quad (X_1, \dots, X_N) \in \mathcal{X}^N, \eta \in [0,1].$$

Note also that condition (4.9.3) is satisfied if some measurable set of parameters $T \in \mathcal{B}'$ is such that $\pi(T) > 0$ and

$$\inf_{x \in \mathcal{X}} q_\theta(x) > 0, \quad \theta \in T.$$

Under these conditions, introducing the abridged notation $X_1^N \overset{\text{def}}{=} (X_1, \dots, X_N)$, we see that

$$\eta \mapsto \mathcal{E}^{X_1^N}(\eta) \overset{\text{def}}{=} \log \mathbb{E}_{\pi(d\theta)}\left[\left(\prod_{n=1}^{N-1} q(\theta, X_n)\right)^{\beta} (q(\theta, X_N))^{\eta}\right] \in \mathcal{C}^2(]0,1]),$$

$$X_1^N \in \mathcal{X}^N.$$

Note that $\mathcal{E}^{X_1^N}(\eta)$ is also defined (according to (4.9.1)) and right continuous at point 0 (apply the monotone convergence theorem on the sets $\{\theta : q(\theta, X_N) \geq 1\}$ and $\{\theta : q(\theta, X_N) < 1\}$), and that it is a convex function on $[0,1]$ (see the expression of its second derivative below). Moreover

$$\frac{\partial}{\partial \eta} \mathcal{E}^{X_1^N}(\eta) = \frac{\mathbb{E}_{\pi(d\theta)}\left[\left(\prod_{n=1}^{N-1} q(\theta, X_n)\right)^{\beta} (q(\theta, X_N))^{\eta} \log q(\theta, X_N)\right]}{\mathbb{E}_{\pi(d\theta)}\left[\left(\prod_{n=1}^{N-1} q(\theta, X_n)\right)^{\beta} (q(\theta, X_N))^{\eta}\right]},$$

$$\frac{\partial^2}{\partial \eta^2} \mathcal{E}^{X_1^N}(\eta) = \frac{\mathbb{E}_{\pi(d\theta)} \left[ \left( \prod_{n=1}^{N-1} q(\theta, X_n) \right)^{\beta} (q(\theta, X_N))^{\eta} \Bigl( \log\bigl(q(\theta, X_N)\bigr) - \frac{\partial}{\partial \eta} \mathcal{E}^{X_1^N}(\eta) \Bigr)^2 \right]}{\mathbb{E}_{\pi(d\theta)} \left[ \left( \prod_{n=1}^{N-1} q(\theta, X_n) \right)^{\beta} (q(\theta, X_N))^{\eta} \right]}$$

Putting

$$\frac{\partial^3}{\partial \eta^3} \mathcal{E}^{X_1^N}(\eta) \stackrel{\text{def}}{=} \frac{\mathbb{E}_{\pi(d\theta)} \left[ \left( \prod_{n=1}^{N-1} q(\theta, X_n) \right)^{\beta} (q(\theta, X_N))^{\eta} \Bigl( \log\bigl(q(\theta, X_N)\bigr) - \frac{\partial}{\partial \eta} \mathcal{E}^{X_1^N}(\eta) \Bigr)^3 \right]}{\mathbb{E}_{\pi(d\theta)} \left[ \left( \prod_{n=1}^{N-1} q(\theta, X_n) \right)^{\beta} (q(\theta, X_N))^{\eta} \right]},$$

and using Fubini's theorem and lemma 4.9.1, we see that $\frac{\partial^2}{\partial \eta^2} \mathcal{E}^{X_1^N}(\eta)$ is absolutely continuous and that

$$\frac{\partial^2}{\partial \eta^2} \mathcal{E}^{X_1^N}(\eta) = \frac{\partial^2}{\partial \eta^2}_{|\eta=\beta} \mathcal{E}^{X_1^N}(\eta) + \int_{\beta}^{\eta} \frac{\partial^3}{\partial \xi^3} \mathcal{E}^{X_1^N}(\xi) d\xi, \qquad \eta \in ]0, 1].$$

(This implies that $\frac{\partial^3}{\partial \eta^3} \mathcal{E}^{X_1^N}(\eta)$ is almost everywhere the derivative of $\frac{\partial^2}{\partial \eta^2} \mathcal{E}^{X_1^N}(\eta)$.)

Now we will need to take expectations with respect to $(X_1, \dots, X_N)$, and to exchange these expectations with integrations with respect to $\eta$. This leads to consider the following set of assumptions:

$$X_1^N \mapsto \frac{\partial}{\partial \eta}_{|\eta=\beta} \mathcal{E}^{X_1^N}(\eta) \in \mathbb{L}^1(\mathcal{X}^N, P_N), \tag{4.9.4}$$

$$(X_1^N, \eta) \mapsto \frac{\mathbb{E}_{\pi(d\theta)}\left[\left(\prod_{n=1}^{N-1} q(\theta, X_n)\right)^{\beta} (q(\theta, X_N))^{\eta} \left|\log(q(\theta, X_N)) - \frac{\partial}{\partial \eta} \mathcal{E}^{X_1^N}(\eta)\right|^3\right]}{\mathbb{E}_{\pi(d\theta)}\left[\left(\prod_{n=1}^{N-1} q(\theta, X_n)\right)^{\beta} (q(\theta, X_N))^{\eta}\right]}$$
$$\in \mathbb{L}^1(\mathcal{X}^N \times [\epsilon, 1], P_N \otimes \lambda), \quad \epsilon > 0. \quad (4.9.5)$$

Note that hypothesis (4.9.5) implies that

$$(X_1^N, \eta) \mapsto \frac{\partial^k}{\partial \eta^k} \mathcal{E}^{X_1^N}(\eta) \in \mathbb{L}^1(\mathcal{X}^N \times [\epsilon, 1], P_N \otimes \lambda), \quad k \in \{2, 3\}, \quad \epsilon > 0.$$

Let $\Theta_1 \subset \Theta$ be the set of parameters for which $q(\theta, X_N)$ is integrable :

$$\Theta_1 \overset{\text{def}}{=} \{\theta \in \Theta : X_1^N \mapsto q(\theta, X_N) \in \mathbb{L}^1(\mathcal{X}^N, P_N)\}. \quad (4.9.6)$$

For any $\theta \in \Theta_1$, let us consider the constant

$$\gamma_\beta(\theta) \overset{\text{def}}{=} \mathbb{E}_{P_N}\left\{\beta \log \prod_{n=1}^{N} q(\theta, X_n) - \frac{\mathbb{E}_{\pi(d\theta')}\left[\left(\prod_{n=1}^{N} q(\theta', X_n)\right)^{\beta} \beta \log\left(\prod_{n=1}^{N} q(\theta', X_n)\right)\right]}{\mathbb{E}_{\pi(d\theta')}\left[\left(\prod_{n=1}^{N} q(\theta', X_n)\right)^{\beta}\right]}\right\}. \quad (4.9.7)$$

Note that $\gamma_\beta(\theta)$ is a well defined real number for any $\theta \in \Theta_1$, because it is from the definition of $\Theta_1$, assumption (4.9.4) and the fact that $P_N$ is exchangeable, the expectation with respect to $P_N$ of the difference of two integrable functions.

Indeed

$$\frac{\mathbb{E}_{\pi(d\theta')}\left(\left(\prod_{n=1}^{N} q(\theta', X_n)\right)^{\beta} \beta \log\left(\prod_{n=1}^{N} q(\theta', X_n)\right)\right)}{\mathbb{E}_{\pi(d\theta')}\left(\left(\prod_{n=1}^{N} q(\theta', X_n)\right)^{\beta}\right)}$$

$$= \sum_{n=1}^{N} \frac{\mathbb{E}_{\pi(d\theta')}\left(\left(\prod_{n=1}^{N} q(\theta', X_n)\right)^{\beta} \beta \log\left(q(\theta', X_n)\right)\right)}{\mathbb{E}_{\pi(d\theta')}\left(\left(\prod_{n=1}^{N} q(\theta', X_n)\right)^{\beta}\right)},$$

and each term in the righthand side has the same distribution as $\beta \frac{\partial}{\partial \eta}_{|\eta=\beta} \mathcal{E}^{X_1^N}(\eta)$.

Assume that the previous integrability and positivity hypotheses hold for some given value of $\beta$. Assume moreover that for some value of $\alpha \in ]\beta, 1]$

$$\chi_{\beta,\alpha} = 0 \vee \operatorname*{ess\,sup}_{\xi \in ]0,\alpha]} \left[ \frac{-\mathbb{E}_{P_N} \frac{\partial^3}{\partial \xi^3} \mathcal{E}^{X_1^N}(\xi)}{\mathbb{E}_{P_N} \frac{\partial^2}{\partial \xi^2} \mathcal{E}^{X_1^N}(\xi)} \mathbb{1}\left(\mathbb{E}_{P_N} \frac{\partial^2}{\partial \xi^2} \mathcal{E}^{X_1^N}(\xi) > 0\right)\right] < +\infty, \tag{4.9.8}$$

and that

$$\frac{\beta^2}{(\alpha - \beta)(2 - \alpha - \beta)} \exp(\alpha \chi_{\beta,\alpha}) \leq 1. \tag{4.9.9}$$

To summarize what has been assumed, we will let $\mathcal{H}(N, \beta, \alpha)$ denote the set of hypotheses (4.9.1) through (4.9.9), which depends on three parameters: N, $\beta$, and $\alpha$.

**Theorem 4.9.1.** *Let us assume that $\mathcal{H}(N, \beta, \alpha)$ holds. In this case*

$$\left(\mathcal{E}^{X_1^N}(1) - \mathcal{E}^{X_1^N}(0)\right) \in \mathbb{L}^1(\mathcal{X}^N, P_N),$$

*and the expectation* $\mathbb{E}_{P_N}\left(\mathcal{E}^{X_1^N}(1) - \mathcal{E}^{X_1^N}(0)\right)$ *is such that*

$$\mathbb{E}_{P_N}\left(\mathcal{E}^{X_1^N}(1) - \mathcal{E}^{X_1^N}(0)\right) \geq \mathbb{E}_{P_N} \frac{\partial}{\partial \eta}_{|\eta=\beta} \mathcal{E}^{X_1^N}(\eta) \tag{4.9.10}$$

$$\geq \sup_{\theta \in \Theta_1} \left(\mathbb{E}_{P_N} \log q(\theta, X_N) - \frac{\gamma_\beta(\theta)}{\beta N}\right). \tag{4.9.11}$$

*Remark* 4.9.1. The hypotheses mentioned in the theorem are all fullfilled when the function $(\theta, x) \mapsto \log\bigl(q(\theta, x)\bigr)$ is bounded.

*Remark* 4.9.2. The condition (4.9.9) on $\beta$ can also be written as

$$\beta \leq \frac{1}{e^{\alpha\chi} - 1}\Bigl(\sqrt{1 + \alpha(e^{\alpha\chi} - 1)(2 - \alpha)} - 1\Bigr),$$

where we have put $\chi_{\beta,\alpha} = \chi$ for short.

When $\chi_{\beta,1}$ is small, we can take $\alpha = 1$ and obtain the sufficient condition

$$\beta \leq (1 + \exp(\chi_{\beta,1}/2))^{-1}.$$

When $\chi_{\beta,1}$ is large, we can take $\alpha = \chi_{\beta,1}^{-1}\log(\chi_{\beta,1})$ and, (reminding that $\alpha \mapsto \chi_{\beta,\alpha}$ is increasing), obtain the sufficient condition

$$\beta \leq \frac{1}{\chi_{\beta,1} - 1}\left(\sqrt{1 + (\chi_{\beta,1} - 1)\left(2 - \frac{\log(\chi_{\beta,1})}{\chi_{\beta,1}}\right)\frac{\log(\chi_{\beta,1})}{\chi_{\beta,1}}} - 1\right)$$
$$\underset{\chi_{\beta,1}\to+\infty}{\sim} \frac{\sqrt{2\log(\chi_{\beta,1})}}{\chi_{\beta,1}}.$$

*Remark* 4.9.3. Obviously,

$$\sup_{\beta,\alpha} \chi_{\beta,\alpha} \leq \sup_{\theta,\theta',x} \log\left[\frac{q(\theta, x)}{q(\theta', x)}\right].$$

This provides a bound when the lefthand side of this inequality is finite, that is when the model $\bigl\{x \mapsto q(\theta, x)\,;\, \theta \in \Theta\bigr\}$ is so to speak "log bounded".

*Remark* 4.9.4. Putting

$$h(\theta, \theta', x_1^N) = \frac{\beta}{N}\log\left(\prod_{n=1}^{N}\frac{q(\theta, x_n)}{q(\theta', x_n)}\right),$$

we can write $\gamma_\beta(\theta)$ in the form

$$\gamma_\beta(\theta) = \mathbb{E}_{P_N}\frac{\mathbb{E}_{\pi(d\theta')}\Bigl[Nh(\theta, \theta', X_1^N)\exp\bigl(-Nh(\theta, \theta', X_1^N)\bigr)\Bigr]}{\mathbb{E}_{\pi(d\theta')}\Bigl[\exp\bigl(-Nh(\theta, \theta', X_1^N)\bigr)\Bigr]},$$

and bound it by

$$\gamma_\beta(\theta) \leq \sup_{x_1^N \in \mathcal{X}^N}\frac{\mathbb{E}_{\pi(d\theta')}\Bigl[Nh(\theta, \theta', x_1^N)\exp\bigl(-Nh(\theta, \theta', x_1^N)\bigr)\Bigr]}{\mathbb{E}_{\pi(d\theta')}\Bigl[\exp\bigl(-Nh(\theta, \theta', x_1^N)\bigr)\Bigr]}.$$

to stress the fact that $\gamma_\beta(\theta)$ can be bounded using some evaluation of Laplace integrals over the parameter space.

When $\Theta$ is countable, and $\pi$ is a finite measure, the following explicit bound is useful:

*Proposition* 4.9.1.

$$\gamma_\beta(\theta) \leq \log \frac{\pi(\Theta)}{\pi(\{\theta\})}. \tag{4.9.12}$$

It is a consequence of the more general lemma 4.1.1.

*Proof of theorem 4.9.1.* For any $\epsilon \in ]0, \beta[$, any $X_1^N \in \mathcal{X}^N$, we can write

$$\begin{aligned}
\mathcal{E}^{X_1^N}(1) - \mathcal{E}^{X_1^N}(\epsilon) &= \int_{\eta=\epsilon}^{1} \frac{\partial}{\partial \eta} \mathcal{E}^{X_1^N}(\eta) d\eta \\
&= (1-\epsilon) \frac{\partial}{\partial \eta}_{|\eta=\beta} \mathcal{E}^{X_1^N}(\eta) + \int_{\eta=\beta}^{1} (1-\eta) \frac{\partial^2}{\partial \eta^2} \mathcal{E}^{X_1^N}(\eta) d\eta \\
&\quad - \int_{\eta=0}^{\beta} \mathbb{1}(\eta \geq \epsilon)(\eta - \epsilon) \frac{\partial^2}{\partial \eta^2} \mathcal{E}^{X_1^N}(\eta) d\eta.
\end{aligned} \tag{4.9.13}$$

From the monotone convergence theorem

$$\lim_{\epsilon \downarrow 0} \int_{\eta=0}^{\beta} \mathbb{1}(\eta \geq \epsilon)(\eta - \epsilon) \frac{\partial^2}{\partial \eta^2} \mathcal{E}^{X_1^N}(\eta) d\eta = \int_{\eta=0}^{\beta} \eta \frac{\partial^2}{\partial \eta^2} \mathcal{E}^{X_1^N}(\eta) d\eta.$$

Thus

$$\begin{aligned}
\mathcal{E}^{X_1^N}(1) - \mathcal{E}^{X_1^N}(0) = \frac{\partial}{\partial \eta}_{|\eta=\beta} \mathcal{E}^{X_1^N}(\eta) + \int_{\eta=\beta}^{1} (1-\eta) \frac{\partial^2}{\partial \eta^2} \mathcal{E}^{X_1^N}(\eta) d\eta \\
- \int_{\eta=0}^{\beta} \eta \frac{\partial^2}{\partial \eta^2} \mathcal{E}^{X_1^N}(\eta) d\eta.
\end{aligned} \tag{4.9.14}$$

The two first terms of the righthand side are in $\mathbb{L}^1(\mathcal{X}^N, P_N)$, according to hypotheses (4.9.4), (4.9.5) and Fubini's theorem. To see that the third one is also in $\mathbb{L}^1(\mathcal{X}^N, P_N)$, we will need to use hypothesis (4.9.8)

Coming back to the expression of $\frac{\partial^2}{\partial \eta^2} \mathcal{E}^{X_1^N}(\eta)$, we see that $\mathbb{E}_{P_N}\left(\frac{\partial^2}{\partial \eta^2} \mathcal{E}^{X_1^N}(\eta)\right) = 0$ if and only if $\theta \mapsto q(\theta, X_N)$ is $\pi$ a.s. constant $P_N$ almost surely, and therefore that either $\mathbb{E}_{P_N} \frac{\partial^2}{\partial \eta^2} \mathcal{E}^{X_1^N}(\eta) > 0$ for all $\eta \in ]0,1]$, or $\frac{\partial^2}{\partial \eta^2} \mathcal{E}^{X_1^N}(\eta) = 0$, $P_N$ a.s. for all $\eta \in ]0,1]$. In the latter case

$$\left(\mathcal{E}^{X_1^N}(1) - \mathcal{E}^{X_1^N}(0)\right) = \frac{\partial}{\partial \eta}_{|\eta=\beta} \mathcal{E}^{X_1^N}(\eta), \qquad P_N \text{ a.s.},$$

and therefore equation (4.9.10) of theorem 4.9.1 holds. In the former case, using Fubini's theorem and lemma 4.9.1, we wee that for any $\xi \in [\beta, \alpha]$ and any $\zeta \in ]0, \beta]$,

$$\log \mathbb{E}_{P_N} \frac{\partial^2}{\partial \eta^2}_{|\eta=\xi} \mathcal{E}^{X_1^N}(\eta) = \log \mathbb{E}_{P_N} \frac{\partial^2}{\partial \eta^2}_{|\eta=\zeta} \mathcal{E}^{X_1^N}(\eta) + \int_{\eta=\zeta}^{\xi} \frac{\mathbb{E}_{P_N} \frac{\partial^3}{\partial \eta^3} \mathcal{E}^{X_1^N}(\eta)}{\mathbb{E}_{P_N} \frac{\partial^2}{\partial \eta^2} \mathcal{E}^{X_1^N}(\eta)} d\eta$$

$$\geq \log \mathbb{E}_{P_N} \frac{\partial^2}{\partial \eta^2}_{|\eta=\zeta} \mathcal{E}^{X_1^N}(\eta) - \alpha \chi_{\beta,\alpha}. \tag{4.9.15}$$

A first consequence of (4.9.15) is that $\eta \mapsto \mathbb{E}_{P_N} \frac{\partial^2}{\partial \eta^2} \mathcal{E}^{X_1^N}(\eta)$ is bounded on $]0, \beta]$ and therefore that (due to Fubini's theorem and (4.9.5))

$$(X_1^N, \eta) \mapsto \frac{\partial^2}{\partial \eta^2} \mathcal{E}^{X_1^N}(\eta) \in \mathbb{L}^1(\mathcal{X}^N \times [0,1], P_N \otimes \lambda). \tag{4.9.16}$$

This establishes the integrability of the third term of the righthand side of (4.9.14), and thus achieves to prove that $\mathcal{E}^{X_1^N}(1) - \mathcal{E}^{X_1^N}(0) \in \mathbb{L}^1(\mathcal{X}^N, P_N)$.

Integrating with respect to $P_N$, we see from Fubini's theorem and (4.9.16) that

$$\begin{aligned} &\mathbb{E}_{P_N} \left[ \int_{\eta=\beta}^1 (1-\eta) \frac{\partial^2}{\partial \eta^2} \mathcal{E}^{X_1^N}(\eta) d\eta - \int_{\eta=0}^\beta \eta \frac{\partial^2}{\partial \eta^2} \mathcal{E}^{X_1^N}(\eta) d\eta \right] \\ &\quad = \int_{\eta=\beta}^1 (1-\eta) \mathbb{E}_{P_N} \left[ \frac{\partial^2}{\partial \eta^2} \mathcal{E}^{X_1^N}(\eta) \right] d\eta - \int_{\eta=0}^\beta \eta \mathbb{E}_{P_N} \left[ \frac{\partial^2}{\partial \eta^2} \mathcal{E}^{X_1^N}(\eta) \right] d\eta. \end{aligned} \tag{4.9.17}$$

We can assume that $\mathbb{E}_{P_N} \left[ \frac{\partial^2}{\partial \eta^2} \mathcal{E}^{X_1^N}(\eta) \right] > 0$ for all $\eta \in ]0,1]$, since in the alternative situation we already proved (4.9.10). Remembering that $\frac{\partial^2}{\partial \eta^2} \mathcal{E}^{X_1^N}(\eta) \geq 0$ and coming back to equation (4.9.17), we see that we can troncate the first integral of the righthand side to obtain

$$\begin{aligned} &\mathbb{E}_{P_N} \left[ \int_{\eta=\beta}^1 (1-\eta) \frac{\partial^2}{\partial \eta^2} \mathcal{E}^{X_1^N}(\eta) d\eta - \int_{\eta=0}^\beta \eta \frac{\partial^2}{\partial \eta^2} \mathcal{E}^{X_1^N}(\eta) d\eta \right] \\ &\quad \geq \int_{\eta=\beta}^\alpha (1-\eta) \mathbb{E}_{P_N} \left[ \frac{\partial^2}{\partial \eta^2} \mathcal{E}^{X_1^N}(\eta) \right] d\eta - \int_{\eta=0}^\beta \eta \mathbb{E}_{P_N} \left[ \frac{\partial^2}{\partial \eta^2} \mathcal{E}^{X_1^N}(\eta) \right] d\eta. \end{aligned} \tag{4.9.18}$$

Using (4.9.15), we see that

$$\inf_{\xi \in [\beta,\alpha]} \mathbb{E}_{P_N} \left[ \frac{\partial^2}{\partial \eta^2}_{|\eta=\xi} \mathcal{E}^{X_1^N}(\eta) \right] \geq \exp(-\alpha \chi_{\beta,\alpha}) \mathbb{E}_{P_N} \left[ \sup_{\zeta \in ]0,\beta]} \frac{\partial^2}{\partial \eta^2}_{|\eta=\zeta} \mathcal{E}^{X_1^N}(\eta) \right]. \tag{4.9.19}$$

Combining equations (4.9.13), (4.9.17), (4.9.18), (4.9.19) and (4.9.9) we obtain

$$\begin{aligned} &\mathbb{E}_{P_N} \left( \mathcal{E}^{X_1^N}(1) - \mathcal{E}^{X_1^N}(0) \right) \\ &\qquad \geq \mathbb{E}_{P_N} \left( \frac{\partial}{\partial \eta}_{|\eta=\beta} \mathcal{E}^{X_1^N}(\eta) \right) \end{aligned}$$

$$+\left[\frac{(\alpha-\beta)(2-\alpha-\beta)}{2}\exp(-\alpha\chi_{\beta,\alpha})-\frac{\beta^2}{2}\right]\sup_{\zeta\in]0,\beta]}\mathbb{E}_{P_N}\left[\frac{\partial^2}{\partial\eta^2}_{|\eta=\zeta}\mathcal{E}^{X_1^N}(\eta)\right]$$
$$\geq \mathbb{E}_{P_N}\left[\frac{\partial}{\partial\eta}_{|\eta=\beta}\mathcal{E}^{X_1^N}(\eta)\right].$$

This ends the proof of (4.9.10). Note that we have not used yet the fact that $P_N$ was supposed to be exchangeable. To prove (4.9.11) from (4.9.10), it is enough to notice that, due to the fact that $P_N$ is exchangeable, for any $\theta \in \Theta_1$

$$\gamma_\beta(\theta) = \mathbb{E}_{P_N}\big(\log q(\theta, X_N)\big) - \beta N \mathbb{E}_{P_N}\left[\frac{\partial}{\partial\eta}_{|\eta=\beta}\mathcal{E}^{X_1^N}(\eta)\right].$$

□

## 4.10 Another type of bound

Theorem 4.9.1 is concerned with values of $\beta$ in the region $]0,1/2[$. Another interesting value of the temperature is $\beta = 1/2$. In this section we will consider a set of assumptions $\mathcal{H}'(N,\beta)$ which is the same as the set of hypotheses $\mathcal{H}(N,\beta,\alpha)$ of appendix 4.9, except that we do not require conditions (4.9.8) and (4.9.9), and that condition (4.9.5) is strengthened to

$$(X_1^N,\eta)\mapsto$$
$$\eta^2\frac{\mathbb{E}_{\pi(d\theta)}\left[\left(\prod_{n=1}^{N-1}q(\theta,X_n)\right)^{\beta}(q(\theta,X_N))^{\eta}\left|\log(q(\theta,X_N))-\frac{\partial}{\partial\eta}\mathcal{E}^{X_1^N}(\eta)\right|^3\right]}{\mathbb{E}_{\pi(d\theta)}\left[\left(\prod_{n=1}^{N-1}q(\theta,X_n)\right)^{\beta}(q(\theta,X_N))^{\eta}\right]}$$
$$\in \mathbb{L}^1(\mathcal{X}^N\times[0,1], P_N\otimes\lambda). \quad (4.10.1)$$

Under this slightly different set of hypotheses, we can write

$$\mathcal{E}^{X_1^N}(1)-\mathcal{E}^{X_1^N}(\epsilon) = (1-\epsilon)\frac{\partial}{\partial\eta}_{|\eta=\beta}\mathcal{E}^{X_1^N}(\eta)$$
$$+\frac{(1-\beta)^2-(\beta-\epsilon)^2}{2}\frac{\partial^2}{\partial\eta^2}_{|\eta=\beta}\mathcal{E}^{X_1^N}(\eta)$$
$$+\int_{\eta=\beta}^{1}\frac{(1-\eta)^2}{2}\frac{\partial^3}{\partial\eta^3}\mathcal{E}^{X_1^N}(\eta)d\eta+\int_{\eta=\epsilon}^{\beta}\frac{(\eta-\epsilon)^2}{2}\frac{\partial^3}{\partial\eta^3}\mathcal{E}^{X_1^N}(\eta)d\eta.$$

Applying Lebesgue's dominated convergence theorem to the righthand side under hypothesis (4.10.1), we see that the poinwise limit of the lefthand side when $\epsilon$ tends to zero, which is equal to $\mathcal{E}^{X_1^N}(1) - \mathcal{E}^{X_1^N}(0)$, is in $\mathbb{L}^1(\mathcal{X}^N, P_N)$. We can then come back to the second order expansion (4.9.13), and integrate up to $\alpha$ only (instead of one), to get the inequality

$$\begin{aligned}\mathcal{E}^{X_1^N}(1) - \mathcal{E}^{X_1^N}(0) \geq \frac{\partial}{\partial \eta}_{|\eta=\beta} & \mathcal{E}^{X_1^N}(\eta) \\ & + \frac{(1-\beta)^2 - (1-\alpha)^2 - \beta^2}{2} \frac{\partial^2}{\partial \eta^2}_{|\eta=\beta} \mathcal{E}^{X_1^N}(\eta) \\ & + \int_{\eta=\beta}^{1} \frac{(1-\eta)^2 - (1-\alpha)^2}{2} \frac{\partial^3}{\partial \eta^3} \mathcal{E}^{X_1^N}(\eta) d\eta \\ & \qquad + \int_{\eta=0}^{\beta} \frac{\eta^2}{2} \frac{\partial^3}{\partial \eta^3} \mathcal{E}^{X_1^N}(\eta) d\eta, \quad P_N \text{ a.s.}\end{aligned}$$

Choosing $\alpha = 1 - \sqrt{1-2\beta}$, to cancel the second order term, and integrating with respect to $P_N$, we obtain the following theorem :

**Theorem 4.10.1.** *Under the set of hypotheses $\mathcal{H}'(N, \beta)$ described above,*

$$\mathcal{E}^{X_1^N}(1) - \mathcal{E}^{X_1^N}(0) \in \mathbb{L}^1(\mathcal{X}^N, P_N)$$

*and*

$$\begin{aligned}\mathbb{E}_{P_N}\Bigl(\mathcal{E}^{X_1^N}(1) - \mathcal{E}^{X_1^N}(0)\Bigr) \geq \sup_{\theta \in \Theta_1} \left( \mathbb{E}_{P_N}\bigl(\log q(\theta, X_N)\bigr) - \frac{\gamma_\beta(\theta)}{\beta N} \right) \\ + \frac{1}{24}\bigl(1 - \sqrt{1-2\beta}\bigr)^3 \bigl(1 + 3\sqrt{1-2\beta}\bigr) \inf_{\eta \in [0, 1-\sqrt{1-2\beta}]} \mathbb{E}_{P_N}\left[\frac{\partial^3}{\partial \eta^3} \mathcal{E}^{X_1^N}(\eta)\right].\end{aligned}$$

*Remark* 4.10.1. This theorem is interesting in cases when $-\inf_\eta \mathbb{E}_{P_N}\left[\frac{\partial^3}{\partial \eta^3} \mathcal{E}^{X_1^N}(\eta)\right]$ turns out to be small when compared to $\frac{\gamma_\beta(\theta)}{\beta N}$.

*Remark* 4.10.2. When $(\theta, x) \mapsto q(\theta, x)$ is bounded on $\Theta \times \mathcal{X}$, then $\mathcal{H}'(N, \beta)$ is satisfied for any $\beta \in [0, 1/2]$ and $\Theta_1 = \Theta$.

# 5

# Randomized estimators and empirical complexity for pattern recognition and least square regression

## 5.1 A pseudo-Bayesian approach to adaptive inference

### 5.1.1 General framework

We consider in this chapter a regression framework. Generalization to other situations should be possible, but beyond the scope we plan to cover here. We will start with a product space $\mathcal{X} \times \mathcal{Y}$, where $(\mathcal{X}, \mathcal{B})$ is a measurable space and where $\mathcal{Y}$ is either a finite set or the real line. Here again, generalizations could easily be conceived, but we prefer to restrict to this case for the sake of simplicity and concreteness. In the case of classification, the set $\mathcal{X}$ is to be thought of as the space of "patterns", and the finite set $\mathcal{Y}$ is to be thought of as a set of labels describing different patterns. In the case of regression, $\mathcal{X}$ may be thought of as a multidimensional space of explanatory variables which we would like to use to predict the value of some real valued random variable. In both cases, we assume that we observe an i.i.d. sample $(X_i, Y_i)_{i=1}^N$ of random variables distributed according to a product probability measure $P^{\otimes N}$, where $P$ is a probability distribution on $\bigl(\mathcal{X} \times \mathcal{Y}, \mathcal{B} \otimes \mathcal{B}'\bigr)$, and where $\mathcal{B}'$ is the Borel sigma algebra when $\mathcal{Y}$ is the real line and is the trivial algebra of all subsets when $\mathcal{Y}$ is finite. Apart from this observation, we also consider a set of regression functions (or classification rules, depending on the context) $\mathcal{R} = \bigl\{f_\theta : \mathcal{X} \to \mathcal{Y}; \theta \in \Theta\bigr\}$.

The aim of all the techniques presented in this chapter is to make the best possible guess about $Y$ given $X$, in the sense of minimizing the *expected risk*

$$R(\theta) \overset{\text{def}}{=} \mathbb{E}_P\bigl[\ell\bigl(Y, f_\theta(X)\bigr)\bigr].$$

The *loss function* $\ell$ will be the Hamming distance $\ell(Y, Y') = \mathbb{1}(Y \neq Y')$ in the case of pattern recognition. Therefore in this case $R(\theta)$ is the *error rate* of the classification rule $f_\theta$, and can also be written as

$$R(\theta) = P\bigl(Y \neq f_\theta(X)\bigr).$$

In the regression framework, we will work with the square loss $\ell(Y, Y') = (Y - Y')^2$. Although other loss functions may be considered, the mean square loss has the distinguished property of being minimized by the conditional expectation of $Y$ given $X$. More precisely, it decomposes into

$$R(\theta) = \mathbb{E}_P\Big\{\big[Y - \mathbb{E}_P(Y \,|\, X)\big]^2\Big\} + \mathbb{E}_P\Big\{\big[\mathbb{E}_P(Y \,|\, X) - f_\theta(X)\big]^2\Big\}.$$

Therefore minimizing the mean square loss is equivalent to minimizing the quadratic distance to the conditional expectation.

We assume that we have no prior information about the distribution $P$ of $(X, Y)$, and that we have to guess it entirely from the observed sample $(X_i, Y_i)_{i=1}^N$.

Working with a *prescribed* set of regression functions $\mathcal{R}$ is crucial in this situation. Indeed, as it is well known, it would be impossible in general to find an estimator $\widehat{f} : \mathcal{X} \times (\mathcal{X}, \mathcal{Y})^N \to \mathcal{Y}$ such that

$$\lim_{N \to +\infty} \sup_{P \in \mathcal{M}_+^1(\mathcal{X} \times \mathcal{Y})} \mathbb{E}_{P^{\otimes (N+1)}}\Big\{\ell\big[Y_{N+1}, \widehat{f}(X_{N+1} \,|\, X_1^N, Y_1^N)\big]\Big\} - \inf_{f : \mathcal{X} \to \mathcal{Y}} \mathbb{E}_P\Big\{\ell\big[Y, f(X)\big]\Big\} = 0,$$

where $f$ ranges in all the measurable functions from $\mathcal{X}$ to $\mathcal{Y}$, and $X_1^N$ is a short notation for $(X_i)_{i=1}^N$. A more realistic requirement is to look for a regression function $\widehat{f}$ such that

$$\lim_{N \to +\infty} \sup_{P \in \mathcal{M}_+^1(\mathcal{X} \times \mathcal{Y})} \mathbb{E}_{P^{\otimes (N+1)}}\Big\{\ell\big[Y_{N+1}, \widehat{f}(X_{N+1} \,|\, X_1^N, Y_1^N)\big]\Big\} - \inf_{\theta \in \Theta} \mathbb{E}_P\Big\{\ell\big[Y, f_\theta(X)\big]\Big\} = 0, \quad (5.1.1)$$

This turns out to be a feasible problem, with a speed of convergence depending on the *complexity* (to be more precisely defined) of the set of regression functions $\mathcal{R} = \{f_\theta : \theta \in \theta\}$.

Another important technical feature of our approach is to dismiss the requirement that $\widehat{f}$ should belong to $\mathcal{R}$. Instead we allow $\widehat{f}$ either to be drawn from $\mathcal{R}$ according to some *posterior* distribution $\rho(d\theta)$ on the parameter set $\Theta$, or to be computed from this posterior distribution in some deterministic way. *Aggregating* estimators is an idea we have already studied in the previous chapters. It is also an active field of research in data compression theory [82] and in computer science probabilistic learning theory [53, 54]. It is possible to show that some aggregation rules have in some situations a lower *mean* risk than any possible selection rule of $\widehat{f}$ in $\mathcal{R}$ (at least in the regression setting, see [22]). When the deviations of the risk are studied, the situation is not completely understood (at least to us) : we will show here how to derive parameter selection rules in the regression setting, based on a control of the oscillations of the expected risk on balls. In the classification setting, it is not clear that the same reasoning could be made, because the oscillations of the expected $L_1$ risk on balls cannot be bounded in the same way. Therefore in our view, aggregation and randomized rules can be expected to bring a more decisive improvement over selection rules in the classification setting than in the least square regression case.

The originality of the results presented here are twofold : on the one hand we derive oracle inequalities for the *deviations* of the risk, instead of its *mean* as in the previous chapters, on the other hand, we express these bounds in term of a new *empirical* measure of complexity in the spirit of PAC-Bayesian bounds given e.g. in [53, 54]. What we call an empirical measure of complexity is a measure which can be computed from the observed sample. This is to be contrasted for instance with the Vapnik entropy or its universal bound in terms of VC dimension, which has to be computed from a theoretical study of the model $\mathcal{R}$.

### 5.1.2 Some pervading ideas

When the parameter set (or more accurately the set of regression functions $\mathcal{R}$) is at the same time "big" and heterogeneous, it is well known that replacing the *expected risk* $R(\theta)$ by the *empirical risk*

$$r(\theta) \overset{\text{def}}{=} \frac{1}{N} \sum_{i=1}^{N} \ell\bigl(Y_i, f_\theta(X_i)\bigr),$$

and solving $\inf_{\theta \in \Theta} r(\theta)$ instead of $\inf_{\theta \in \Theta} R(\theta)$ is not consistent in the sense of equation (5.1.1). Indeed, the discrepancy $\sup_\theta \bigl(R(\theta) - r(\theta)\bigr)$ may be large enough to spoil the game. An intensively studied way to get some control on this situation is to add a penalty term $\gamma_N(\theta)$ and to solve the *penalized* minimization problem

$$\inf_{\theta \in \Theta} r(\theta) + \gamma_N(\theta).$$

A different path is followed here : instead of adding a penalty term to $r(\theta)$, we use the *quantiles* of the distribution of $r(\theta)$ under some prior distribution $\pi$ on the parameter set $\Theta$. In order to make this meaningful we assume that the parameter set $(\Theta, \mathcal{F})$ is a measurable space and that $(\theta, x) \mapsto f_\theta(x)$ is a measurable function on $\bigl(\Theta \times \mathcal{X}, \mathcal{F} \otimes \mathcal{B}\bigr)$. Under these assumptions $\theta \mapsto r(\theta)$ and $\theta \mapsto R(\theta)$ are measurable functions (by Fubini's theorem). The *$\alpha$-quantile* of $r(\theta)$ under $\pi$ can then be defined as

$$q(\alpha) = \inf\bigl\{\eta \,|\, \pi\bigl(r(\theta) \leq \eta\bigr) > \alpha\bigr\}.$$

As $q(0) = \operatorname{essinf}_{\theta \in \Theta} r(\theta)$, considering $q(\alpha)$ for $\alpha > 0$ may be regarded as a generalization of empirical risk minimization. Its main advantage is that, unless the "extreme" quantile $q(0)$, the $\alpha$-quantile $q(\alpha)$ with $\alpha > 0$ can fluctuate in a way which is controlled by $\alpha$, whatever $\Theta$ may be.

A simple classification rule $\widehat{f}$ related to the quantiles of $r(\theta)$ under $\pi$ is obtained by drawing $\theta$ according to $\pi\bigl[d\theta \,|\, r(\theta) \leq q(\alpha)\bigr]$, the prior probability distribution $\pi$ conditioned by the set of positive measure $\{\theta : r(\theta) \leq q(\alpha)\}$. We will explain how $\alpha$ can be chosen to make this "randomized" classification rule efficient. To achieve this, we will show how $\alpha \mapsto q(\alpha)$ can be used to obtain a local empirical measure of the "complexity" of the set of classification rules $\mathcal{R}$.

Another kind of posterior distributions worthy of some interest are the *Gibbs posterior distributions* introduced in [22], and defined as a function of some real positive *inverse temperature* parameter $\beta$ by

$$\rho(d\theta) \overset{\text{def}}{=} \frac{\exp\bigl(-\beta r(\theta)\bigr)}{\mathbb{E}_{\pi(d\theta')}\Bigl[\exp\bigl(-\beta r(\theta')\bigr)\Bigr]}\pi(d\theta).$$

We derived in [22] oracle inequalities for the mean risk of regression rules based on Gibbs posterior distributions for properly chosen values of the parameter $\beta$. We will continue this study here with oracle inequalities concerned with the deviations of their risk.

We start with the pattern recognition framework.

## 5.2 A randomized rule for pattern recognition

A detailed formulation of Bernstein's inequality will be useful. It can be found in McDiarmid's monograph [55, p 203-204].

**Theorem 5.2.1.** *Let $(\sigma_1, \dots, \sigma_N)$ be independent real valued random variables such that*

$$\sigma_i - \mathbb{E}(\sigma_i) \leq b, \qquad i = 1, \dots, N.$$

*Let*

$$S = \frac{1}{N}\sum_{i=1}^{N} \sigma_i$$

*be their normalized sum,*

$$m = \mathbb{E}(S) = \frac{1}{N}\sum_{i=1}^{N} \mathbb{E}(\sigma_i)$$

*its expectation and*

$$V = N\mathbb{E}\Bigl[\bigl(S - \mathbb{E}(S)\bigr)^2\Bigr] = \frac{1}{N}\sum_{i=1}^{N} \mathbb{E}\Bigl[\bigl(\sigma_i - \mathbb{E}(\sigma_i)\bigr)^2\Bigr]$$

*its renormalized variance. Let*

$$g(x) = \frac{1}{x^2}(e^x - 1 - x).$$

*The deviations of $S$ are bounded, for any $\lambda \in \mathbb{R}_+$, any $\eta \in \mathbb{R}_+$, by*

$$\begin{aligned} P(S - m \geq \eta) &\leq \mathbb{E}\Big[\exp\Big(-\lambda\eta + \lambda(S-m)\Big)\Big] \\ &\leq \exp\left(-\eta\lambda + g\left(\frac{b\lambda}{N}\right)\frac{V}{N}\lambda^2\right), \end{aligned}$$

*moreover when $\lambda$ is chosen to be*

$$\lambda = \frac{N}{b}\log\left(1 + \frac{b\eta}{V}\right),$$

*the right-hand side of the previous equation is itself bounded by*

$$\exp\left(-\eta\lambda + g\left(\frac{b\lambda}{N}\right)\frac{V}{N}\lambda^2\right) \leq \exp\left(-\frac{3N\eta^2}{6V + 2b\eta}\right).$$

Applying Bernstein's theorem to

$$\sigma_i \overset{\text{def}}{=} -\mathbb{1}\big(Y_i \neq f_\theta(X_i)\big)$$

and integrating with respect to $\pi(d\theta)$, we obtain a "learning" lemma which improves on the PAC-Bayesian bounds derived in [53, 54] from Hoeffding's inequality and can be used to learn a posterior distribution on the parameters from a prior one :

Let us remind the definition of the Kullback Leibler divergence $\mathcal{K}(\rho, \pi)$ between two probability distributions $\rho$ and $\pi$:

$$\mathcal{K}(\rho, \pi) \overset{\text{def}}{=} \begin{cases} \displaystyle\int_\theta \log\left[\frac{\rho}{\pi}(\theta)\right]\rho(d\theta) & \text{if } \rho \ll \pi, \\ +\infty & \text{otherwise.} \end{cases}$$

It will be useful in the following to compute the Legendre transform of the convex function $\rho \mapsto \mathcal{K}(\rho, \pi)$. It is given by the following formula: for any measurable function $h : \Theta \to \mathbb{R}$,

$$\log\Big\{\mathbb{E}_{\pi(d\theta)}\Big\{\exp\big[h(\theta)\big]\Big\}\Big\} = \sup_{\rho \in \mathcal{M}_+^1(\Theta)} \mathbb{E}_{\rho(d\theta)}\big[h(\theta)\big] - \mathcal{K}(\rho, \pi), \qquad (5.2.1)$$

where the value of $\mathbb{E}_{\rho(d\theta)}\big[h(\theta)\big]$ is defined by convention as

$$\mathbb{E}_{\rho(d\theta)}\big[h(\theta)\big] \overset{\text{def}}{=} \sup_{B \in \mathbb{R}} \mathbb{E}\Big[\min\{B, h(\theta)\}\Big], \qquad (5.2.2)$$

and where it is also understood that

$$\infty - \infty = \sup_{B \in \mathbb{R}} (B) - \infty = \sup_{B \in \mathbb{R}} (B - \infty) = -\infty. \tag{5.2.3}$$

In other words a priority is given to $-\infty$ in ambiguous cases : the expectation of a function whose negative part is not integrable will be assumed to be $-\infty$, even when its positive part integrates to $+\infty$.

Let us give for the sake of completeness a short proof of this well known result. In the case when $h$ is upperbounded, consider

$$\nu(d\theta) = \frac{\exp\bigl[h(\theta)\bigr]}{\mathbb{E}_{\pi(d\theta)}\Bigl\{\exp\bigl[h(\theta)\bigr]\Bigr\}}\pi(d\theta),$$

and remark that

$$\log\Bigl\{\mathbb{E}_{\pi(d\theta)}\Bigl\{\exp\bigl[h(\theta)\bigr]\Bigr\}\Bigr\} + \mathcal{K}(\rho,\pi) - \mathbb{E}_{\rho}\bigl[h(\theta)\bigr] = \mathcal{K}(\rho,\nu),$$

where both members may be finite or equal to $+\infty$. This proves equation (5.2.1) in this case, showing moreover that the maximum in $\rho$ is attained at the point $\rho = \nu$. In the general case, using the notation $\min\{B, h(\theta)\} = B \wedge h(\theta)$, we get

$$\begin{aligned}
\log\Bigl\{\mathbb{E}_{\pi(d\theta)}\Bigl\{\exp\bigl[h(\theta)\bigr]\Bigr\}\Bigr\} &= \sup_{B\in\mathbb{R}} \log\Bigl\{\mathbb{E}_{\pi(d\theta)}\Bigl\{\exp\bigl[B \wedge h(\theta)\bigr]\Bigr\}\Bigr\} \\
&= \sup_{B\in\mathbb{R}} \sup_{\rho\in\mathcal{M}_+^1(\Theta)} \Bigl\{\mathbb{E}_{\rho(d\theta)}\bigl[B\wedge h(\theta)\bigr] - \mathcal{K}(\rho,\pi)\Bigr\} \\
&= \sup_{\rho\in\mathcal{M}_+^1(\Theta)} \sup_{B\in\mathbb{R}} \Bigl\{\mathbb{E}_{\rho(d\theta)}\bigl[B\wedge h(\theta)\bigr] - \mathcal{K}(\rho,\pi)\Bigr\} \\
&= \sup_{\rho\in\mathcal{M}_+^1(\Theta)} \sup_{B\in\mathbb{R}} \Bigl\{\mathbb{E}_{\rho(d\theta)}\bigl[B\wedge h(\theta)\bigr]\Bigr\} - \mathcal{K}(\rho,\pi) \\
&= \sup_{\rho\in\mathcal{M}_+^1(\Theta)} \mathbb{E}_{\rho(d\theta)}\bigl[h(\theta)\bigr] - \mathcal{K}(\rho,\pi).
\end{aligned}$$

**Lemma 5.2.1.** *Let $\eta : \Theta \to \mathbb{R}_+$ and $\lambda : \Theta \to \mathbb{R}_+$ be two given non negative real valued measurable functions defined on the parameter set $(\Theta, \mathcal{F})$. Let $\pi \in \mathcal{M}_+^1(\Theta,\mathcal{F})$ be a probability distribution on the parameters. Then*

$$\begin{aligned}
&(P^{\otimes N})\Bigl\{\sup_{\rho\in\mathcal{M}_+^1(\Theta)}\Bigl\{\mathbb{E}_{\rho(d\theta)}\Bigl[\lambda(\theta)\bigl[R(\theta)-r(\theta)-\eta(\theta)\bigr]\Bigr] - \mathcal{K}(\rho,\pi)\Bigr\} \leq 0\Bigr\} \\
&\geq 1 - \mathbb{E}_{\pi(d\theta)}\left\{\exp\left[-\eta(\theta)\lambda(\theta) + g\left(\frac{R(\theta)\lambda(\theta)}{N}\right)\frac{R(\theta)\bigl(1-R(\theta)\bigr)}{N}\lambda(\theta)^2\right]\right\},
\end{aligned} \tag{5.2.4}$$

*with the conventions defined by equations* (5.2.2) *and* (5.2.3). *As a consequence*

$$\left(P^{\otimes N}\right)_* \left\{ \sup_{\Lambda \in \mathcal{F}} \pi(\Lambda) \inf_{\theta \in \Lambda} \left\{ \exp\left[\lambda(\theta)\Big(R(\theta) - r(\theta) - \eta(\theta)\Big)\right]\right\} \leq 1 \right\}$$
$$\geq 1 - \mathbb{E}_{\pi(d\theta)} \exp\left[-\eta(\theta)\lambda(\theta) + g\left(\frac{R(\theta)\lambda(\theta)}{N}\right)\frac{R(\theta)\big(1-R(\theta)\big)}{N}\lambda(\theta)^2\right], \tag{5.2.5}$$

*where* $\left(P^{\otimes N}\right)_*(A) = \sup\Big\{P^{\otimes N}(B) : B \in (\mathcal{B}\otimes\mathcal{B}')^{\otimes N},\, B \subset A\Big\}$ *is the* interior measure *associated with* $P^{\otimes N}$.

*Remark* 5.2.1. For example if we choose

$$\lambda(\theta) = \frac{N}{R(\theta)} \log\left[1 + \frac{\eta(\theta)}{1 - R(\theta)}\right],$$

we get

$$\left(P^{\otimes N}\right)_* \left\{ \sup_{\Lambda \in \mathcal{F}} \pi(\Lambda) \inf_{\theta \in \Lambda} \left\{ \exp\left[\lambda(\theta)\Big(R(\theta) - r(\theta) - \eta(\theta)\Big)\right]\right\} \leq 1 \right\}$$
$$\geq 1 - \mathbb{E}_{\pi(d\theta)} \exp\left[-\frac{3N\eta(\theta)^2}{R(\theta)\big[6\big(1-R(\theta)\big) + 2\eta(\theta)\big]}\right].$$

*Proof.* In this application of theorem 5.2.1, $b = R(\theta)$ and $V = R(\theta)\big(1 - R(\theta)\big)$. Integrating with respect to $\pi(d\theta)$ and using Fubini's theorem, we get

$$P^{\otimes N}\left\{\mathbb{E}_{\pi(d\theta)}\left[\exp\Big\{\lambda(\theta)\big[R(\theta) - r(\theta) - \eta(\theta)\big]\Big\}\right] \geq 1\right\}$$
$$\leq \mathbb{E}_{P^{\otimes N}}\left\{\mathbb{E}_{\pi(d\theta)}\left[\exp\Big\{\lambda(\theta)\big[R(\theta) - r(\theta) - \eta(\theta)\big]\Big\}\right]\right\}$$
$$= \mathbb{E}_{\pi(d\theta)}\left\{\mathbb{E}_{P^{\otimes N}}\left[\exp\Big\{\lambda(\theta)\big[R(\theta) - r(\theta) - \eta(\theta)\big]\Big\}\right]\right\}$$
$$\leq \mathbb{E}_{\pi(d\theta)}\left\{\exp\left[-\eta(\theta)\lambda(\theta) + g\left(\frac{R(\theta)\lambda(\theta)}{N}\right)\frac{R(\theta)\big(1-R(\theta)\big)}{N}\lambda(\theta)^2\right]\right\}.$$

We obtain equation (5.2.4) of lemma 5.2.1 using the identity (5.2.1), which reads in this case

$$\log\Big\{\mathbb{E}_{\pi(d\theta)}\Big\{\exp\Big[\lambda(\theta)\big[R(\theta) - r(\theta) - \eta(\theta)\big]\Big]\Big\}\Big\}$$
$$= \sup_{\rho \in \mathcal{M}_+^1(\Theta)} \left\{\mathbb{E}_{\rho(d\theta)}\Big[\lambda(\theta)\big[R(\theta) - r(\theta) - \eta(\theta)\big]\Big] - \mathcal{K}(\rho, \pi)\right\}. \tag{5.2.6}$$

The second part of lemma 5.2.1, equation (5.2.5), is obtained by considering the special case $\rho(d\theta) = \pi(d\theta|\Lambda)$, (the probability measure $\pi$ conditioned by the set $\Lambda$, namely $\pi(d\theta|\Lambda) = \mathbb{1}(\theta \in \Lambda)\pi(d\theta)/\pi(\Lambda)$), and noticing that $\mathcal{K}\big[\pi(d\theta|\Lambda), \pi(d\theta)\big] = -\log\big[\pi(\Lambda)\big]$. □

**Theorem 5.2.2.** *For any constant $\gamma \in \mathbb{R}_+$, with $P^{\otimes N}$ probability at least $1-\epsilon$, where*

$$\epsilon = \exp\left\{-\gamma\left[1 - g\left(\sqrt{\frac{\gamma}{N}}\right)\right]\right\},$$

*for any measurable subset of parameters $\Lambda \in \mathfrak{F}$,*

$$\mathbb{E}_{\pi(d\theta \,|\, \Lambda)}\big[R(\theta)\big] \leq \mathbb{E}_{\pi(d\theta \,|\, \Lambda)}\big[r(\theta)\big] + \int_{\xi=0}^{1} \min\left\{1, \pi(\Lambda)^{-1} \exp\left[-\sqrt{\gamma N}\left(\frac{\xi}{\sqrt{\xi + \sup_{\theta\in\Lambda} r(\theta)}} - \sqrt{\frac{\gamma}{N}}\right)\right]\right\} d\xi.$$

*This bound is computable from the observations, with the help of a numerical integration scheme, and the choice of $\Lambda$ can be optimized. However, it is also possible to weaken it slightly to obtain the more explicit following inequality : with probability at least $1-\epsilon$,*

$$\mathbb{E}_{\pi(d\theta \,|\, \Lambda)}\big[R(\theta)\big] \leq \mathbb{E}_{\pi(d\theta \,|\, \Lambda)}\big[r(\theta)\big] + \begin{cases} \sqrt{\dfrac{2\sup_{\theta\in\Lambda} r(\theta)}{\gamma N}}\Big(3 + \gamma - \log\big[\pi(\Lambda)\big]\Big) \\ \qquad + \dfrac{4}{\gamma N}, \text{ when } \sup\limits_{\theta} r(\theta) \geq \dfrac{2}{\gamma N}\Big(\gamma - \log\big[\pi(\Lambda)\big]\Big)^2, \\ \dfrac{2}{\gamma N}\Big\{\Big(1 + \gamma - \log\big[\pi(\Lambda)\big]\Big)^2 + 1\Big\}, \text{ otherwise.} \end{cases} \tag{5.2.7}$$

*We can also bound the deviations with respect to the posterior distribution on $\theta$, by stating that with $P^{\otimes N}$ probability at least $1-\epsilon$ (where $\epsilon$ is as above), for any $\Lambda \in \mathfrak{F}$, for any $\xi \in \mathbb{R}_+$, with $\pi(d\theta \,|\, \Lambda)$ probability at least*

$$1 - \pi(\Lambda)^{-1} \exp\left[-\sqrt{\gamma N}\left(\frac{\xi}{\sqrt{\xi + \sup_{\theta\in\Lambda} r(\theta)}} - \sqrt{\frac{\gamma}{N}}\right)\right], \tag{5.2.8}$$

*the expected risk is bounded by*

$$R(\theta) \leq r(\theta) + \xi.$$

*Remark* 5.2.2. The upper bound in this theorem is minimized when the parameter set $\Lambda$ is chosen from the level sets of the empirical risk $\Lambda = \{\theta \in \Theta : r(\theta) \leq \mu\}$, however different choices of $\Lambda$ could also be interesting in some situations. For instance, if $\Theta = \bigsqcup_{k\in K} \Theta_k$ is a disjoint union of different parameter sets $\Theta_k$ standing for different possible choices of classification models, one may want to perform a model selection by considering parameter subsets $\Lambda_{k,\mu} = \{\theta \in \Theta_k : r(\theta) \leq \mu\}$.

*Remark* 5.2.3. A sequence $(\gamma_k)_{k\in\mathbb{N}}$ of values of $\gamma$ being chosen, with probability at least $1-\epsilon$ where

$$\epsilon = \sum_{k\in\mathbb{N}} \exp\left\{-\gamma_k\left[1-g\left(\sqrt{\frac{\gamma_k}{N}}\right)\right]\right\},$$

we can also take the infimum over $\gamma \in \{\gamma_k\}$ of the right-hand side of (5.2.7), or of (5.2.8).

*Proof.* Let us choose

$$\eta(\theta) = \sqrt{\frac{\gamma R(\theta)}{N}},$$
$$\lambda(\theta) = \frac{N\eta(\theta)}{R(\theta)} = \sqrt{\frac{\gamma N}{R(\theta)}}$$
$$\epsilon = \exp\left\{-\gamma\left[1-g\left(\sqrt{\frac{\gamma}{N}}\right)\right]\right\}.$$

With $P^{\otimes N}$ probability $1-\epsilon$,

$$\sup_{\Lambda\subset\Theta} \pi(\Lambda) \inf_{\theta\in\Lambda}\left\{\exp\left[\sqrt{\frac{\gamma N}{R(\theta)}}\left(R(\theta)-r(\theta)-\sqrt{\frac{\gamma R(\theta)}{N}}\right)\right]\right\} \leq 1. \quad (5.2.9)$$

All the following inequalities hold simultaneously with probability $1-\epsilon$ for any value of $\Lambda$, the notation $\mu$ being a short hand for $\mu = \sup_{\theta\in\Lambda} r(\theta)$ :

$$\mathbb{E}_{\pi(d\theta\,|\,\Lambda)}\big[R(\theta)\big] \leq \mathbb{E}_{\pi(d\theta\,|\,\Lambda)}\big[r(\theta)\big] + \int_{\xi=0}^{1} \pi\big[R(\theta)-r(\theta)\geq \xi\,|\,\Lambda\big]d\xi. \quad (5.2.10)$$

$$\begin{aligned}
\pi\Big[R(\theta)-r(\theta)\geq\xi \text{ and } \theta\in\Lambda\Big] \\
&\leq \sup\left\{\exp\left[-\sqrt{\gamma N}\left(\sqrt{R(\theta)}-\frac{r(\theta)}{\sqrt{R(\theta)}}-\sqrt{\frac{\gamma}{N}}\right)\right]\right. \\
&\qquad\qquad\left. : R(\theta)\geq r(\theta)+\xi \text{ and } r(\theta)\leq\mu\right\} \\
&\leq \exp\left[-\sqrt{\gamma N}\left(\frac{\xi}{\sqrt{\xi+\mu}}-\sqrt{\frac{\gamma}{N}}\right)\right]. \quad (5.2.11)
\end{aligned}$$

Taking into account the trivial bound

$$\pi\big[R(\theta)\geq r(\theta)+\xi \text{ and } \theta\in\Lambda\big] \leq \pi(\Lambda) \quad (5.2.12)$$

and coming back to equation (5.2.10), gives the first part of the theorem. The more explicit upper bound (5.2.7) is then obtained by applying lemma 5.10.2. □

Another choice of constants can be made when the optimal value of $\sup_{\theta \in \Lambda} r(\theta)$ is expected to be of order $1/N$ or less. It leads to the following theorem :

**Theorem 5.2.3.** *With $P^{\otimes N}$ probability at least $1-\epsilon$, where*

$$\epsilon = \exp(-\gamma)\mathbb{E}_{\pi(d\theta)}\left\{\exp\left[-\frac{N}{4}R(\theta)\left[1-g\left(\tfrac{1}{2}\right)\right]\right]\right\},$$

*for any measurable parameter subset $\Lambda \in \mathcal{F}$*

$$\mathbb{E}_{\pi(d\theta \mid \Lambda)}\bigl[R(\theta)\bigr] \leq 2\mathbb{E}_{\pi(d\theta \mid \Lambda)}\bigl[r(\theta)\bigr] - \frac{4}{N}\log\bigl[\pi(\Lambda)\bigr] + \frac{4\gamma}{N}. \tag{5.2.13}$$

*We can also bound the deviations with respect to the posterior distribution on $\theta$ by stating that with $P^{\otimes N}$ probability at least $1-\epsilon$, where $\epsilon$ is as above, for any $\Lambda \in \mathcal{F}$, for any $\xi \in \mathbb{R}_+$, with $\pi(d\theta \mid \Lambda)$ probability at least*

$$1 - \pi(\Lambda)^{-1}\exp\left[-\frac{N}{4}\left(\xi - \frac{4\gamma}{N}\right)\right],$$

*the expected risk $R(\theta)$ is bounded by*

$$R(\theta) \leq 2r(\theta) + \xi.$$

*Remark* 5.2.4. This theorem is sharper than theorem 5.2.2 in the case when $\mathbb{E}_{\pi(d\theta|\Lambda)}\bigl[r(\theta)\bigr]$ is of order $1/N$, but it is weaker when this quantity is larger.

*Remark* 5.2.5. Equation (5.2.13) delivers a clearcut message : choose $\Lambda$ to minimize

$$\mathbb{E}_{\pi(d\theta|\Lambda)}\bigl[r(\theta)\bigr] - \frac{2}{N}\log\bigl[\pi(\Lambda)\bigr].$$

One nice feature is that this criterion is independent of $\gamma$, and therefore of the desired level of confidence $\epsilon$. Note that (5.2.13) can also be written as

$$\begin{aligned}\mathbb{E}_{\pi(d\theta|\Lambda)}\bigl[R(\theta)\bigr] &\leq 2\mathbb{E}_{\pi(d\theta|\Lambda)}\bigl[r(\theta)\bigr] \\ &- \frac{4}{N}\log\bigl[\pi(\Lambda)\bigr] + \frac{4}{N}\log\left\{\epsilon^{-1}\mathbb{E}_{\pi(d\theta)}\left[\exp\left(-\frac{N}{4}R(\theta)\bigl[1-g(\tfrac{1}{2})\bigr]\right)\right]\right\}.\end{aligned}$$

*Proof.* Let us take

$$\begin{aligned}\eta(\theta) &= \frac{1}{2}R(\theta) + \frac{2\gamma}{N}, \\ \lambda(\theta) &= \frac{N}{2}.\end{aligned}$$

In the case when $\lambda(\theta)$ is chosen to be a constant, as it is here, we can apply equation (5.2.4) of lemma 5.2.1 to $\rho = \pi(d\theta|\Lambda)$, for which $\mathcal{K}(\rho,\pi) = -\log\bigl[\pi(\Lambda)\bigr]$. This proves the first part of theorem 5.2.3.

Using equation (5.2.5) of lemma 5.2.1, we then obtain that with probability at least $1-\epsilon$, for any parameter set $\Lambda \in \mathcal{F}$, for any $\xi \geq 0$,

$$\pi\Big[\theta \in \Lambda \text{ and } R(\theta) \geq 2r(\theta) + \xi\Big] \leq \exp\left[-\frac{N}{4}\left(\xi - \frac{4\gamma}{N}\right)\right],$$

which proves the second part of theorem 5.2.3.

Note that we could have proved from the second part of the theorem and equation (5.2.10) that

$$\begin{aligned}\mathbb{E}_{\pi(d\theta\,|\,\Lambda)}\big[R(\theta)\big] &\leq \mathbb{E}_{\pi(d\theta\,|\,\Lambda)}\big[r(\theta)\big] \\ &\quad + \int_{\xi=0}^{1} 1 \wedge \pi(\Lambda)^{-1}\exp\left[-\frac{N}{4}\left(\xi - \mu - \frac{4\gamma}{N}\right)\right] d\xi \\ &\leq \mathbb{E}_{\pi(d\theta\,|\,\Lambda)}\big[r(\theta)\big] + \mu - \frac{4}{N}\log\big[\pi(\Lambda)\big] + \frac{4\gamma}{N} + \frac{4}{N}.\end{aligned}$$

It is weaker, but still of the same order of magnitude as (5.2.13). □

## 5.3 Generalizations of theorem 5.2.3

More generally, we can choose

$$\lambda(\theta) = \lambda,$$
$$\eta(\theta) = \frac{\gamma}{\lambda} + \frac{R(\theta)}{2}.$$

We obtain from lemma 5.2.1 that with probability at least $1-\epsilon$ with respect to $P^{\otimes N}$, where

$$\epsilon = \exp(-\gamma)\mathbb{E}_{\pi(d\theta)}\left[\exp\left\{-\frac{R(\theta)\lambda}{2}\left[1 - \tfrac{2\lambda}{N} g\left(\tfrac{\lambda}{N}\right)\right]\right\}\right],$$

for any posterior distribution $\rho \in \mathcal{M}_+^1(\Theta, \mathcal{F})$,

$$\mathbb{E}_{\rho(d\theta)}\big[R(\theta)\big] \leq 2\left\{\mathbb{E}_{\rho(d\theta)}\big[r(\theta)\big] + \frac{1}{\lambda}\mathcal{K}(\rho,\pi) + \frac{\gamma}{\lambda}\right\}.$$

We can now remark that the righthand side of this last equation is minimized when the posterior distribution $\rho$ is chosen to be a *Gibbs* distribution with energy function equal to $r(\theta)$ and inverse temperature equal to $\lambda$, namely when

$$\rho(d\theta) = \frac{\exp\big[-\lambda r(\theta)\big]}{\mathbb{E}_{\pi(d\theta')}\Big\{\exp\big[-\lambda r(\theta')\big]\Big\}}\pi(d\theta) \stackrel{\text{def}}{=} \rho_\lambda(d\theta).$$

In this case we obtain with $P^{\otimes N}$ probability at least $1-\epsilon$

$$\begin{aligned}\mathbb{E}_{\rho_\lambda(d\theta)}\bigl[R(\theta)\bigr] \leq -\frac{2}{\lambda}\log\Bigl[\mathbb{E}_{\pi(d\theta)}\Bigl\{\exp\bigl(-\lambda r(\theta)\bigr)\Bigr\}\Bigr] \\ + \frac{2}{\lambda}\log\biggl\{\mathbb{E}_{\pi(d\theta)}\biggl[\exp\biggl\{-\frac{R(\theta)\lambda}{2}\biggl[1-\frac{2\lambda}{N}g\Bigl(\frac{\lambda}{N}\Bigr)\biggr]\biggr\}\biggr]\biggr\} - \frac{2\log(\epsilon)}{\lambda} \\ \leq -\frac{2}{\lambda}\log\Bigl[\mathbb{E}_{\pi(d\theta)}\Bigl\{\exp\bigl(-\lambda r(\theta)\bigr)\Bigr\}\Bigr] \\ - \biggl[1-\frac{2\lambda}{N}g\Bigl(\frac{\lambda}{N}\Bigr)\biggr]\inf_{\theta\in\Theta} R(\theta) - \frac{2\log(\epsilon)}{\lambda} \\ \leq \inf_{\Lambda\in\mathfrak{F}} 2\mathbb{E}_{\pi(d\theta|\Lambda)}\bigl[r(\theta)\bigr] - \frac{2}{\lambda}\log\bigl[\pi(\Lambda)\bigr] \\ - \biggl[1-\frac{2\lambda}{N}g\Bigl(\frac{\lambda}{N}\Bigr)\biggr]\inf_{\theta\in\Theta} R(\theta) - \frac{2\log(\epsilon)}{\lambda}.\end{aligned}$$

In the same way, with $P^{\otimes N}$ probability at least $1-\epsilon$, for any $\Lambda \in \mathfrak{F}$,

$$\begin{aligned}\mathbb{E}_{\pi(d\theta|\Lambda)}\bigl[R(\theta)\bigr] \leq 2\mathbb{E}_{\pi(d\theta|\Lambda)}\bigl[r(\theta)\bigr] - \frac{2}{\lambda}\log\bigl[\pi(\Lambda)\bigr] \\ + \frac{2}{\lambda}\log\biggl\{\mathbb{E}_{\pi(d\theta)}\biggl[\exp\biggl\{-\frac{R(\theta)\lambda}{2}\biggl[1-\frac{2\lambda}{N}g\Bigl(\frac{\lambda}{N}\Bigr)\biggr]\biggr\}\biggr]\biggr\} - \frac{2\log(\epsilon)}{\lambda} \\ \leq 2\mathbb{E}_{\pi(d\theta|\Lambda)}\bigl[r(\theta)\bigr] - \frac{2}{\lambda}\log\bigl[\pi(\Lambda)\bigr] \\ - \biggl[1-\frac{2\lambda}{N}g\Bigl(\frac{\lambda}{N}\Bigr)\biggr]\inf_{\theta\in\Theta} R(\theta) - \frac{2\log(\epsilon)}{\lambda}.\end{aligned}$$

Let us notice that the correction term $-\Bigl[1-\frac{2\lambda}{N}g\Bigl(\frac{\lambda}{N}\Bigr)\Bigr]\inf_{\theta\in\Theta} R(\theta)$ makes up for the factor 2 in front of $\mathbb{E}_{\pi(d\theta|\Lambda)}\bigl[r(\theta)\bigr]$. Indeed, assuming for simplicity that $\inf_{\theta\in\Theta} R(\theta)$ is reached and is equal to $R(\widetilde{\theta})$, we get from a new application of theorem 5.2.1 to $\sigma_i = \mathbb{1}\bigl(f_{\widetilde{\theta}}(X_i) \neq Y_i\bigr)$ that with probability $1-\epsilon$, where

$$\epsilon = \exp\biggl\{-\eta\lambda + g\Bigl(\frac{\lambda}{N}\Bigr)\frac{R(\widetilde{\theta})}{N}\lambda^2\biggr\},$$

we have

$$R(\widetilde{\theta}) \geq r(\widetilde{\theta}) - \eta.$$

This proves that with $P^{\otimes N}$ probability at least $1-2\epsilon$,

$$\begin{aligned}\mathbb{E}_{\pi(d\theta|\Lambda)}\bigl[R(\theta)\bigr] \leq & 2\mathbb{E}_{\pi(d\theta|\Lambda)}\bigl[r(\theta)\bigr] - \biggl[1-\frac{2\lambda}{N}g\Bigl(\frac{\lambda}{N}\Bigr)\biggr]\inf_{\theta\in\Theta} r(\theta) \\ & - \frac{2}{\lambda}\log\bigl[\pi(\Lambda)\bigr] \\ & + \biggl[1-\frac{2\lambda}{N}g\Bigl(\frac{\lambda}{N}\Bigr)\biggr]\biggl[\frac{\lambda R(\widetilde{\theta})}{N}g\Bigl(\frac{\lambda}{N}\Bigr) - \frac{\log(\epsilon)}{\lambda}\biggr] - \frac{2}{\lambda}\log(\epsilon).\end{aligned}$$

For instance, restricting to the parameter subsets

$$\Lambda = \{\theta : r(\theta) \leq \inf_{\theta \in \Theta} r(\theta) + \mu\}$$

and using the fact that $g(0.5) \leq 0.6$, we get for any $\lambda \leq N/2$ with $P^{\otimes N}$ probability at least $1 - \epsilon$, for any $\mu \in \mathbb{R}_+$,

$$\mathbb{E}_{\pi[d\theta | r(\theta) \leq \inf_\theta r(\theta) + \mu]} \bigl[R(\theta)\bigr] \leq \left[1 + \frac{1.2\,\lambda}{N}\right] \inf_{\theta \in \Theta} r(\theta) + 2\mu \\ - \frac{2}{\lambda} \log\Bigl\{\pi\bigl[r(\theta) \leq \inf_\theta r(\theta) + \mu\bigr]\Bigr\} + \frac{0.6\,\lambda}{N} - \frac{3}{\lambda}\log(\epsilon/2).$$

## 5.4 The non-ambiguous case

The results of the last section are satisfactory when the level of noise is low, meaning that $\inf_{\theta \in \Theta} R(\theta)$ is of order $1/N$. In the case when $\inf_{\theta \in \Theta} R(\theta)$ is not small, two different situations may occur, depending on the level of *ambiguity* of the classification problem. A precise definition of ambiguity will be stated. Let us say for the moment that the problem is ambiguous if no rule is everywhere better than the others, and the choice has to be done by comparing the performances of different classification rules on different sets of patterns, or if the error rate of the best rule is for some patterns close to the error rate of the second best rule. To give a precise meaning to the notion of ambiguity (or rather to the absence of ambiguity), let us consider, in the same setting as in the previous section, some fixed distinguished classification rule $f_{\widetilde{\theta}} \in \mathcal{R}$ belonging to our reference family of rules. The most interesting case is when $R(\widetilde{\theta}) = \inf_{\theta \in \Theta} R(\theta)$, but we will not require it explicitly in the following. The case when $\widetilde{\theta} \notin \Theta$ can also be covered by considering $\Theta \cup \{\widetilde{\theta}\}$ and a distribution $\pi \in \mathcal{M}_+^1\bigl(\Theta \cup \{\widetilde{\theta}\}\bigr)$ such that $\pi(\{\widetilde{\theta}\}) = 0$. Needless to say that we do not assume that $\widetilde{\theta}$ is known to the statistician.

Let us introduce some notations :

$$\begin{aligned}
\overline{R}(\theta) &= P\bigl(Y \neq f_\theta(X)\bigr) - P\bigl(Y \neq f_{\widetilde{\theta}}(X)\bigr),\\
\overline{r}(\theta) &= \frac{1}{N}\sum_{i=1}^{N} \mathbb{1}\bigl(Y \neq f_\theta(X)\bigr) - \mathbb{1}\bigl(Y \neq f_{\widetilde{\theta}}(X)\bigr),\\
\overline{V}(\theta) &= \operatorname{Var}\Bigl[\mathbb{1}\bigl(Y \neq f_\theta(X)\bigr) - \mathbb{1}\bigl(Y \neq f_{\widetilde{\theta}}(X)\bigr)\Bigr],\\
\overline{R}(\theta \,|\, X) &= P\bigl(Y \neq f_\theta(X) \,|\, X\bigr) - P\bigl(Y \neq f_{\widetilde{\theta}}(X) \,|\, X\bigr),\\
\overline{V}(\theta \,|\, X) &= \operatorname{Var}\Bigl[\mathbb{1}\bigl(Y \neq f_\theta(X)\bigr) - \mathbb{1}\bigl(Y \neq f_{\widetilde{\theta}}(X)\bigr) \,|\, X\Bigr].
\end{aligned}$$

To improve on the results of the previous section, it is useful to introduce the margin (of non ambiguity) at point $x \in \mathcal{X}$, defined as

$$\alpha(x) = \min\Bigl\{\overline{R}(\theta \,|\, x) : \theta \in \Theta, f_\theta(x) \neq f_{\widetilde{\theta}}(x)\Bigr\}.$$

(In this formula, we assume that some realization of the conditional expectations involved has been chosen once for all.) The margin $\alpha(x)$ tells us about the increase in the classification error rate of pattern $x$ when the label $f_{\widetilde{\theta}}(x)$ given by the rule $f_{\widetilde{\theta}}$ is replaced with another label. It can be negative when $f_{\widetilde{\theta}}(x)$ is not the most probable label for pattern $x$.

Associated with this notion of margin at point $x$ is the notion of $\alpha$-ambiguity set $\Omega_\alpha$, where $\alpha$ is an arbitrary positive constant :

$$\Omega_\alpha \overset{\text{def}}{=} \{x \in \mathcal{X} \,|\, \alpha(x) < \alpha\}.$$

The interest of this notion is that we can control the variance by the mean in the following way :

$$\begin{aligned}\overline{V}(\theta \,|\, x) &\leq \mathbb{1}\bigl[f_\theta(X) \neq f_{\widetilde{\theta}}(X)\bigr] \\ &\leq \frac{\overline{R}(\theta \,|\, x)}{\alpha}, \qquad x \notin \Omega_\alpha.\end{aligned}$$

Thus

$$\begin{aligned}\overline{V}(\theta) &= \mathbb{E}_P\Bigl[\overline{V}(\theta \,|\, X)\Bigr] + \mathrm{Var}\bigl[\overline{R}(\theta \,|\, X)\bigr] \\ &\leq \mathbb{E}_P\left(\frac{\overline{R}(\theta \,|\, X)}{\alpha}\mathbb{1}(X \notin \Omega_\alpha) + \mathbb{1}(\Omega_\alpha)\right) + \mathbb{E}\bigl[\overline{R}(\theta \,|\, X)^2\bigr] \\ &\leq \frac{1}{\alpha}\Bigl\{\overline{R}(\theta) + P(\Omega_0)\Bigr\} + P(\Omega_\alpha) + \overline{R}(\theta) + 2P(\Omega_0) \\ &= \left(\frac{1}{\alpha} + 1\right)\overline{R}(\theta) + \left(\frac{1}{\alpha} + 2\right)P(\Omega_0) + P(\Omega_\alpha) \\ &\overset{\text{def}}{=} \overline{R}^*(\theta).\end{aligned}$$

Notice that we took the opportunity of the previous equation to introduce the short notation $\overline{R}^*(\theta)$ which will be useful in the following.

Applying this and theorem 5.2.1, we get a lemma analogous to lemma 5.2.1:

**Lemma 5.4.1.** *For any given measurable functions $\eta : \Theta \to \mathbb{R}_+$ and $\lambda : \Theta \to \mathbb{R}_+$, we have*

$$\begin{aligned}(P^{\otimes N})&\left\{\sup_{\rho \in \mathcal{M}_+^1(\Theta)} \Bigl\{\mathbb{E}_{\rho(d\theta)}\Bigl[\lambda(\theta)\Bigl(\overline{R}(\theta) - \overline{r}(\theta) - \eta(\theta)\Bigr)\Bigr] - \mathcal{K}(\rho, \pi)\Bigr\} \leq 0\right\} \\ &\geq 1 - \mathbb{E}_{\pi(d\theta)} \exp\left[-\eta(\theta)\lambda(\theta) + g\left(\frac{(1 + \overline{R}(\theta))\lambda(\theta)}{N}\right)\frac{\overline{R}^*(\theta)}{N}\bigl[\lambda(\theta)\bigr]^2\right].\end{aligned}$$

As a consequence :

**Lemma 5.4.2.** *For any given measurable functions* $\eta : \Theta \rightarrow \mathbb{R}_+$ *and* $\lambda : \Theta \rightarrow \mathbb{R}_+$, *we have*

$$\left(P^{\otimes N}\right)_* \left\{ \sup_{\Lambda \in \mathfrak{F}} \pi(\Lambda) \inf_{\theta \in \Lambda} \left\{ \exp\left[\lambda(\theta)\left(\overline{R}(\theta) - \overline{r}(\theta) - \eta(\theta)\right)\right]\right\} \leq 1 \right\}$$
$$\geq 1 - \mathbb{E}_{\pi(d\theta)} \exp\left[ -\eta(\theta)\lambda(\theta) + g\left(\frac{(1+\overline{R}(\theta))\lambda(\theta)}{N}\right) \frac{\overline{R}^*(\theta)}{N} \left[\lambda(\theta)\right]^2 \right].$$

Using this last lemma and appropriate choices of $\lambda(\theta)$ and $\eta(\theta)$ yields the following theorem.

**Theorem 5.4.1.** *For any positive constant* $\gamma$, *with* $P^{\otimes N}$ *probability at least* $1 - \epsilon$, *where*

$$\epsilon = \exp\Bigl( - \left(1 - \tfrac{e}{4}\right) \gamma \Bigr),$$

*for any measurable parameter set* $\Lambda \in \mathfrak{F}$, *putting* $\mu = \sup_{\theta \in \Lambda} r(\theta)$, *we have*

$$\begin{aligned}
\mathbb{E}_{\pi(d\theta \,|\, \Lambda)}\bigl[R(\theta)\bigr] \leq R(\widetilde{\theta}) + \mathbb{E}_{\pi(d\theta \,|\, \Lambda)}\bigl[r(\theta)\bigr] - r(\widetilde{\theta}) \\
+ \int_{\xi=0}^{1} \min \Biggl\{ 1, \\
\pi(\Lambda)^{-1} \exp\left[ -\frac{1}{2}\sqrt{\gamma N} \left( \frac{\xi}{\sqrt{\max\{\gamma/N, (1+\frac{1}{\alpha})(\mu + \xi - r(\widetilde{\theta})) + C_\alpha\}}} \right.\right. \\
\left.\left. - \sqrt{\frac{\gamma}{N}} \right)\right] \Biggr\} d\xi, \qquad (5.4.1)
\end{aligned}$$

*for any* $C_\alpha \geq (2 + 1/\alpha) P(\Omega_0) + P(\Omega_\alpha)$.

*The deviations with respect to the posterior distribution on* $\theta$ *are moreover thus bounded : with* $P^{\otimes N}$ *probability at least* $1-\epsilon$, *for any parameter set* $\Lambda \in \mathfrak{F}$, *putting* $\mu = \sup_{\theta \in \Lambda} r(\theta)$, *for any* $\xi \geq 0$, *with* $\pi\bigl(d\theta \,|\, \Lambda\bigr)$ *probability at least*

$$\begin{aligned}
1 - \pi(\Lambda)^{-1} \exp\left[ -\frac{1}{2}\sqrt{\gamma N} \left( \frac{\xi}{\sqrt{\max\{\gamma/N, (1+\frac{1}{\alpha})(\mu + \xi - r(\widetilde{\theta})) + C_\alpha\}}} \right.\right. \\
\left.\left. - \sqrt{\frac{\gamma}{N}} \right)\right],
\end{aligned}$$

*the expected risk* $R(\theta)$ *is bounded by*

$$R(\theta) \leq R(\widetilde{\theta}) + r(\theta) - r(\widetilde{\theta}) + \xi.$$

*Remark* 5.4.1. Note that the right-hand side of (5.4.1) is not an observed quantity, but that it can be upper bounded by

$$R(\widetilde{\theta}) + \mu - \inf_{\theta \in \Theta} r(\theta) + \int_0^1 \min \Biggl\{ 1, \\ \pi(\Lambda)^{-1} \exp \Biggl[ -\frac{1}{2}\sqrt{\gamma N} \Biggl( \frac{\xi}{\sqrt{(1+\frac{1}{\alpha})(\xi + \mu - \inf_{\theta\in\Theta} r(\theta)) + C_\alpha + \frac{\gamma}{N}}} \\ - \sqrt{\frac{\gamma}{N}} \Biggr) \Biggr] \Biggr\} d\xi,$$

which can be optimized in $\mu$ without knowing $\widetilde{\theta}$.

*Remark* 5.4.2. Note also that the theorem still holds when $\widetilde{\theta} \not\in \Theta$. To see this, we can simply apply it to $\Theta \cup \{\widetilde{\theta}\}$ and a prior $\pi$ which puts no mass on $\widetilde{\theta}$.

*Remark* 5.4.3. A more explicit bound can be obtained using lemma 5.10.2. Namely

$$\mathbb{E}_{\pi(d\theta \mid \Lambda)}\bigl[R(\theta)\bigr] \leq R(\widetilde{\theta}) + \mathbb{E}_{\pi(d\theta \mid \Lambda)}\bigl[r(\theta)\bigr] - r(\widetilde{\theta}) \\ + \begin{cases} 2\sqrt{2}\sqrt{\dfrac{(1+1/\alpha)(\mu - r(\widetilde{\theta})) + C_\alpha + \gamma/N}{\gamma N}} \\ \qquad \times \Bigl(3 + \gamma/2 - \log\bigl[\pi(\Lambda)\bigr]\Bigr) + \dfrac{16(1+1/\alpha)}{\gamma N}, \\ \qquad \text{when } \mu - r(\widetilde{\theta}) + \dfrac{C_\alpha + \gamma/N}{(1+1/\alpha)} \\ \qquad\qquad \geq (1+1/\alpha)\dfrac{8}{\gamma N}\Bigl(\gamma/2 - \log\bigl[\pi(\Lambda)\bigr]\Bigr)^2, \\ \dfrac{8(1+1/\alpha)}{\gamma N}\Bigl[1 + \Bigl(1 + \gamma/2 - \log \pi(\Lambda)\Bigr)^2\Bigr], \text{otherwise.} \end{cases}$$

Here again, $r(\widetilde{\theta})$ can be lower bounded by the observed quantity $\inf_{\theta\in\Theta} r(\theta)$.

*Remark* 5.4.4. A case of special interest is when there exists $\widetilde{\theta} \in \Theta$ such that $P(\Omega_\alpha) = 0$ for some $\alpha > 0$. In this case $f_{\widetilde{\theta}}$ is optimal among all the classification rules in $\mathcal{R}$, $C_\alpha = 0$ and the proposed estimator has an error rate which approaches the optimal rate $R(\widetilde{\theta})$ at speed of order $1/N$ when $\Theta$ is finite. Note that the fluctuations of $\mathbb{E}_{\pi(d\theta \mid \Lambda)}\bigl[r(\theta)\bigr]$ are of order $1/\sqrt{N}$, but that they are compensated by the fluctuations of $\widetilde{r}(\theta)$ (or of $\inf_{\theta\in\Theta} r(\theta)$).

*Proof.* Let us write

$$\mathbb{E}_{\pi(d\theta\,|\,\Lambda)}\bigl[R(\theta)\bigr] \leq R(\widetilde{\theta}) - r(\widetilde{\theta}) + \mathbb{E}_{\pi(d\theta\,|\,\Lambda)}\bigl[r(\theta)\bigr] + \int_0^1 \pi\bigl(\overline{r}(\theta) \geq \xi \,|\, \Lambda\bigr) d\xi.$$

Let us choose

$$\lambda(\theta) = \frac{1}{2}\sqrt{\frac{\gamma N}{\max\{\gamma/N, \overline{R}^*(\theta)\}}}$$

$$\eta(\theta) = \sqrt{\frac{\gamma \max\{\gamma/N, \overline{R}^*(\theta)\}}{N}}.$$

Applying lemma 5.4.2, we see that, with $P^{\otimes N}$ probability at least $1-\epsilon$,

$$\pi\Bigl[\overline{R}(\theta) - \overline{r}(\theta) \geq \xi \text{ and } \theta \in \Lambda\Bigr]$$
$$\leq \exp\left\{-\frac{1}{2}\sqrt{\gamma N}\left(\frac{\xi}{\sqrt{\max\{\gamma/N, (1+1/\alpha)(\mu + \xi - r(\widetilde{\theta})) + C_\alpha\}}}\right) - \sqrt{\frac{\gamma}{N}}\right\}.$$

□

A second somewhat simpler variant of the previous theorem can be obtained by a different choice of constants.

**Theorem 5.4.2.** *Let us put $a = 1 + 1/\alpha$. For any*

$$\gamma \geq \frac{N}{2+2/\alpha}\Bigl[(2+1/\alpha)P(\Omega_0) + P(\Omega_\alpha)\Bigr] \qquad (5.4.2)$$

*with $P^{\otimes N}$ probability at least $1-\epsilon$, where*

$$\epsilon = \mathbb{E}_{\pi(d\theta)}\left\{\exp\left[-\frac{1}{8a}\left(\gamma + \frac{N\overline{R}(\theta)}{2}\right)\left(2 - g\left(\frac{1}{2a}\right)\right)\right]\right\},$$

*for any measurable parameter subset $\Lambda \in \mathfrak{F}$, we have*

$$\mathbb{E}_{\pi(d\theta|\Lambda)}\bigl[R(\theta)\bigr] \leq R(\widetilde{\theta}) + 2\Bigl[\mathbb{E}_{\pi(d\theta|\Lambda)}\bigl[r(\theta)\bigr] - r(\widetilde{\theta})\Bigr] + \frac{2\gamma}{N} - \frac{8a}{N}\log\bigl[\pi(\Lambda)\bigr]. \qquad (5.4.3)$$

*Moreover, the deviations of $\theta$ under the posterior $\pi(d\theta\,|\,\Lambda)$ are thus bounded : with $P^{\otimes N}$ probability at least $1-\epsilon$, where $\epsilon$ is as above, for any parameter subset $\Lambda \in \mathfrak{F}$, for any $\xi \in \mathbb{R}_+$, with $\pi(d\theta\,|\,\Lambda)$ probability at least*

$$1 - \pi(\Lambda)^{-1}\exp\left(-\frac{N\xi}{8a}\right)$$

*the expected risk $R(\theta)$ is bounded by*

$$R(\theta) \leq R(\widetilde{\theta}) + 2\,\overline{r}(\theta) + \frac{2\gamma}{N} + \xi$$
$$\leq R(\widetilde{\theta}) + 2\,r(\theta) - 2\inf_{\theta\in\Theta} r(\theta) + \frac{2\gamma}{N} + \xi.$$

*Remark* 5.4.5. Here again the message delivered by equation (5.4.3) is clear : whatever level of confidence is sought, minimize in $\Lambda$ the quantity

$$\mathbb{E}_{\pi(d\theta|\Lambda)}\big[r(\theta)\big] - \frac{4a}{N}\log\big[\pi(\Lambda)\big].$$

It is to be remarked that the criterion to minimize is of the same type as in theorem 5.2.3, the only change being in the value of the constant in front of the penalization term.

*Remark* 5.4.6. The second part of the theorem can also be stated in the following way : For any $\epsilon$ such that

$$\epsilon \leq \exp\left\{-\frac{N}{16a^2}\Big[(2+1/\alpha)P(\Omega_0)+P(\Omega_\alpha)\Big]\right\} \times \mathbb{E}_{\pi(d\theta)}\left\{\exp\left[-\frac{N\overline{R}(\theta)}{16a}\big(2-g(\tfrac{1}{2a})\big)\right]\right\},$$

with $P^{\otimes N}$ probability at least $1-\epsilon$, for any $\Lambda \in \mathcal{F}$, with $\pi(d\theta|\Lambda)$ at least $1-\eta$,

$$\begin{aligned} R(\theta) \leq R(\widetilde{\theta}) + 2\,\overline{r}(\theta) - \frac{8a}{N}\log\big[\pi(\Lambda)\big] \\ + \frac{16a}{N}\log\left\{\mathbb{E}_{\pi(d\theta)}\left[\exp\left(-\frac{N\overline{R}(\theta)}{16a}\big(2-g(\tfrac{1}{2a})\big)\right)\right]\right\} \\ - \frac{16a}{N}\log(\epsilon) - \frac{8a}{N}\log(\eta) \end{aligned}$$

*Proof.* Let us put to simplify notations $b = 2a\gamma/N$, and notice that according to hypothesis (5.4.2) we have $a\overline{R}(\theta)+b \geq \overline{R}^*(\theta)$. Let us apply lemma 5.4.1 while choosing

$$\begin{aligned} \eta(\theta) &= \frac{\overline{R}(\theta)}{2} + \frac{\gamma}{N}, \\ \lambda(\theta) &= \frac{N\eta(\theta)}{2(a\overline{R}(\theta)+b)} \\ &= \frac{N}{4a}. \end{aligned}$$

We get that with probability at least $1-\epsilon$ where

$$\begin{aligned} \epsilon &= \mathbb{E}_{\pi(d\theta)}\left\{\exp\left[-\left(\frac{1}{2}-\frac{1}{4}g\left(\frac{2\lambda(\theta)}{N}\right)\right)\frac{N\eta(\theta)^2}{a\overline{R}(\theta)+b}\right]\right\}, \\ &= \mathbb{E}_{\pi(d\theta)}\left\{\exp\left[-\frac{1}{8a}\left(\gamma+\frac{N\overline{R}(\theta)}{2}\right)\left(2-g\left(\frac{1}{2a}\right)\right)\right]\right\}, \end{aligned}$$

for any parameter subset $\Lambda \in \mathcal{F}$,

$$\mathbb{E}_{\pi(d\theta|\Lambda)}\left\{\frac{N}{4a}\left[\tfrac{1}{2}\overline{R}(\theta)-\overline{r}(\theta)-\frac{\gamma}{N}\right]\right\}\leq-\log\bigl[\pi(\Lambda)\bigr],$$

from which the first part of the theorem follows.

Lemma 5.4.2 then shows that with $P^{\otimes N}$ probability at least $1-\epsilon$, for any measurable parameter subset $\Lambda\in\mathfrak{F}$, the following inequality holds:

$$\pi\Bigl\{\overline{R}(\theta)\geq 2\,\overline{r}(\theta)+\xi \text{ and } \theta\in\Lambda\Bigr\}\leq\exp\left\{-\frac{N}{4a}\left(\frac{\xi}{2}-\frac{\gamma}{N}\right)\right\}.$$

□

## 5.5 Empirical complexity bounds for the Gibbs estimator

We have introduced Gibbs estimators and studied their mean expected risk in previous papers [22, 23] (see also our students' works on related subjects: [14, 15, 16] and [78, 79, 80]). We would like to show briefly in this section how to derive empirical deviation bounds for the Gibbs estimator for pattern recognition from the previous theorems of this chapter. Let us recall first that the Gibbs estimator for pattern recognition is a "soft threshold" version of the thresholded posterior $\pi\bigl(d\theta\,|\,r(\theta)\leq\mu\bigr)$ depending on an inverse temperature parameter $\beta\in\mathbb{R}_+$ : it is defined as

$$G_\beta(d\theta)\stackrel{\text{def}}{=}\bigl[Z(\beta)\bigr]^{-1}\exp\bigl[-\beta r(\theta)\bigr]\pi(d\theta),$$

where $Z(\beta)=\mathbb{E}_{\pi(d\theta)}\bigl\{\exp\bigl[-\beta r(\theta)\bigr]\bigr\}$.

Building on what has been done so far, it is easy to bound the randomized risk of the Gibbs estimator, remarking that

$$\begin{aligned}\mathbb{E}_{G_\beta(d\theta)}\bigl[R(\theta)\bigr]&=\mathbb{E}_{G_\beta(d\theta)}\bigl[r(\theta)\bigr]\\&+\bigl[Z(\beta)\bigr]^{-1}\int_{\mu=0}^{+\infty}\beta\exp(-\beta\mu)\pi\bigl[r(\theta)\leq\mu\bigr]\mathbb{E}_{\pi(d\theta|r(\theta)\leq\mu)}\bigl[R(\theta)-r(\theta)\bigr]d\mu,\end{aligned}\tag{5.5.1}$$

and using the fact that theorems 5.2.2, 5.2.3, 5.4.1 and 5.4.2 give bounds that hold with $P^{\otimes N}$ probability $1-\epsilon$ uniformly for any value of $\mu\geq\inf_\theta r(\theta)$. The deviations with respect to $\theta$ under $G_\beta(d\theta)$ can also be bounded from the equality

$$\begin{aligned}&G_\beta\bigl[R(\theta)\geq r(\theta)+\xi\bigr]\\&=\bigl[Z(\beta)\bigr]^{-1}\int_{\mu=0}^{+\infty}\beta\exp(-\beta\mu)\pi\bigl[r(\theta)\leq\mu\text{ and }R(\theta)\geq r(\theta)+\xi\bigr]d\mu.\end{aligned}\tag{5.5.2}$$

Moreover, the first part of theorems 5.2.3 and 5.4.2 can be generalized to *any* posterior distributions (including of course the family of distributions

$\{G_\beta(d\theta) : \theta \in \Theta\}$). The generalization can easily be figured out from a mere inspection of their proofs, where $\pi(d\theta|\Lambda)$ can be replaced *mutatis mutandis* by any posterior $\rho \in \mathcal{M}_+^1(\Theta, \mathcal{F})$. It is the following : For any $\gamma$, with $P^{\otimes N}$ probability $1-\epsilon$, where $\epsilon$ takes the value indicated in theorem 5.2.3, for any posterior distribution $\rho \in \mathcal{M}_+^1(\Theta, \mathcal{F})$,

$$\mathbb{E}_{\rho(d\theta)}\big[R(\theta)\big] \leq R(\widetilde{\theta}) + 2\Big[\mathbb{E}_{\rho(d\theta)}\big[r(\theta)\big] - r(\widetilde{\theta})\Big] + \frac{2\gamma}{N} + \frac{8a}{N}\mathcal{K}(\rho, \pi). \quad (5.5.3)$$

In the same way, for any $\gamma$, with $P^{\otimes N}$ probability $1-\epsilon$, where $\epsilon$ takes the value stated in theorem 5.4.2, for any posterior distribution $\rho \in \mathcal{M}_+^1(\Theta, \mathcal{F})$,

$$\mathbb{E}_{\rho(d\theta)}\big[R(\theta)\big] \leq 2\mathbb{E}_{\rho(d\theta)}\big[r(\theta)\big] + \frac{4\gamma}{N} + \frac{4}{N}\mathcal{K}(\rho, \pi). \quad (5.5.4)$$

The bounds obtained from equations (5.5.1) and (5.5.2) hold with $P^{\otimes N}$ probability $1-\epsilon$ uniformly for all values of $\beta$, therefore they can be optimized in $\beta$, this gives an empirical way to choose the inverse temperature parameter $\beta$.

As for equations (5.5.3) and (5.5.4), their right-hand side is minimized when $\rho = G_{N/4a}$ and $\rho = G_{N/2}$ respectively (as it can be seen from the Legendre transform equation (5.2.1)). To obtain bounds for other values of $\beta$, we can generalize theorems 5.2.3 and 5.4.2 to more general choices of auxiliary distributions $\lambda(\theta)$ and $\eta(\theta)$ in the following way : take for $\lambda(\theta)$ any constant value $\lambda$ and put $\eta(\theta) = \xi R(\theta) + \gamma/\lambda$ in the case of theorem 5.2.3 and $\eta(\theta) = \xi \overline{R}(\theta) + \gamma/\lambda$ in the case of theorem 5.4.2, where $\xi$ is a positive parameter smaller than one. This leads to

**Theorem 5.5.1.** *For any $\lambda > 0$ and any $\xi \in ]0,1[$, with $P^{\otimes N}$ probability at least $1-\epsilon$, for any $\rho \in \mathcal{M}_+^1(\Theta)$,*

$$\begin{aligned}\mathbb{E}_{\rho(d\theta)}\big[R(\theta)\big] \leq (1-\xi)^{-1}\Bigg\{&\mathbb{E}_{\rho(d\theta)}\big[r(\theta)\big] + \frac{1}{\lambda}\mathcal{K}(\rho, \pi) \\ &+ \frac{\log(\epsilon^{-1})}{\lambda} + \frac{1}{\lambda}\log\Bigg\{\mathbb{E}_{\pi(d\theta)}\Bigg[\exp\Bigg\{-\lambda R(\theta)\Bigg(\xi - \frac{\lambda}{N}g\Big(\frac{\lambda}{N}\Big)\Bigg)\Bigg\}\Bigg]\Bigg\}\Bigg\}. \quad (5.5.5)\end{aligned}$$

*Moreover, in the non ambiguous setting, putting $a = 1 + 1/\alpha$ and $b = (2+1/\alpha)P(\Omega_0) + P(\Omega_\alpha)$, we get with $P^{\otimes N}$ probability at least $1-\epsilon$*

$$\begin{aligned}\mathbb{E}_{\rho(d\theta)}\big[R(\theta)\big] \leq R(\widetilde{\theta}) + (1-\xi)^{-1}\Bigg\{&\Big[\mathbb{E}_{\rho(d\theta)}\big[r(\theta)\big] - r(\widetilde{\theta})\Big] + \frac{1}{\lambda}\mathcal{K}(\rho, \pi) \\ &+ \frac{\log(\epsilon^{-1})}{\lambda} + \frac{1}{\lambda}\log\Bigg\{\mathbb{E}_{\pi(d\theta)}\Bigg[\exp\Bigg\{-\lambda \overline{R}(\theta)\Bigg(\beta - \frac{a\lambda}{N}g\Big(\frac{2\lambda}{N}\Big)\Bigg)\Bigg\}\Bigg]\Bigg\} \\ &+ \frac{\lambda b}{N}g\Big(\frac{2\lambda}{N}\Big)\Bigg\}. \quad (5.5.6)\end{aligned}$$

*Remark* 5.5.1. The right-hand sides of the two inequalities (5.5.5) and (5.5.6) are minimized by choosing $\rho(d\theta) = G_\lambda(d\theta)$, as a consequence of equation (5.2.1).

*Remark* 5.5.2. By choosing $\xi = g(\frac{\lambda}{N})\frac{\lambda}{N}$ in equation (5.5.5) and $\xi = g(\frac{2\lambda}{N})\frac{a\lambda}{N}$ in equation (5.5.6), we get a simpler (but not necessarily sharper) right-hand side.

*Remark* 5.5.3. The value of $\lambda$ can be optimized through the following scheme. Apply theorem 5.5.1 to each $\lambda_{m,k} = \exp(k2^{-m})$ and $\epsilon_{m,k} = 2^{-(m+1)}\alpha_m(1-\alpha_m)^k\epsilon$, where $\alpha_m = 2^{-(m+1)}\log(N)^{-1}$, and use the union bound to get a result holding with probability $1-\epsilon$ for any value of $\lambda \in \{\lambda_{m,k}; m,k \in \mathbb{N}\}$.

Putting the three previous remarks together gives the following corollary of theorem 5.5.1 :

**Corollary 5.5.1.** *With $P^{\otimes N}$ probability at least $1-\epsilon$, for any $(m,k) \in \mathbb{N}^2$, putting $\lambda_{m,k} = \exp(k2^{-m})$,*

$$\begin{aligned}
\mathbb{E}_{G_{\lambda_{m,k}}(d\theta)}\bigl[R(\theta)\bigr] \leq {} & \frac{1}{\lambda_{m,k}\bigl[1 - g(\frac{\lambda_{m,k}}{N})\frac{\lambda_{m,k}}{N}\bigr]} \\
& \times \Biggl\{ -\log\Biggl\{ \mathbb{E}_{\pi(d\theta)}\Bigl[\exp\bigl[-\lambda_{m,k} r(\theta)\bigr]\Bigr]\Biggr\} \\
& + \log(\epsilon^{-1}) + 2(m+1)\log(2) + \log\bigl[\log(N)\bigr] + \log(\lambda_{m,k})\log(N)^{-1}\Biggr\} \\
\leq {} & \frac{1}{1 - g(\frac{\lambda_{m,k}}{N})\frac{\lambda_{m,k}}{N}} \Biggl\{ \inf_{\Lambda \in \mathfrak{F}} \mathbb{E}_{\pi(d\theta|\Lambda)}\bigl[r(\theta)\bigr] - \frac{\log\bigl[\pi(\Lambda)\bigr]}{\lambda_{m,k}} \\
& + \frac{\log(\epsilon^{-1}) + 2(m+1)\log(2) + \log\bigl[\log(N)\bigr] + \log(\lambda_{m,k})\log(N)^{-1}}{\lambda_{m,k}} \Biggr\}.
\end{aligned}$$

*Moreover*

$$\begin{aligned}
\mathbb{E}_{G_{\lambda_{m,k}}(d\theta)}\bigl[R(\theta)\bigr] \leq {} & R(\widetilde{\theta}) \\
& + \frac{1}{\lambda_{m,k}\bigl[1 - \frac{a\lambda_{m,k}}{N} g(\frac{2\lambda_{m,k}}{N})\bigr]} \Biggl\{ -\log\Biggl\{ \mathbb{E}_{\pi(d\theta)}\Bigl[\exp\bigl[-\lambda_{m,k} r(\theta)\bigr]\Bigr]\Biggr\} - \lambda_{m,k} r(\widetilde{\theta}) \\
& + \log(\epsilon^{-1}) + 2(m+1)\log(2) + \log\bigl[\log(N)\bigr] \\
& + \log(\lambda_{m,k})\log(N)^{-1} + \frac{b\lambda_{m,k}^2}{N} g\left(\frac{2\lambda_{m,k}}{N}\right)\Biggr\} \\
\leq {} & R(\widetilde{\theta}) \\
& + \frac{1}{\bigl[1 - \frac{a\lambda_{m,k}}{N} g(\frac{2\lambda_{m,k}}{N})\bigr]} \Biggl\{ \inf_{\Lambda \in \mathfrak{F}} \mathbb{E}_{\pi(d\theta|\Lambda)}\bigl[r(\theta)\bigr] - \frac{\log\bigl[\pi(\Lambda)\bigr]}{\lambda_{m,k}} - \inf_{\theta \in \Theta} r(\theta)
\end{aligned}$$

$$+\frac{\log(\epsilon^{-1})+2(m+1)\log(2)+\log\big[\log(N)\big]}{\lambda_{m,k}}$$
$$+\frac{\log(\lambda_{m,k})\log(N)^{-1}}{\lambda_{m},k}+\frac{b\lambda_{m,k}}{N}g\left(\frac{2\lambda_{m,k}}{N}\right)\Bigg\}.$$

## 5.6 Non randomized classification rules

Once a posterior distribution $\rho(d\theta)$ has been chosen, as in the previous sections, it is also possible to use it to make a weighted majority vote between the classifiers. This is an alternative to choosing $\theta$ at random according to the posterior $\rho(d\theta)$. It is possible to deduce the performance of the voting rule from the performance of the randomized rule in the following way.

Let us consider, in the pattern recognition framework described in the previous sections, a posterior distribution $\rho(d\theta)$ on the parameter set $\Theta$. Let us define the classification rule

$$\widehat{f}(x)=\arg\max_{y\in\mathcal{Y}}\mathbb{E}_{\rho(d\theta)}\Big\{\mathbb{1}\big[y=f_\theta(x)\big]\Big\}.$$

We can bound the error rate of $\widehat{f}(x)$ from the randomized error rate of $\rho(d\theta)$ using the following inequality :

$$\mathbb{1}\big(\widehat{f}(x)\neq y\big)\leq 2\,\mathbb{E}_{\rho(d\theta)}\big[\mathbb{1}\big(f_\theta(x)\neq y\big)\big]. \tag{5.6.1}$$

It comes from the fact that any value of $y$ which is not the mode of the distribution of $f_\theta(x)$ under $\rho(d\theta)$, cannot have a probability larger than $1/2$.

In the non ambiguous case, we can instead write the following equations : for any $x\in\mathcal{X}$ such that $\alpha(x)>\alpha>0$ (that is for any $x\notin\Omega_\alpha$),

$$\begin{aligned}
&P\big[Y\neq\widehat{f}(X)\,|\,X=x\big]-P\big[Y\neq f_{\widetilde{\theta}}(X)\,|\,X=x\big]\\
&\quad\leq\frac{1+\alpha(x)}{2}\,\mathbb{1}\Big[\widehat{f}(x)\neq\widetilde{f}(x)\Big]\\
&\quad\leq\big[1+\alpha(x)\big]\,\mathbb{E}_{\rho(d\theta)}\Big\{\mathbb{1}\big[f_\theta(x)\neq f_{\widetilde{\theta}}(x)\big]\Big\}\\
&\quad\leq\frac{1+\alpha(x)}{\alpha(x)}\,\mathbb{E}_{\rho(d\theta)}\Big\{P\big[Y\neq f_\theta(X)\,|\,X=x\big]-P\big[Y\neq f_{\widetilde{\theta}}(X)\,|\,X=x\big]\Big\}\\
&\quad\leq\left(1+\frac{1}{\alpha}\right)\mathbb{E}_{\rho(d\theta)}\Big\{P\big[Y\neq f_\theta(X)\,|\,X=x\big]-P\big[Y\neq f_{\widetilde{\theta}}(X)\,|\,X=x\big]\Big\}.
\end{aligned}$$

The first of the above chain of inequalities comes from the fact that

$$P\big[Y=f_{\widetilde{\theta}}(X)\,|\,X=x\big]\leq\frac{1+\alpha(x)}{2}$$

and the obvious identity

$$P\big[Y \neq \widehat{f}(X) \,|\, X = x\big] - P\big[Y \neq f_{\widetilde{\theta}}(X) \,|\, X = x\big] \\ = P\big[Y = f_{\widetilde{\theta}}(X) \,|\, X = x\big] - P\big[Y = \widehat{f}(X) \,|\, X = x\big].$$

Putting this together with the general case (5.6.1), we obtain that

$$P\big[Y \neq \widehat{f}(X)\big] - P\big[Y \neq f_{\widetilde{\theta}}(X)\big] \\ \leq \left(1 + \frac{1}{\alpha}\right) \mathbb{E}_{\rho(d\theta)} \Big\{ P\big[Y \neq f_{\theta}(X)\big] - P\big[Y \neq f_{\widetilde{\theta}}(X)\big] \Big\} + \frac{1}{\alpha} P(\Omega_{\alpha}). \quad (5.6.2)$$

From inequalities (5.6.1) and (5.6.2), one can deduce results for $\widehat{f}$ from results for $\rho(d\theta)$, both in the general and non ambiguous cases of classification problems.

## 5.7 Application to classification trees

We are going in this section to illustrate the use of a simple classification bound, namely the generalization of theorem 5.2.3 given by equation (5.5.5) of theorem 5.5.1, which says that with $P^{\otimes N}$ probability at least $1 - \epsilon$,

$$\mathbb{E}_{\pi(d\theta | \Lambda)} \big[R(\theta)\big] \leq \left(1 - \tfrac{\lambda}{N} g\left(\tfrac{\lambda}{N}\right)\right)^{-1} \left\{ \mathbb{E}_{\pi(d\theta|\Lambda)} \big[r(\theta)\big] - \frac{1}{\lambda} \log\big[\pi(\Lambda)\big] - \frac{\log(\epsilon)}{\lambda} \right\}.$$

Let us consider some measurable space $\mathcal{X}$ and some finite set of labels $\mathcal{Y}$. Assume that we can make on $\mathcal{X}$ a pool of $L$ possible real valued normalized measurements $\{f_k : \mathcal{X} \to [0,1]; k = 1, \ldots, L\}$.

**Description of a set of classification rules.** We study the case when a classification rule is deduced from comparing the measurements with threshold values. We allow for an adaptive choice of the next question to ask, depending on the answers to previous questions. More formally, we build a sequence of questions $q_1(X), \ldots, q_T(X)$, where $T$ is a stopping time for the filtration generated by $\big(q_k(X)\big)_{k \in \mathbb{N}}$. To be realistic, we will consider that we can afford to ask at most $D$ questions, in other terms that $T \leq D$ a.s.. Our definition of question $q_k(X)$ will more precisely be

$$q_k(X) = \mathbb{1}\Big[f_{j(q_1, \ldots, q_{k-1})}(X) \geq \tau(q_1, \ldots, q_{k-1})\Big],$$

allowing for an adaptive choice of the measurement through $j(q_1, \ldots, q_{k-1})$ and of the threshold, through $\tau(q_1, \ldots, q_{k-1})$. Once we have collected the answers $(q_1, \ldots, q_T) \in \{0,1\}^T$, where $T$ is the time when we decided to stop, we have to take a decision on the label of pattern $X$, using some decision function

$$\varphi : \bigcup_{k \in \mathbb{N}} \{0,1\}^k \to \mathcal{Y}.$$

The resulting decision function can thus be described as $g(X) = \varphi(q_1, \ldots, q_T)$.

**Explicit description of the parameter set.** The choice of a stopping time $T$ is equivalent to the choice of a *complete prefix set* $S \in \bigcup_{k=0}^{D} \{0,1\}^k$. We will let $\mathcal{S}$ be the set of all the complete prefix subsets of $\bigcup_{k=0}^{D} \{0,1\}^k$. We will also let $\overset{\circ}{S}$ be the set of all the strict prefixes of the binary words of $S$. The definition of $T$ from $S$ is

$$T = \inf\{k : (q_1, \ldots, q_k) \in S\}.$$

On the other hand, $S$ may be retrieved from $T$ as the range of $(q_1, \ldots, q_T)$. In this description through complete prefix sets, the choice of the next measurement $j$ can be any mapping of $\overset{\circ}{S}$ to $\{1, \ldots, L\}$ :

$$j : \overset{\circ}{S} \to \{1, \ldots, L\},$$

and the choice of the next threshold $\tau$, any mapping from $\overset{\circ}{S}$ to $[0, 1]$ :

$$\tau : \overset{\circ}{S} \to [0, 1].$$

Thus, each possible decision rule $g(X)$ is described by the vector of parameters $\theta = (S, j, \tau, \varphi)$, where $S \in \mathcal{S}$, $j \in \{1, \ldots, L\}^{\overset{\circ}{S}}$, $\tau \in [0, 1]^{\overset{\circ}{S}}$ and $\varphi \in \mathcal{Y}^S$, through the formula

$$g(S, j, \tau, \phi, X) = \varphi(q_1, \ldots, q_T).$$

**Choice of a prior.** A simple choice of a prior distribution $\pi(d\theta)$ on the parameter set is

$$\begin{aligned} \pi(S) &= 2^{-|\overset{\circ}{S}|} 2^{-|S \setminus \{0,1\}^D|} \qquad \text{(critical branching process)} \\ \pi(j|S) &= L^{-|\overset{\circ}{S}|} \\ \pi(d\tau|S, j) &= \text{ Lebesgue measure on } [0, 1]^{\overset{\circ}{S}} \\ \pi(\varphi|S, j, \tau) &= |\mathcal{Y}|^{-|S|}. \end{aligned}$$

**Choice of a subfamily of parameter subsets :**

A natural way to choose a family of parameter subsets is to define some finite partition of $\Theta$, such that the component $\Lambda(\theta)$ of this partition into which a given value of the parameter $\theta$ falls satisfies

$$\Lambda(\theta) \subset \{\theta' : r(\theta') = r(\theta)\}.$$

To this purpose, let us put, for any $s \in \overset{\circ}{S}$,

$$m_s(\tau) \stackrel{\text{def}}{=} \max\{f_{j(s)}(X_i) : (q_1(X_i), \dots, q_{l(s)}(X_i)) = s \\ \text{and } f_{j(s)}(X_i) < \tau(s)\},$$
$$M_s(\tau) \stackrel{\text{def}}{=} \min\{f_{j(s)}(X_i) : (q_1(X_i), \dots, q_{l(s)}(X_i)) = s \\ \text{and } f_{j(s)}(X_i) \geq \tau(s)\},$$
$$\Lambda(S, j, \tau, \varphi) \stackrel{\text{def}}{=} \{S\} \times \{j\} \times \prod_{s \in \overset{\circ}{S}} ]m_s(\tau), M_s(\tau)] \times \{\varphi\}.$$

As

$$\pi\bigl[\Lambda(S, j, \tau, \varphi)\bigr] = 2^{-|S|-1+|S\setminus\{0,1\}^D|} L^{-(|S|-1)} |\mathcal{Y}|^{-|S|} \times \prod_{s \in \overset{\circ}{S}} \bigl[M_s(\tau) - m_s(\tau)\bigr],$$

the size of the cell $\Lambda(S, j, \tau, \varphi)$ is reasonably big, in the sense that $-\log\bigl[\pi\bigl(\Lambda(S, j, \tau, \varphi)\bigr)\bigr]$ grows at most linearly with $|S|$. Indeed, the parameter set can be viewed as a union of submodels, indexed by the choice of $S$, the "dimension" of each submodel being related to $|S|$.

**Selection of $\theta = (S, j, \tau, \varphi)$ :** Let us put

$$a_s^y \stackrel{\text{def}}{=} \sum_{i=1}^N \mathbb{1}\bigl[Y_i = y, (q_1, \dots, q_{\ell(s)}) = s\bigr], \qquad s \in \{\varnothing\} \cup \bigcup_{k=1}^D \{0,1\}^k, y \in \mathcal{Y},$$
$$b_s \stackrel{\text{def}}{=} \sum_{y \in \mathcal{Y}} a_s^y.$$

The empirical error rate can be expressed with the help of these notations as

$$r(S, j, \tau, \varphi) = \frac{1}{N} \sum_{s \in S} (b_s - a_s^{\varphi(s)}).$$

Let us put moreover

$$B(\theta, \lambda) \stackrel{\text{def}}{=} \frac{1}{1 - \frac{\lambda}{N} g\left(\frac{\lambda}{N}\right)} \Biggl\{ \frac{1}{\lambda} \sum_{s \in \overset{\circ}{S}} \Bigl\{ \log(2) + \log(L) - \log\bigl[M_s(\tau) - m_s(\tau)\bigr] \Bigr\} \\ + \frac{1}{\lambda} \sum_{s \in S} \Bigl\{ \log(2) \mathbb{1}\left[\ell(s) < D\right] + \log\bigl(|\mathcal{Y}|\bigr) \Bigr\} \\ + \frac{1}{N} \sum_{s \in S} (b_s - a_s^{\varphi(s)}) - \frac{\log(\epsilon)}{\lambda} \Biggr\}.$$

With these notations, for any $\lambda \in \mathbb{R}_+$ such that $\frac{\lambda}{N} g\left(\frac{\lambda}{N}\right) < 1$, with $P^{\otimes N}$ probability at least $1 - \epsilon$, for any $\widehat{\theta} \in \Theta$,

$$\mathbb{E}_{\pi(d\theta|\Lambda(\widehat{\theta}))}\big[R(\widehat{\theta})\big] \leq B(\widehat{\theta},\lambda). \tag{5.7.1}$$

Ideally, we would like to choose $\widehat{\theta} = \arg\min_\theta B(\theta,\lambda)$. (We assume for simplicity in this discussion that some fixed value of $\lambda$ has been chosen beforehand.) However, minimizing $B(\theta,\lambda)$ in $\theta$ maybe difficult to achieve when $\Theta$ is large. Some part of this optimization can be performed efficiently : indeed, once $j : \bigcup_{k=0}^{D-1}\{0,1\}^k \rightarrow \{1,\dots,L\}$ and $\tau : \bigcup_{k=0}^{D-1}\{0,1\}^k \rightarrow [0,1]$ have been chosen, one can put $\widehat{\varphi}(s) = \arg\max_{y\in\mathcal{Y}} a_s^y$ and choose $\widehat{S}$ (that is $T$) by dynamic programming on $\{0,1\}^D$. It is on the other hand not obvious to find a way of minimizing $B(\theta,\lambda)$ in $j$ and $\tau$ otherwise than by an exhaustive inspection of the $L^{2^D-1}$ values that $j$ may take and the at most $(N+1)^{2^D-1}$ intervals of sampled values in which $\tau$ may fall. Therefore, except for very small values of $D$, only some partial optimization in $j$ and $\tau$, using some heuristic scheme, will presumably be feasible. This will not prevent us from obtaining a rigorous upper bound for the deviations of the error rate, since the bound (5.7.1) holds uniformly for any value of $\widehat{\theta}$, which may depend in any given way on the observed sample.

The bound $B(\widehat{\theta},\lambda)$, suitably modified to this purpose, can furthermore be optimized in $\lambda$ as explained in remark 5.5.3.

We would like to stress the fact that the empirical complexity approach as led us to a criterion $B(\theta,\lambda)$ which is a mix of supervised and unsupervised learning : the empirical error rate which requires a labelled sample (supervised learning) is balanced with some measure of *empirical margin* $\sum_{s\in\overset{\circ}{S}} -\log\big[M_s(\tau)-m_s(\tau)\big]$, which depends only on the way in which the observed patterns $(X_1,\dots,X_N)$ are clustured (unsupervised learning).

**Computing the law of $g(\theta,x)$ under $\pi\big[d\theta|\Lambda(\widehat{\theta})\big]$ :**

The computation of the probability of each label can be made explicitly as follows:

$$\mathbb{E}_{\pi(d\theta|\Lambda(\widehat{\theta}))}\big\{\mathbb{1}\big[g(\theta,x)=y\big]\big\} =$$
$$\sum_{s\in S}\mathbb{1}\big[\varphi(s)=y\big]\prod_{k=1}^{\ell(s)} 0\vee 1\wedge\left\{\left[\frac{\big(f_{j(s_1^{k-1})}(x)-m_{s_1^{k-1}}(\tau)\big)}{M_{s_1^{k-1}}(\tau)-m_{s_1^{k-1}}(\tau)}\right](2s_k-1)+(1-s_k)\right\},$$

where $\widehat{\theta}=(S,j,\tau,\varphi)$. Note that this can be factorized as a recursive computation on $\overset{\circ}{S}$ with $2|S|-1$ steps.

**Comparison with Vapnik's entropy :** An application of aggregation rules of the previous chapters would have led to

$$\mathbb{E}_{P^{\otimes N}}\big[R(\widehat{f})\big] \leq \inf_{S,j,\varphi}\ \inf_{\theta\in\Theta_{S,j,\varphi}} CR(\theta)$$

$$+\frac{C'}{N}\mathbb{E}_{P^{\otimes N}}\biggl\{-\log\bigl[\pi(S,j,\varphi)\bigr]+\underbrace{\log\biggl(\Bigl|\Bigl\{\bigl[g(\theta,X_i)\bigr]_{i=1}^N:\theta\in\Theta_{S,j,\varphi}\Bigr\}\Bigr|\biggr)}_{\leq(|S|-1)\log(N)}\biggr\},$$

where $\Theta_{S,j,\varphi}=\bigl\{(S,j,\tau,\varphi):\tau\in[0,1]^{\overset{\circ}{S}}\bigr\}$. Note that an exact computation of the shattering number $\log\biggl(\Bigl|\Bigl\{\bigl[g(\theta,X_i)\bigr]_{i=1}^N:\theta\in\Theta_{S,j,\varphi}\Bigr\}\Bigr|\biggr)$ would not be so easy.

## 5.8 The regression setting

Let us assume now that we observe a sample $(X_i,Y_i)_{i=1}^N$ drawn from a product distribution $P^{\otimes N}$ on $\Bigl(\bigl(\mathcal{X}\times\mathbb{R}\bigr)^N,\bigl(\mathcal{B}\otimes\mathcal{B}'\bigr)^{\otimes N}\Bigr)$, where $(\mathcal{X},\mathcal{B})$ is a measurable space and $\mathcal{B}'$ is the Borel sigma algebra.

Assume that a set of regression functions $\{f_\theta:\mathcal{X}\to\mathbb{R}\,;\,\theta\in\Theta\}$ is given, and that $(\theta,x)\mapsto f_\theta(x)$ is measurable and bounded.

Let us put $f_{\widetilde{\theta}}(X)=\mathbb{E}_P(Y\,|\,X)$, and let us assume without loss of generality that $\widetilde{\theta}\in\Theta$ (if it is not the case, replace $\Theta$ by $\Theta\cup\{\widetilde{\theta}\}$).

Let us introduce the following notations :

$$\begin{aligned}
R(\theta)&=\mathbb{E}_P\bigl\{\bigl[Y-f_\theta(X)\bigr]^2\bigr\},\\
r(\theta)&=\frac{1}{N}\sum_{i=1}^N\bigl[Y_i-f_\theta(X_i)\bigr]^2,\\
\overline{R}(\theta)&=R(\theta)-R(\widetilde{\theta})\\
&=\mathbb{E}_P\bigl\{\bigl[f_\theta(X)-f_{\widetilde{\theta}}(X)\bigr]^2\bigr\},\\
\overline{r}(\theta)&=r(\theta)-r(\widetilde{\theta})\\
&=\frac{1}{N}\sum_{i=1}^N 2\bigl[Y_i-f_{\widetilde{\theta}}(X_i)\bigr]\bigl[f_{\widetilde{\theta}}(X_i)-f_\theta(X_i)\bigr]+\bigl[f_\theta(X_i)-f_{\widetilde{\theta}}(X_i)\bigr]^2.
\end{aligned}$$

Let us assume that

$$\sup_{x\in\mathcal{X},\theta\in\Theta}|f_\theta(x)-f_{\widetilde{\theta}}(x)|\leq B<+\infty.$$

Let us also assume that there is some regular version of the conditional probabilities $P(dY\,|\,X)$ (which we will use in the following without further mention), such that for some positive real constants $M$ and $\alpha$,

$$\sup_{x\in\mathcal{X}}\mathbb{E}_P\Bigl[\exp\Bigl(2\alpha B\bigl|Y-f_{\widetilde{\theta}}(X)\bigr|\Bigr)\,\Big|\,X=x\Bigr]\leq M. \tag{5.8.1}$$

This assumption implies that for any $\beta<\alpha$

$$\sup_{x \in \mathcal{X}} \mathbb{E}_P \Big[ \exp\Big( 2\beta B \big| Y - f_{\widetilde{\theta}}(X) \big| \Big) \,|\, X = x \Big] \leq M^{\beta/\alpha}.$$

Let us put

$$\begin{aligned} Z &\overset{\text{def}}{=} -\big[Y - f_\theta(X)\big]^2 + \big[Y - f_{\widetilde{\theta}}(X)\big]^2 \\ &= \big[f_\theta(X) - f_{\widetilde{\theta}}(X)\big]\big[2Y - f_\theta(X) - f_{\widetilde{\theta}}(X)\big]. \end{aligned} \tag{5.8.2}$$

Our first concern will be to get a deviation inequality for this random variable. Let us introduce the notation

$$T \overset{\text{def}}{=} B|2Y - f_\theta(X) - f_{\widetilde{\theta}}(X)|. \tag{5.8.3}$$

Assumption (5.8.1) implies that for any positive $\beta \leq \alpha$,

$$\mathbb{E}_P \big[ \exp(\beta T) \,|\, X = x \big] \leq M^{\beta/\alpha} \exp(\beta B^2).$$

We are going to prove the following lemma :

**Lemma 5.8.1.** *Let $g : \mathbb{R}_+ \to \mathbb{R}_+$ be the function*

$$g(\beta) \overset{\text{def}}{=} \frac{\exp(\beta B^2)}{B^2} \sup_{x \in \mathcal{X}} \mathbb{E} \Big[ T^2 \exp(\beta T) \,|\, X = x \Big].$$

*It satisfies the following properties :*

$$g(0) \leq B^2 + 4 \sup_{x \in \mathcal{X}} \mathbb{E}_P \Big[ \big(Y - f_{\widetilde{\theta}}(X)\big)^2 \,|\, X = x \Big], \tag{5.8.4}$$

$$\begin{aligned} g(\beta) - g(0) &\leq \Big[ \exp(\beta B^2) - 1 \Big] g(0) \\ &+ M^{\beta/\alpha} \exp(2\beta B^2) \frac{\beta}{B^2} \inf_{\gamma \in [0, \frac{\alpha - \beta}{3}]} (e\gamma)^{-3} \exp\left[ 3\gamma \left( B^2 + \frac{\log(M)}{\alpha} \right) \right], \end{aligned} \tag{5.8.5}$$

$$\log \Bigg\{ \mathbb{E}_P \Bigg[ \exp \Bigg\{ \beta \Big[ \overline{R}(\theta) - \big(Y - f_\theta(X)\big)^2 + \big(Y - f_{\widetilde{\theta}}(X)\big)^2 \Big] \Bigg\} \Bigg] \Bigg\} \leq \frac{\beta^2}{2} \overline{R}(\theta) g(\beta). \tag{5.8.6}$$

*Remark* 5.8.1. In the case when the distribution of $Y$ knowing $X$ is a Gaussian distribution with mean $f_{\widetilde{\theta}}(X)$ and constant standard deviation $\sigma$, that is when $P(dY \,|\, X) = \mathcal{N}(f_{\widetilde{\theta}}(X), \sigma)$, one can take

$$g(\beta) = 4\sigma^2 + 2(2\beta\sigma^2 - 1)^2 B^2 \psi\big(\beta|2\beta\sigma^2 - 1|B^2\big)$$

with

$$\psi(s) = \frac{\exp(s) - 1 - s}{s^2}.$$

*Proof.* Let us consider the order 2 Taylor expansion of the log-Laplace transform of $Z$:

$$\begin{aligned}
&\log \mathbb{E}_P\bigl[\exp(\beta Z)\bigr] - \beta \mathbb{E}_P(Z) \\
&\quad = \int_{\gamma=0}^{\beta} (\beta-\gamma) \left[ \frac{\mathbb{E}_P\bigl[Z^2 \exp(\gamma Z)\bigr]}{\mathbb{E}_P\bigl[\exp(\gamma Z)\bigr]} - \left( \frac{\mathbb{E}_P\bigl[Z \exp(\gamma Z)\bigr]}{\mathbb{E}_P\bigl[\exp(\gamma Z)\bigr]} \right)^2 \right] d\gamma \\
&\quad \leq \int_{\gamma=0}^{\beta} (\beta-\gamma) \mathbb{E}_P\bigl[Z^2 \exp(\gamma Z)\bigr] \exp\bigl[-\gamma \mathbb{E}_P(Z)\bigr] d\gamma \\
&\quad \leq \frac{\beta^2}{2} \overline{R}(\theta) \exp\bigl[\beta \overline{R}(\theta)\bigr] \sup_{x \in \mathcal{X}} \mathbb{E}_P \left\{ \frac{T^2}{B^2} \exp(\beta T) \,|\, X = x \right\} \\
&\quad \leq \frac{\beta^2}{2} \overline{R}(\theta) g(\beta).
\end{aligned}$$

Moreover

$$\begin{aligned}
g(0) &= \sup_{x \in \mathcal{X}} \mathbb{E}_P \left\{ \frac{T^2}{B^2} \,|\, X = x \right\} \\
&= \sup_{x \in \mathcal{X}} \bigl[ f_\theta(x) - f_{\widetilde{\theta}}(x) \bigr]^2 + 4\, \mathbb{E}_P \left\{ \bigl[ Y - f_{\widetilde{\theta}}(X) \bigr]^2 \,|\, X = x \right\} \\
&\leq B^2 + \sup_{x \in \mathcal{X}} 4\, \mathbb{E}_P \left\{ \bigl[ Y - f_{\widetilde{\theta}}(X) \bigr]^2 \,|\, X = x \right\}.
\end{aligned}$$

As for $g(\beta)$,

$$\begin{aligned}
g(\beta) - g(0) &= \bigl[ \exp(\beta B^2) - 1 \bigr] g(0) \\
&\qquad + \exp(\beta B^2) \sup_{x \in \mathcal{X}} \mathbb{E}_P \left[ \left( \frac{T}{B} \right)^2 \bigl[ \exp(\beta T) - 1 \bigr] \,|\, X = x \right],
\end{aligned}$$

and for any positive real number $\gamma$

$$\begin{aligned}
\left( \frac{T}{B} \right)^2 \bigl[ \exp(\beta T) - 1 \bigr] &\leq \frac{\beta}{B^2} T^3 \exp(\beta T) \\
&\leq \frac{\beta}{\gamma^3 B^2} \exp\bigl[ \beta T + 3 \log(\gamma T) \bigr] \\
&\leq \frac{\beta}{(e\gamma)^3 B^2} \exp\bigl[ (\beta + 3\gamma) T \bigr],
\end{aligned}$$

whence

$$\begin{aligned}
&\mathbb{E}_P \left[ \left( \frac{T}{B} \right)^2 \bigl[ \exp(\beta T) - 1 \bigr] \,|\, X = x \right] \\
&\qquad \leq \inf_{\gamma \in [0, \frac{\alpha-\beta}{3}]} \frac{\beta}{(e\gamma)^3 B^2} \exp\bigl[ (\beta + 3\gamma) B^2 \bigr] M^{(\beta+3\gamma)/\alpha}.
\end{aligned}$$

The bound for $g(\beta) - g(0)$ stated in the lemma follows immediately. □

We deduce from lemma 5.8.1 the following analogous of lemma 5.2.1 for the regression setting:

**Lemma 5.8.2.** *Let $\eta : \Theta \rightarrow \mathbb{R}_+$ and $\lambda : \Theta \rightarrow \mathbb{R}_+$ be given measurable functions defined on the measurable parameter set $(\Theta, \mathcal{F})$. Let $\pi \in \mathcal{M}_+^1(\Theta, \mathcal{F})$ be a prior probability distribution.*

$$P^{\otimes N}\left(\sup_{\rho \in \mathcal{M}_+^1(\Theta)} \mathbb{E}_{\rho(d\theta)}\Bigl\{\lambda(\theta)\Bigl[\overline{R}(\theta) - \overline{r}(\theta) - \eta(\theta)\Bigr]\Bigr\} - \mathcal{K}(\rho, \pi) \leq 0\right)$$
$$\geq 1 - \mathbb{E}_{\pi(d\theta)}\left\{\exp\left[-\eta(\theta)\lambda(\theta) + g\left(\frac{\lambda(\theta)}{N}\right)\overline{R}(\theta)\frac{\lambda(\theta)^2}{2N}\right]\right\},$$

*where conventions (5.2.2) and (5.2.3) are assumed. Consequently,*

$$\left(P^{\otimes N}\right)_*\left\{\sup_{\Lambda \in \mathcal{F}} \pi(\Lambda) \inf_{\theta \in \Lambda} \exp\Bigl\{\lambda(\theta)\Bigl[\overline{R}(\theta) - \overline{r}(\theta) - \eta(\theta)\Bigr]\Bigr\} \leq 1\right\}$$
$$\geq 1 - \mathbb{E}_{\pi(d\theta)}\left\{\exp\left[-\eta(\theta)\lambda(\theta) + g\left(\frac{\lambda(\theta)}{N}\right)\overline{R}(\theta)\frac{\lambda(\theta)^2}{2N}\right]\right\}.$$

The same reasoning as in the pattern recognition case then leads to

**Theorem 5.8.1.** *For any positive constants $\beta$ and $\gamma$, with $P^{\otimes N}$ probability at least $1 - \epsilon$, where*

$$\epsilon = \exp\left\{-\gamma\left[1 - \frac{g(\beta)}{2g(0)}\right]\right\}$$

*for any measurable parameter subset $\Lambda \in \mathcal{F}$, putting $\mu = \sup_{\theta \in \Lambda} r(\theta)$,*

$$\mathbb{E}_{\pi(d\theta \,|\, \Lambda)}\bigl[R(\theta)\bigr] \leq R(\widetilde{\theta}) - r(\widetilde{\theta}) + \mathbb{E}_{\pi(d\theta \,|\, \Lambda)}\bigl[r(\theta)\bigr]$$
$$+ \int_{\xi=0}^{1} \min\left\{1, \pi(\Lambda)^{-1} \exp\left[-\sqrt{\gamma N}\left(\frac{\xi}{\sqrt{g(0)\bigl[\xi + \mu - r(\widetilde{\theta})\bigr] + \gamma/(\beta^2 N)}} - \sqrt{\frac{\gamma}{N}}\right)\right]\right\} d\xi$$

$$\leq R(\widetilde{\theta}) - r(\widetilde{\theta}) + \mathbb{E}_{\pi(d\theta \,|\, \Lambda)}\bigl[r(\theta)\bigr]$$
$$+ \begin{cases} \sqrt{\dfrac{g(0)}{\gamma N}\bigl[\mu - r(\widetilde{\theta})\bigr] + \dfrac{1}{\beta^2 N^2}\Bigl\{3 + \gamma - \log\bigl[\pi(\Lambda)\bigr]\Bigr\}} + \dfrac{4g(0)}{\gamma N}, \\ \quad \text{when } -\log\bigl[\pi(\Lambda)\bigr] \leq \sqrt{\gamma N}\left(\sqrt{\dfrac{\mu - r(\widetilde{\theta})}{2g(0)} + \dfrac{\gamma}{2g(0)^2\beta^2 N}}\right), \\ \dfrac{2g(0)}{\gamma N}\Bigl[\Bigl(1 + \gamma - \log\bigl[\pi(\Lambda)\bigr]\Bigr)^2 + 1\Bigr], \quad \text{otherwise.} \end{cases}$$

— *Let us recall that* $g(0) \leq B^2 + 4 \sup_{x \in \mathcal{X}} \mathbb{E}_P \Bigl\{ \bigl[ Y - f_{\widetilde{\theta}}(X) \bigr]^2 \,|\, X = x \Bigr\}$ *and that* $g(\beta) - g(0)$ *is explicitely bounded as stated in lemma 5.8.1. Moreover with* $\pi\bigl(d\theta \,|\, \Lambda\bigr)$ *probability at least*

$$1 - \pi\bigl(\Lambda\bigr)^{-1} \exp \left\{ -\sqrt{\gamma N} \left( \frac{\xi}{\sqrt{g(0)\bigl(\xi + \mu - r(\widetilde{\theta})\bigr) + \frac{\gamma}{\beta^2 N}}} - \sqrt{\frac{\gamma}{N}} \right) \right\}$$

*the expected risk* $R(\theta)$ *is bounded by*

$$R(\theta) \leq R(\widetilde{\theta}) - r(\widetilde{\theta}) + r(\theta) + \xi.$$

*As in the pattern recognition case, note that* $-r(\widetilde{\theta})$ *can be upper bounded in the right-hand side of the two previous inequalities by the observable quantity* $-\inf_{\theta \in \Theta} r(\theta)$.

*Proof.* Put

$$\eta(\theta) = \sqrt{\frac{\gamma\bigl[g(0)\overline{R}(\theta) + \gamma/(\beta^2 N)\bigr]}{N}},$$

$$\lambda(\theta) = \sqrt{\frac{\gamma N}{g(0)\overline{R}(\theta) + \gamma/(\beta^2 N)}},$$

and use lemma 5.8.2. The more explicit upper bound is obtained by applying lemma 5.10.2 with $a = \sqrt{\gamma N}$, $b = g(0)$, $d = \mu - r(\widetilde{\theta}) + \dfrac{\gamma}{g(0)\beta^2 N}$, and $e = \sqrt{\dfrac{\gamma}{N}}$. □

As in the pattern recognition case, another choice of the functions $\eta(\theta)$ and $\lambda(\theta)$ is also of interest and gives a different kind of result. Let us put

$$\eta(\theta) = \frac{\overline{R}(\theta)}{2} + \gamma \frac{2g(0)}{\beta N},$$

$$\lambda(\theta) = \frac{\beta N}{2g(0)},$$

using lemma 5.8.2 again, we get :

**Theorem 5.8.2.** *For any positive constant* $\beta$ *and real constant* $\gamma$*, with* $P^{\otimes N}$ *probability at least* $(1 - \epsilon)$*, where*

$$\epsilon = \exp(-\gamma) \mathbb{E}_{\pi(d\theta)} \left\{ \exp \left[ -\beta N \frac{\overline{R}(\theta)}{4g(0)} \left( 1 - \frac{\beta g\left(\frac{\beta}{2g(0)}\right)}{2g(0)} \right) \right] \right\},$$

*for any posterior probability distribution* $\rho \in \mathcal{M}_+^1(\Theta)$,

$$\mathbb{E}_{\rho(d\theta)}\big[\overline{R}(\theta)\big] \leq 2\Big\{\mathbb{E}_{\rho(d\theta)}\big[r(\theta)\big] - r(\widetilde{\theta})\Big\} + \frac{4g(0)}{\beta N}\Big[\gamma + \mathcal{K}(\rho, \pi)\Big]. \quad (5.8.7)$$

*As a consequence, for any measurable parameter subset* $\Lambda \in \mathcal{F}$,

$$\mathbb{E}_{\pi(d\theta|\Lambda)}\big[\overline{R}(\theta)\big] \leq 2\Big\{\mathbb{E}_{\pi(d\theta|\Lambda)}\big[r(\theta)\big] - r(\widetilde{\theta})\Big\} + \frac{4g(0)}{\beta N}\Big[\gamma - \log\big[\pi(\Lambda)\big]\Big]. \quad (5.8.8)$$

*Moreover for any* $\xi$, *with* $\pi\big(d\theta \,|\, \Lambda\big)$ *probability at least*

$$1 - \pi\big(\Lambda\big)^{-1} \exp\left(-\frac{\beta N \xi}{4g(0)}\right),$$

*the expected risk* $R(\theta)$ *is bounded by*

$$R(\theta) \leq R(\widetilde{\theta}) + 2r(\theta) - 2r(\widetilde{\theta}) + \frac{4\gamma g(0)}{\beta N} + \xi.$$

*Here again, note that* $-r(\widetilde{\theta})$ *can be upper bounded in the right-hand side of the previous inequalities by the observable quantity* $-\inf_{\theta\in\Theta} r(\theta)$.

*Remark* 5.8.2. It should be noted that the *randomized* regression estimators presented in this section, in which $f_\theta$ is drawn from some posterior distribution $\rho(d\theta)$, can be replaced by *aggregated* estimators of the type $\mathbb{E}_{\rho(d\theta)}(f_\theta)$. Indeed, the convexity of the quadratic loss function implies that

$$\begin{aligned}
&\mathbb{E}_{P(dX,dY)}\Big\{\Big[Y - \mathbb{E}_{\rho(d\theta)}\big[f_\theta(X)\big]\Big]^2\Big\} \\
&\qquad \leq \mathbb{E}_{\rho(d\theta)}\bigg\{\mathbb{E}_{P(dX,dY)}\Big[\big[Y - f_\theta(X)\big]^2\Big]\bigg\} = \mathbb{E}_{\rho(d\theta)}\big[R(\theta)\big]. \quad (5.8.9)
\end{aligned}$$

## 5.9 Links with penalized least square regression

### 5.9.1 The general case

Unlike in the classification setting, it is possible in the least square regression framework to control the oscillations of the expected risk $R(\theta)$ in the following way: for any $(\theta, \widehat{\theta}) \in \Theta^2$,

$$\begin{aligned}
\overline{R}(\widehat{\theta}) - \overline{R}(\theta) &= \mathbb{E}_P\Big\{\big[2f_{\widetilde{\theta}}(X) - f_\theta(X) - f_{\widehat{\theta}}(X)\big]\big[f_\theta(X) - f_{\widehat{\theta}}(X)\big]\Big\} \\
&\leq \sqrt{\mathbb{E}_P\Big\{\big[2f_{\widetilde{\theta}}(X) - f_\theta(X) - f_{\widehat{\theta}}(X)\big]^2\Big\}}\sqrt{\mathbb{E}_P\Big\{\big[f_{\widehat{\theta}}(X) - f_\theta(X)\big]^2\Big\}} \\
&\leq \Big(\sqrt{\overline{R}(\theta)} + \sqrt{\overline{R}(\widehat{\theta})}\Big)\, \|f_\theta - f_{\widehat{\theta}}\|_\infty.
\end{aligned} \quad (5.9.1)$$

This opens the possibility to derive a deterministic parameter selection rule. To achieve this we will consider different types of parameter subsets. In the third case below, we will assume that the following bound is available to the statistician :

$$\mathbb{E}_P\Bigl\{\bigl[f_\theta(X) - f_{\widehat{\theta}}(X)\bigr]^2\Bigr\} \leq d(\theta, \widehat{\theta}),$$

through a known measurable function $d : \Theta \times \Theta \rightarrow \mathbb{R}_+$.

$$\Lambda_\infty(\widehat{\theta}, \delta) \overset{\text{def}}{=} \Bigl\{\theta \in \Theta : \|f_\theta - f_{\widehat{\theta}}\|_\infty \leq \delta\Bigr\}, \qquad \widehat{\theta} \in \Theta, \delta \in \mathbb{R}_+^*,$$

$$\Lambda_2(\widehat{\theta}, \delta) \overset{\text{def}}{=} \Bigl\{\theta \in \Theta : \mathbb{E}\Bigl(\bigl[f_\theta(X) - f_{\widehat{\theta}}(X)\bigr]^2\Bigr) \leq \delta^2\Bigr\}, \qquad \widehat{\theta} \in \Theta, \delta \in \mathbb{R}_+^*,$$

$$\Lambda_d(\widehat{\theta}, \delta) \overset{\text{def}}{=} \Bigl\{\theta \in \Theta : d(\theta, \widehat{\theta}) \leq \delta^2\Bigr\} \subset \Lambda_2(\widehat{\theta}, \delta), \qquad \widehat{\theta} \in \Theta, \delta \in \mathbb{R}_+^*.$$

All these sets being measurable, we derive from equation (5.9.1) that

$$\sqrt{\overline{R}(\widehat{\theta})}\left[\sqrt{\overline{R}(\widehat{\theta})} - \delta\right] \leq \mathbb{E}_{\pi[d\theta \,|\, \Lambda(\widehat{\theta},\delta)]}\bigl[\overline{R}(\theta)\bigr] + \delta\sqrt{\mathbb{E}_{\pi[d\theta \,|\, \Lambda(\widehat{\theta},\delta)]}\bigl[\overline{R}(\theta)\bigr]}, \quad (5.9.2)$$

where we have used the fact that $x \mapsto \sqrt{x}$ is a concave function, and where $\Lambda(\widehat{\theta}, \delta)$ may be either $\Lambda_\infty(\widehat{\theta}, \delta)$, $\Lambda_2(\widehat{\theta}, \delta)$, or $\Lambda_d(\widehat{\theta}, \delta)$. Using equation (5.9.2) and theorem 5.8.1 or 5.8.2, and optimizing the resulting inequalities in $\delta$ and $\widehat{\theta}$ gives an oracle deviation inequality for a deterministic (non randomized) estimator $\widehat{\theta}$.

### 5.9.2 Adaptive linear least square regression estimation

To illustrate this, let us see what we get in the case of a family of linear regression models. Let $\mathcal{X}$ be a bounded subset of $\mathbb{R}^D$ and let $\overline{\Theta} \subset \mathbb{R}^D$ be a bounded set of parameters. Let $H_j$, $j \in J$, be a countable family of linear subspaces of $\mathbb{R}^D$ (we do not assume that it contains necessarily $\mathbb{R}^D$ itself). Let $\dim(H_j) = d_j$. Put $\Theta_j = \overline{\Theta} \cap H_j$ and

$$\Theta = \bigsqcup_{j \in J} \Theta_j,$$

where the union is meant to be disjoint. This is a way to introduce some "structure" in the set of parameters. Consider the family of regression functions $\{f_\theta, \theta \in \Theta\}$ defined by

$$f_\theta(X) = \langle \theta, X \rangle, \qquad \theta \in \Theta.$$

Note that this setting covers (through a straightforward change of notations) the basis selection problem, where the regression functions are of the form

$$f(X) = \sum_{k=1}^{D} \alpha_k \phi_k(X),$$

and the problem is to select an appropriate subfamily of basis functions $\phi_k$ (assumed here to be bounded functions). Let

$$\begin{aligned}
C &\overset{\text{def}}{=} \mathbb{E}_P\bigl[XX'\bigr],\\
C_j &\overset{\text{def}}{=} C_{|H_j},\\
\widehat{C} &\overset{\text{def}}{=} \frac{1}{N}\sum_{i=1}^N X_i X_i',\\
\widehat{C}_j &\overset{\text{def}}{=} \widehat{C}_{|H_j},\\
\widetilde{\theta} &\overset{\text{def}}{=} \arg\min_{\theta\in\mathbb{R}^D} R(\theta) = C^{-1}\mathbb{E}_P(YX),\\
\widetilde{\theta}_j &\overset{\text{def}}{=} \arg\min_{\theta\in H_j} R(\theta) = C_j^{-1}\mathbb{E}_P\bigl(YX_{|H_j}\bigr),\\
\widehat{\theta}_j &\overset{\text{def}}{=} \arg\min_{\theta\in H_j} r(\theta) = \frac{1}{N}\sum_{i=1}^N Y_i \widehat{C}_j^{-1} X_{i|H_j},
\end{aligned}$$

where $X_{|H_j}$ is the orthogonal projection of $X$ on the linear subspace $H_j$ of $\mathbb{R}^D$. Let us assume that for any $j\in J$, $\widetilde{\theta}_j \subset \Theta$ and that $\widetilde{\theta}\subset\overline{\Theta}$.

Let us consider a prior distribution $\pi\in\mathcal{M}_+^1(\Theta)$ of the form

$$d\pi(\theta) = \sum_{j\in J}\mathbb{1}(\theta\in\Theta_j)\mu_j d_j(\theta),$$

where $d_j(\theta)$ is a notation for the $d_j$ dimensional Lebesgue measure on $H_j$. Define the following parameter subsets :

$$\Lambda_j(\widehat{\theta}_j,\delta) \overset{\text{def}}{=} \{\theta\in H_j : (\theta-\widehat{\theta}_j)'\widehat{C}_j(\theta-\widehat{\theta}_j)\le\delta^2\},\qquad \delta\in\mathbb{R}_+.$$

A careful look at the proof of lemma 5.8.1, and consequently of theorems 5.8.1 and 5.8.2 reveals that the assumption that $f_{\widetilde{\theta}}(X) = \mathbb{E}_P(Y\,|\,X)$ was used only once, to prove that $\overline{R}(\theta) = E_P\Bigl\{\bigl[f_\theta(X) - f_{\widetilde{\theta}}(X)\bigr]^2\Bigr\}$. As

$$\mathbb{E}_P\Bigl\{\bigl(Y - \langle\widetilde{\theta},X\rangle\bigr)\langle\theta-\widetilde{\theta},X\rangle\Bigr\} = 0$$

we still have here that

$$\overline{R}(\theta) = \mathbb{E}_P\Bigl[\langle\theta-\widetilde{\theta},X\rangle^2\Bigr],$$

therefore theorems 5.8.1 and 5.8.2 hold true in the present setting. Let us apply more specifically theorem 5.8.2. Instead of using (5.9.2), let us also use (5.8.9), and take advantage of the fact that, as soon as $\Lambda_j(\widehat{\theta}_j,\delta)\subset\Theta$,

$$\mathbb{E}_{\pi[d\theta|\Lambda_j(\widehat{\theta}_j,\delta)]}\bigl(\langle\theta,X\rangle\bigr) = \langle\widehat{\theta}_j,X\rangle,$$

and therefore

$$\overline{R}(\widehat{\theta}_j) \leq \mathbb{E}_{\pi[d\theta | \Lambda_j(\widehat{\theta}_j)]} \bigl[\overline{R}(\theta)\bigr].$$

As

$$\pi\bigl[\Lambda_j(\widehat{\theta}_j, \delta)\bigr] = \mu_j \frac{2\pi^{d_j/2}\delta^{d_j}}{d_j \Gamma(d_j/2) \det(\widehat{C}_j)^{1/2}},$$

we get the following corollary :

**Corollary 5.9.1.** *Let* $A = \frac{\beta}{2g(0)}\left(1 - \frac{\beta g\left(\frac{\beta}{2g(0)}\right)}{2g(0)}\right)$. *With* $P^{\otimes N}$ *probability at least* $1 - \epsilon$, *where*

$$\begin{aligned}\epsilon &= \exp(-\gamma) \sum_{j \in J} \mu_j \int_{\theta \in \Theta_j} \exp\left[-\frac{AN}{2}(\theta - \widetilde{\theta})' C (\theta - \widetilde{\theta})\right] d_j\theta \\ &\leq \exp(-\gamma) \sum_{j \in J} \mu_j \det(C_j)^{-1/2} (2\pi)^{d_j/2} (AN)^{-d_j/2} \exp\left[-\frac{AN}{2}\overline{R}(\widetilde{\theta}_j)\right],\end{aligned}$$

*for any* $j \in J$, *any* $\delta \in \mathbb{R}_+^*$, *as soon as* $\Lambda_j(\widehat{\theta}_j, \delta) \subset \Theta$,

$$\begin{aligned}&\overline{R}(\widehat{\theta}_j) \leq \mathbb{E}_{\pi[d\theta|\Lambda_j(\widehat{\theta}_j,\delta)]}\bigl[\overline{R}(\theta)\bigr] \leq 2\bigl[r(\widehat{\theta}_j) - r(\widetilde{\theta})\bigr] + 2\delta^2 \\ &+ \frac{4g(0)}{\beta N}\left\{\gamma - \log(\mu_j) - \log\left(\frac{2\pi^{d_j/2}}{d_j\Gamma(d_j/2)}\right) + \tfrac{1}{2}\log\bigl[\det(\widehat{C}_j)\bigr] - \frac{d_j}{2}\log(\delta^2)\right\}.\end{aligned}$$

Let us remark that the normalizing constant of the prior distribution $\pi$ can be absorbed in this corollary in the constant $\gamma$. In other words, we can choose for $\pi$ any finite positive measure. We will do so to simplify notations. Let us apply this corollary with

$$\delta_j = \sqrt{\frac{g(0)d_j}{\beta N}}, \tag{5.9.3}$$

$$\mu_j = \nu_j \det(C_j)^{1/2} (2\pi)^{-d_j/2} (AN)^{d_j/2} \exp\left[\frac{AN}{2}\overline{R}(\widetilde{\theta}_j)\right], \qquad j \in J, \tag{5.9.4}$$

$$\gamma = \log(\epsilon^{-1}), \tag{5.9.5}$$

where $\nu_j$, $j \in J$, is a probability distribution on $J$, that is a collection of non negative weights such that $\sum_{j\in J} \nu_j = 1$. With this choice of parameters, we get with probability at least $1-\epsilon$, as soon as $\Lambda_j\left(\widehat{\theta}_j, \sqrt{\frac{g(0)d_j}{\beta N}}\right) \subset \Theta$,

$$\overline{R}(\widehat{\theta}_j) \leq 2\Bigl[r(\widehat{\theta}_j) - r(\widetilde{\theta})\Bigr] - \frac{2Ag(0)}{\beta}\overline{R}(\widetilde{\theta}_j) + \frac{2g(0)d_j}{\beta N}$$

$$+\frac{4g(0)}{\beta N}\left\{\log(\epsilon^{-1})-\log(\nu_j)+\frac{1}{2}\log\left(\frac{\det(\widehat{C}_j)}{\det(C_j)}\right)\right.$$
$$\left.+\log\left[\frac{d_j}{2}\Gamma(d_j/2)\right]-\frac{d_j}{2}\log\left(\frac{Ag(0)d_j}{2\beta}\right)\right\}. \quad (5.9.6)$$

Eventually, we can take into account the following upper bound of the $\Gamma$ function:

$$\log\bigl[\Gamma(d/2)\bigr]\leq\left(\tfrac{d}{2}\right)\log\left(\tfrac{d}{2}\right)-\tfrac{d}{2}-\tfrac{1}{2}\log\left(\tfrac{d}{2}\right)+\tfrac{1}{2}\log(2\pi)+\tfrac{1}{6d}. \quad (5.9.7)$$

It is a consequence of the well known bound (see [81, p. 253])

$$\tfrac{1}{12t}-\tfrac{1}{360t^3}\leq\log\bigl[\Gamma(t)\bigr]-\Bigl\{\left(t-\tfrac{1}{2}\right)\log(t)-t+\tfrac{1}{2}\log(2\pi)\Bigr\}\leq\tfrac{1}{12t},\qquad t\in\mathbb{R}_+.$$

Combining (5.9.6) and (5.9.7) proves the following proposition:

**Proposition 5.9.1.** *Let* $A=\frac{\beta}{2g(0)}\left(1-\frac{\beta g\left(\frac{\beta}{2g(0)}\right)}{2g(0)}\right)$. *With* $P^{\otimes N}$ *probability at least* $1-\epsilon$, *for any* $j\in J$ *such that*

$$\Lambda_j\left(\widehat{\theta}_j,\sqrt{\frac{g(0)d_j}{\beta N}}\right)\subset\Theta,$$

$$\overline{R}(\widehat{\theta}_j)\leq 2\Bigl[r(\widehat{\theta}_j)-r(\widetilde{\theta})\Bigr]-\left[1-\frac{\beta g\left(\frac{\beta}{2g(0)}\right)}{2g(0)}\right]\overline{R}(\widetilde{\theta}_j)$$
$$+\frac{2g(0)}{\beta N}\left\{d_j\log\left(\frac{2\beta}{Ag(0)}\right)+\log(\pi d_j)\right.$$
$$\left.+\frac{1}{3d_j}+\log\left(\frac{\det(\widehat{C}_j)}{\det(C_j)}\right)-2\log(\nu_j)-2\log(\epsilon)\right\}. \quad (5.9.8)$$

*Remark* 5.9.1. The simplest thing to do is to drop the term $-\left[1-\frac{\beta g\left(\frac{\beta}{2g(0)}\right)}{2g(0)}\right]\overline{R}(\widetilde{\theta}_j)$ which is not observable and to optimize the sum of the remaining terms of the right-hand side in $j$. We kept this refinement to show that the factor 2 in front of $r(\widehat{\theta}_j)-r(\widetilde{\theta})$ is partially compensated, at least when $\beta$ is small enough. Indeed, let

$$\chi(j)=2r(\widehat{\theta}_j)+\frac{2g(0)}{\beta N}\left\{d_j\log\left(\frac{2\beta}{Ag(0)}\right)\right.$$

$$+ \log(d_j) + \frac{1}{3d_j} + \log\left(\frac{\det(\widehat{C}_j)}{\det(C_j)}\right) - 2\log(\nu_j)\Bigg\}.$$

Let

$$\widehat{J} = \left\{ j \in J : \Lambda_j\left(\widehat{\theta}_j, \sqrt{\frac{g(0)d_j}{\beta N}}\right) \subset \Theta \right\}.$$

Whenever $\widehat{J} \neq \varnothing$, let let $\widehat{\jmath} \in \widehat{J}$ be such that $\chi(\widehat{\jmath}) = \min_{j \in \widehat{J}} \chi(j)$. Let moreover $\widetilde{\jmath}$ be such that $R(\theta_{\widetilde{\jmath}}) = \min_{j \in J} R(\theta_j)$. Then with probability at least $1 - \epsilon$, whenever $\widetilde{\jmath} \in \widehat{J}$,

$$\overline{R}(\widehat{\theta}_{\widehat{\jmath}}) \leq 2\overline{r}(\widehat{\theta}_{\widehat{\jmath}}) - \left[1 - \frac{\beta g\left(\frac{\beta}{2g(0)}\right)}{2g(0)}\right] \overline{R}(\widetilde{\theta}_{\widetilde{\jmath}}) \\ + \frac{2g(0)}{\beta N}\Bigg\{ d_{\widetilde{\jmath}} \log\left(\frac{2\beta}{Ag(0)}\right) + \log(\pi d_{\widetilde{\jmath}}) \\ + \frac{1}{3d_{\widetilde{\jmath}}} + \log\left(\frac{\det(\widehat{C}_{\widetilde{\jmath}})}{\det(C_{\widetilde{\jmath}})}\right) - 2\log(\nu_{\widetilde{\jmath}}) - 2\log(\epsilon)\Bigg\}.$$

One can then notice that by construction $\overline{r}(\widehat{\theta}_{\widehat{\jmath}}) \leq \overline{r}(\widetilde{\theta}_{\widetilde{\jmath}})$. This leads to a right-hand side where all quantities depending on $\theta$ are evaluated at the fixed parameter value $\widetilde{\theta}_{\widetilde{\jmath}}$. One can then easily prove that for any value of $\theta \in \Theta$

$$\mathbb{E}\Big\{\exp\Big[\lambda\big[\overline{r}(\theta) - \overline{R}(\theta) - \eta\big]\Big]\Big\} \leq -\eta\lambda + \frac{\lambda^2}{2N}\overline{R}(\theta) g\left(\frac{\lambda}{N}\right).$$

Taking

$$\eta = \frac{\beta \overline{R}(\theta)}{2} + \log(\epsilon^{-1}) \frac{g\big[\beta/g(0)\big]}{\beta N}$$
$$\lambda = \frac{\beta N}{g\big[\beta/g(0)\big]},$$

we obtain that with probability at least $1 - \epsilon$,

$$\overline{r}(\theta) \leq \overline{R}(\theta)(1 + \beta/2) - \log(\epsilon)\frac{g\big[\beta/g(0)\big]}{\beta N}.$$

Applying this to the value $\theta = \widetilde{\theta}_{\widetilde{\jmath}}$, we get that with probability at least $1 - 2\epsilon$, as soon as

$$\Lambda_j\left(\widehat{\theta}_{\widetilde{\jmath}}, \sqrt{\frac{g(0)d_{\widetilde{\jmath}}}{\beta N}}\right) \subset \Theta,$$

$$\overline{R}(\widehat{\theta}_{\widetilde{\jmath}}) \leq \overline{R}(\widetilde{\theta}_{\widetilde{\jmath}}) \left[ 1 + \beta + \frac{\beta g\left(\frac{\beta}{2g(0)}\right)}{2g(0)} \right] - \frac{2g\left[\beta/g(0)\right] + 4g(0)}{\beta N} \log(\epsilon)$$
$$+ \frac{2g(0)}{\beta N} \left\{ d_{\widetilde{\jmath}} \log\left(\frac{2\beta}{Ag(0)}\right) + \log(\pi d_{\widetilde{\jmath}}) \right.$$
$$\left. + \frac{1}{3d_{\widetilde{\jmath}}} + \log\left(\frac{\det(\widehat{C}_{\widetilde{\jmath}})}{\det(C_{\widetilde{\jmath}})}\right) - 2\log(\nu_{\widetilde{\jmath}}) \right\}.$$

When $\overline{R}(\widetilde{\theta}_{\widetilde{\jmath}})$ is of order one, it is advisable to take $\beta$ of order $\sqrt{N}$, and we get a convergence at speed $1/\sqrt{N}$, as expected.

*Remark* 5.9.2. Let us remind that $\lim_{N \to +\infty} \widehat{C}_j = C_j$ almost surely when $N$ tends to infinity, and therefore that $\log\left(\frac{\det(\widehat{C}_j)}{\det(C_j)}\right)$ should be most of the time small when $N$ is large when compared to $d_j$.

*Remark* 5.9.3. Let us notice that we get a penalized risk criterion, with a penalty proportional to the dimension over the sample size, with an explicit constant. We avoided a worse speed of order $\frac{\log(N)}{N}$ because we localized the computation of $\epsilon$ around $\widetilde{\theta}$. As opposed to the more classical approach to penalized empirical risk minimization via deviation inequalities (see e.g. [12, 13] for recent developments), we were able to express directly our upper bound on the risk in terms of the penalized empirical risk.

*Remark* 5.9.4. When $\overline{\Theta}$ is a large enough ball when compared with $\|\widetilde{\theta}\|$, the probability that $\Lambda_j(\widehat{\theta}_j, \delta_j) \not\subset \Theta$ tends to zero exponentially fast when $N$ tends to infinity, for any given $j$, and therefore for $\widetilde{\jmath}$. A more detailed non asymptotic treatment of the case $\Lambda_j(\widehat{\theta}_j, \delta_j) \not\subset \Theta$ is given in appendix : by some modification of the definition of $\widehat{\theta}_j$ and of $\Lambda_j(\widehat{\theta}_j, \delta_j)$ we can simply rule out this possibility.

*Remark* 5.9.5. The simpler case of a non random design can be treated in the same way. If $x_1^N \in \mathfrak{X}^N$ are fixed points and $Y_1^N$ are independent variables with distributions $P(Y_i \,|\, X_i = x_i)$, then the risk

$$R(\theta) = \frac{1}{N} \sum_{i=1}^{N} P\left[Y_i \neq f_\theta(X_i) \,|\, X_i = x_i\right]$$

and the empirical risk

$$r(\theta) = \frac{1}{N} \sum_{i=1}^{N} \mathbb{1}\left[Y_i \neq f_\theta(x_i)\right],$$

satisfy equation (5.9.8) with $\widehat{C}_j = C_j$ (and therefore with $\log\left(\frac{\det(\widehat{C}_j)}{\det(C_j)}\right)$ replaced by zero).

*Remark* 5.9.6. Let us eventually point out that the choice of a prior probability distribution made in equation (5.9.4) is definitely not inspired by a Bayesian approach, *since it depends on the sample size N*. In other words, the prior distributions on the parameter space used in this chapter should in no way be interpreted as a stochastic modeling of the parameters by random variables.

## 5.10 Some elementary bounds

**Lemma 5.10.1.**

$$\int_g^{+\infty} \exp\Bigl(-h\sqrt{\xi}\Bigr) d\xi = \frac{2}{h^2}\Bigl(h\sqrt{g}+1\Bigr)\exp\Bigl(-h\sqrt{g}\Bigr).$$

*Proof.* Put $\zeta = \sqrt{\xi}$ and integrate by parts. □

**Lemma 5.10.2.**

$$\int_0^{+\infty} \min\left\{1, p^{-1}\exp\left[-a\left(\frac{\xi}{\sqrt{b(\xi+d)}} - e\right)\right]\right\} d\xi$$
$$\leq \begin{cases} \dfrac{\sqrt{2bd}}{a}\Bigl(1+ae-\log(p)\Bigr) \\ \qquad + \left(\dfrac{2\sqrt{2bd}}{a} + \dfrac{4b}{a^2}\right) \underbrace{\exp\left[a\left(e - \sqrt{\dfrac{d}{2b}}\right) - \log(p)\right]}_{\leq 1}, \\ \qquad\qquad when \ -\log(p) \leq a\left(\sqrt{\dfrac{d}{2b}} - e\right), \\ \dfrac{2b}{a^2}\left[\Bigl(1+ae-\log(p)\Bigr)^2 + 1\right], \ otherwise. \end{cases} \tag{5.10.1}$$

*Proof.* Let

$$I = \int_0^{+\infty} \min\left\{1, p^{-1}\exp\left[-a\left(\frac{\xi}{\sqrt{b(\xi+d)}} - e\right)\right]\right\} d\xi$$
$$s = \frac{1}{a}\Bigl(ae - \log(p)\Bigr).$$

Notice that

$$\frac{\xi}{\sqrt{b(\xi+d)}} \geq \begin{cases} \dfrac{\xi}{\sqrt{2bd}} & \text{when } \xi \leq d, \\ \sqrt{\dfrac{\xi}{2b}}, & \text{when } \xi \geq d. \end{cases}$$

Assume first that $s \geq \sqrt{\dfrac{d}{2b}}$. Then

$$
\begin{aligned}
I &\leq 2bs^2 + p^{-1} \int_{2bs^2}^{+\infty} \exp\left\{ -a \left( \sqrt{\frac{\xi}{2b}} - e \right) \right\} d\xi \\
&\leq 2bs^2 + p^{-1} \frac{4b}{a^2} \left( \frac{a}{\sqrt{2b}} s\sqrt{2b} + 1 \right) \underbrace{\exp(-as)\exp(ae)}_{=p} \\
&\leq 2bs^2 + \frac{4b}{a^2}(1+as) = \frac{2b}{a^2}(a^2s^2 + 2as + 2) \\
&= \frac{2b}{a^2}\Big[ (as+1)^2 + 1 \Big] = \frac{2b}{a^2}\Big[ \big(1 + ae - \log(p)\big)^2 + 1 \Big].
\end{aligned}
$$

Assume now that $s \leq \sqrt{\frac{d}{2b}}$. Then

$$
\begin{aligned}
I &\leq s\sqrt{2bd} + \int_{s\sqrt{2bd}}^{d} p^{-1} \exp\left\{ -a \left( \frac{\xi}{\sqrt{2bd}} - e \right) \right\} d\xi \\
&\quad + \int_{d}^{+\infty} p^{-1} \exp\left\{ -a \left( \sqrt{\frac{\xi}{2b}} - e \right) \right\} d\xi \\
&\leq s\sqrt{2bd} + \frac{\sqrt{2bd}}{a} + p^{-1} \frac{4b}{a^2} \left( \frac{a}{\sqrt{2b}} \sqrt{d} + 1 \right) \exp\left( -\frac{a}{\sqrt{2b}} \sqrt{d} \right) \exp(ae) \\
&\leq \frac{\sqrt{2bd}}{a}\Big( ae - \log(p) + 1 \Big) + \left( \frac{2\sqrt{2bd}}{a} + \frac{4b}{a^2} \right) \exp\left[ a \left( e - \sqrt{\frac{d}{2b}} \right) - \log(p) \right].
\end{aligned}
$$

□

## 5.11 Some refinements about the linear regression case

In this appendix, we will explain how to deal more accurately with the case when $\widehat{\Lambda}_j(\theta_j, \delta_j) \not\subset \Theta$ in section 5.9.2. Let us assume that for any $j \in J$, $\|\widetilde{\theta}_j\| \leq K_1$ (the constant $K_1$ being known to us). Let $K_1 < K_2 < K_3$ be two other constants. Let us put $\overline{\Theta} = \{\theta \in \mathbb{R}^D : \|\theta\| \leq K_3\}$. Let also

$$
\begin{aligned}
\delta_j &= \sqrt{\frac{g(0)d_j}{\beta N}}, \\
\sigma_j &= \frac{\delta_j^2}{(K_3 - K_2)^2}.
\end{aligned}
$$

For any symmetric matrix $C$ of size $d \times d$, and any real number $\sigma$, let $C \vee \sigma$ be obtained from $C$ in the following way : let $C = U^{-1} \operatorname{Diag}(\xi_i : 1 \leq i \leq d) U$ be the diagonal decomposition of $C$ (where $U'U = I$, $\xi_i$, $i = 1, \ldots, d$ is the spectrum of $C$, and $\operatorname{Diag}(\xi_i)$ is the diagonal matrix with diagonal entries $\xi_i$); define $C \vee \sigma$ by

$$C \vee \sigma = U^{-1} \operatorname{Diag}(\max\{\sigma, \xi_i\} : 1 \leq i \leq d) U.$$

Let

$$\widehat{J} = \{ j \in J : \|\widehat{\theta}_j\| \leq K_2 \}.$$

When $j \in \widehat{J}$, let us put

$$\Lambda_j = \Big\{ \theta \in H_j : (\theta - \widehat{\theta}_j)'(\widehat{C}_j \vee \sigma_j)(\theta - \widehat{\theta}_j) \leq \delta_j^2 \Big\}.$$

Then for any $j \in \widehat{J}$, $\Lambda_j \subset \Theta$. Indeed for any $\theta \in \Lambda_j$,

$$\|\theta\| \leq \|\widehat{\theta}_j\| + \sqrt{\frac{(\theta - \widehat{\theta}_j)'(\widehat{C}_j \vee \sigma_j)(\theta - \widehat{\theta}_j)}{\sigma_j}} \leq K_2 + \frac{\delta_j}{\sqrt{\sigma_j}} \leq K_3.$$

Moreover, for any $\theta \in \Lambda_j$,

$$r(\theta) \leq r(\widehat{\theta}_j) + (\theta - \widehat{\theta}_j)'\widehat{C}_j(\theta - \widehat{\theta}_j) \leq r(\widehat{\theta}_j) + \delta_j^2.$$

An easy adaptation of proposition 5.9.1 then shows that

**Proposition 5.11.1.** *With probability at least $1 - \epsilon$, for any $j \in \widehat{J}$,*

$$\begin{aligned} \overline{R}(\widehat{\theta}_j) \leq 2\Big[ r(\widehat{\theta}_j) - r(\widetilde{\theta}) \Big] - \left[ 1 - \frac{\beta g\left(\frac{\beta}{2g(0)}\right)}{2g(0)} \right] \overline{R}(\widetilde{\theta}_j) \\ + \frac{2g(0)}{\beta N} \Bigg\{ d_j \log\left( \frac{2\beta}{A g(0)} \right) + \log(\pi d_j) \\ + \frac{1}{3d_j} + \log\left( \frac{\det(\widehat{C}_j \vee \sigma_j)}{\det(C_j)} \right) - 2\log(\nu_j) - 2\log(\epsilon) \Bigg\}. \qquad (5.11.1) \end{aligned}$$

The practical way to use this proposition is to minimize in $j \in \widehat{J}$ the right-hand side of equation (5.11.1). Therefore we would like to bound the probability of any given (deterministic) value of $j$ *not* to be in $\widehat{J}$, knowing that the value of $j \in J$ we are really interested in is the one minimizing the *expectation* of the right-hand side of equation (5.11.1). Therefore, let us consider in the following discussion a *fixed* value of $j \in J$, and let us show that $\|\widehat{\theta}_j\| \leq K_2$ with high probability.

To this purpose, whenever $\|\widehat{\theta}_j\| \geq K_2$, let us put

$$\begin{aligned} \overline{\theta}_j &= \arg\min_{\theta \in H_j} r(\theta) + K_4 \|\theta\|^2, \\ &= (\widehat{C}_j + K_4 I)^{-1} \frac{1}{N} \sum_{n=1}^N Y_n X_{n|H_j}. \end{aligned}$$

and let us take $K_4$ such that $\|\overline{\theta}_j\| = K_2$ (which is obviously possible because $K_4 \mapsto \|\overline{\theta}_j\|$ is continuous and tends to zero at infinity.) As it is easily seen that

$$\|\overline{\theta}_j\| \leq K_4^{-1} \left( \frac{1}{N} \sum_{n=1}^{N} |Y_n| \right) \sup_{x \in \mathcal{X}} \|x_{|H_j}\|,$$

we can take

$$K_4 \leq K_2^{-1} \left( \frac{1}{N} \sum_{n=1}^{N} |Y_n| \right) \sup_{x \in \mathcal{X}} \|x_{|H_j}\|.$$

Then we can define the random set

$$\Lambda_j = \Big\{ \theta \in H_j : (\theta - \overline{\theta}_j)' \big[ (\widehat{C}_j + K_4 I) \vee \sigma_j \big] (\theta - \overline{\theta}_j) \leq \delta_j^2 \Big\}.$$

It is easily seen as previously that $\Lambda_j \subset \Theta_j$. We can then apply theorem 5.8.2 to the parameter set $\Theta_j$, to obtain that with probability at least $1 - \epsilon$, where the value of $\epsilon$ is indicated in the theorem,

$$\mathbb{E}_{\pi(d\theta|\Lambda_j)} \big[ R(\theta) - R(\widetilde{\theta}_j) \big] \leq 2 \Big\{ \mathbb{E}_{\pi(d\theta|\Lambda_j)} \big[ r(\theta) \big] \Big\} - r(\widetilde{\theta}_j) + \frac{4g(0)}{\beta N} \Big[ \gamma - \log \big[ \pi(\Lambda_j) \big] \Big].$$

Adding $2K_4 \mathbb{E}_{\pi(d\theta|\Lambda_j)} \big[ \|\theta\|^2 \big]$ to both members of this inequality and using the convexity of $\theta \mapsto R(\theta) + 2K_4 \|\theta\|^2$ to write

$$R(\overline{\theta}_j) + 2K_4 \|\overline{\theta}_j\|^2 \leq \mathbb{E}_{\pi(d\theta|\Lambda_j)} \Big[ R(\theta) + 2K_4 \|\theta\|^2 \Big],$$

and the fact that

$$\sup_{\theta \in \Lambda_j} r(\theta) + K_4 \|\theta\|^2 \leq r(\overline{\theta}_j) + K_4 \|\overline{\theta}_j\|^2 + \delta_j^2,$$

we get that with probability at least $1 - \epsilon$, whenever $\|\widehat{\theta}_j\| \geq K_2$,

$$\begin{aligned} R(\overline{\theta}_j) - R(\widetilde{\theta}_j) \leq 2 \Big[ & r(\overline{\theta}_j) - r(\widetilde{\theta}_j) \Big] \\ & + \frac{2g(0)}{\beta N} \left\{ d_j \log \left( \frac{2\beta}{A g(0)} \right) + \log(\pi d_j) \right. \\ & \left. + \frac{1}{3 d_j} + \log \left( \frac{\det \big[ (\widehat{C}_j + K_4 I) \vee \sigma_j \big]}{\det(C_j)} \right) - 2 \log(\epsilon) \right\}. \end{aligned} \quad (5.11.2)$$

But now we can remark that

$$r(\overline{\theta}_j) \leq r(\widetilde{\theta}_j) + K_4 \Big[ \|\widetilde{\theta}_j\|^2 - \|\overline{\theta}_j\|^2 \Big] \leq r(\widetilde{\theta}_j),$$

and that

$$R(\overline{\theta}_j) = R(\widetilde{\theta}_j) + (\overline{\theta}_j - \widetilde{\theta}_j)' C_j (\overline{\theta}_j - \widetilde{\theta}_j).$$

Let $\xi_j$ be the smallest eigenvalue of $C_j$. We have established that with probability $1-\epsilon$, whenever $\|\widehat{\theta}_j\| \geq K_2$,

$$(K_2 - K_1)^2 \xi_j \leq \frac{2g(0)}{\beta N} \biggl\{ d_j \log\left(\frac{2\beta}{Ag(0)}\right) + \log(\pi d_j) + \frac{1}{3d_j} + \log\left(\frac{\det\bigl[(\widehat{C}_j + K_4 I) \vee \sigma_j\bigr]}{\det(C_j)}\right) - 2\log(\epsilon) \biggr\}.$$

Let us notice that the spectrum of $\widehat{C}_j$ is bounded by

$$\sup_{\theta \in H_j : \|\theta\| = 1} \frac{1}{N} \sum_{n=1}^{N} \langle \theta, X_n \rangle^2 \leq R_j^2,$$

where $R_j = \sup_{x \in \mathcal{X}} \|x_{|H_j}\|$. Thus

$$\log\Bigl\{\det\bigl[(\widehat{C}_j + K_4 I) \vee \sigma_j\bigr]\Bigr\} \leq d_j \log\bigl[(R_j^2 + K_4) \vee \sigma_j\bigr].$$

All this put together gives the following bound :

$$P(\|\widehat{\theta}_j\| \geq K_2) \leq \inf_{\epsilon} \Bigl[\epsilon + P^{\otimes N}(\Omega_\epsilon)\Bigr],$$

where $\Omega_\epsilon$ is the event

$$\left(R_j^2 + \frac{R_j}{K_2} \frac{1}{N} \sum_{n=1}^{N} |Y_i|\right) \vee \frac{g(0) d_j}{(K_3 - K_2)^2 \beta N} \geq \frac{Ag(0)}{2\beta} \left(\frac{1}{d_j \pi}\right)^{1/d_j} \det(C_j)^{1/d_j} \epsilon^{2/d_j} \times \exp\left[\frac{N\beta}{2g(0)d_j}(K_2 - K_1)^2 \xi_j - \frac{1}{3d_j^2}\right].$$

It is clear from these last two lines that, as soon as we assume, as we did, that $Y_i$ has an exponential moment, it is possible to find an explicit constant $K_5$, (depending on $j$), such that $P^{\otimes N}\bigl(\|\widehat{\theta}_j\| \geq K_2\bigr) \leq \exp\bigl(-K_5 N\bigr)$.

# 6

# Laplace transform estimates and deviation inequalities

## Introduction

We are going to give "almost Gaussian" finite sample bounds for the log-Laplace transform of some functions of the type $f(X_1, \dots, X_N)$, where the random variables $(X_1, \dots, X_N)$ are assumed to be independent, or to form a Markov chain.

We will use throughout this chapter a normalisation that parallels the classical case of the sum

$$\frac{1}{\sqrt{N}} \sum_{i=1}^{N} X_i$$

of real valued random variables. As is usual in these matters, we will deduce from upper log-Laplace estimates finite sample almost sub-Gaussian deviation inequalities for $f(X_1, \dots, X_N)$. We will also obtain that the central limit theorem holds when $N$ goes to infinity. Although limit laws are not the main subject of this chapter, they will be a guide for using a relevant normalisation of constants.

To make sure that this is feasible, it is necessary to make assumptions not only on the first order partial derivatives (or more generally first order partial finite differences) of $f$, but also on the second order partial derivatives (or more generally on the second order partial finite differences) of $f$. Indeed, the simple example of

$$f(X_1, \dots, X_N) = g\left(\frac{1}{\sqrt{N}} \sum_{i=1}^{N} X_i\right),$$

should immediately convince the reader that Lipschitz conditions are not enough to enforce a Gaussian limit.

A proper normalisation being chosen, we will be interested in expansions of $Z \overset{\text{def}}{=} f(X_1, \dots, X_N) - \mathbb{E}\big(f(X_1, \dots, X_N)\big)$ of the type

$$\log \mathbb{E}\big(\exp(\lambda Z)\big) = \frac{\lambda^2}{2} \operatorname{\mathbb{V}ar} Z + \cdots ,$$

where $\lambda$ is "of order one", and the remaining terms are small when $N$ is large.

Our line of proof will be a combination of the martingale difference sequence approach initiated by Hoeffding [43] and Yurinskii [88] and the statistical mechanics philosophy we already used in [22]. Reviews about the martingale approach to deviation inequalities are also to be found in McDiarmid [55] and Ledoux and Talagrand [48, page 30]. More precisely, we will decompose $Z$ into its martingale difference sequence $Z = \sum_{i=1}^N F_i$ and we will take appropriate partial derivatives of the "free energy" function

$$(\lambda_1, \dots, \lambda_N) \longmapsto \log \mathbb{E} \exp \left( \sum_{i=1}^N \lambda_i F_i \right) .$$

We will consider first the case of *independent* random variables $X_1, \dots, X_N$, ranging in some product of probability spaces $\bigotimes_{i=1}^N (\mathfrak{X}_i, \mathfrak{B}_i, \mu_i)$. In the first section, we will assume that the partial finite differences of $f(x_1, \dots, x_N)$ of order one and two are bounded. In the second section, we will assume that they have exponential moments instead, and in the third section, we will study the case of Markov chains.

## 6.1 Bounded range functionals of independent variables

Let the collection of random variables $(X_1, \dots, X_N)$ take its values in some product of measurable spaces $\bigotimes_{i=1}^N (\mathfrak{X}_i, \mathfrak{B}_i)$. We will assume in the following that $(X_1, \dots, X_N)$ is the canonical process. Let $\mathbb{P} = \bigotimes_{i=1}^N \mu_i$ be a product probability measure on $\bigotimes_{i=1}^N (\mathfrak{X}_i, \mathfrak{B}_i)$.

If $W$ is some (suitably integrable) measurable function of $(X_1, \dots, X_N)$, we will consider the modified probability distribution $\mathbb{P}_W$ defined by

$$\mathbb{P}_W = \frac{\exp(W)}{\mathbb{E} \exp(W)} \mathbb{P},$$

and we will use the notation $\mathbb{E}_W$ for the expectation operator with respect to $\mathbb{P}_W$.

On the other hand, if $\mathfrak{F}$ is some sub-sigma algebra of $\bigotimes_{i=1}^N \mathfrak{B}_i$, then $\mathbb{E}^{\mathfrak{F}}$ will be used as a short notation for the conditional expectation with respect to $\mathfrak{F}$.

As $(\mathbb{E}^{\mathfrak{F}})_W = (\mathbb{E}_W)^{\mathfrak{F}}$, we will simply write $\mathbb{E}_W^{\mathfrak{F}}$ for the conditional expectation with respect to the conditional measure

$$\frac{\exp(W)}{\mathbb{E}\big(\exp(W) \,|\, \mathfrak{F}\big)} \mathbb{P}(\cdot \,|\, \mathfrak{F}).$$

Note that we have $\mathbb{E}_W(\mathbb{E}_W^{\mathfrak{F}}) = \mathbb{E}_W$, whereas $\mathbb{E}(\mathbb{E}_W^{\mathfrak{F}}) \neq \mathbb{E}_W$ and $\mathbb{E}_W\left(E^{\mathfrak{F}}\right) \neq \mathbb{E}_W$ in general.

For each $i = 1, \dots, N$, let $\mathfrak{F}_i$ be the sigma algebra generated by $(X_1, \dots, X_i)$ and let $\mathfrak{G}_i$ be the sigma algebra generated by $(X_1, \dots, X_{i-1}, X_{i+1}, X_N)$.

To stress the role of the independence assumption, we will put the superscript "i" on the equalities and inequalities requiring this assumption.

Let us introduce some notations linked with the martingale differences of a random variable $W$ measurable with respect to $\mathfrak{F}_N$. We will put

$$\begin{aligned} G_i(W) &= W - \mathbb{E}^{\mathfrak{G}_i}(W), \\ F_i(W) &= \mathbb{E}^{\mathfrak{F}_i}(W) - \mathbb{E}^{\mathfrak{F}_{i-1}}(W) \\ &\overset{\text{i}}{=} \mathbb{E}^{\mathfrak{F}_i}(G_i). \end{aligned}$$

As we explained in the introduction, we will study the log-Laplace transform of

$$Z = f(X) - \mathbb{E}\big(f(X)\big).$$

We will decompose $Z$ into the sum of its martingale differences

$$Z = \sum_{i=1}^{N} F_i(Z),$$

and use the short notation $Z_i = \mathbb{E}^{\mathfrak{F}_i}(Z)$.

In this context, it is natural to assume that for any $(x_1, \dots, x_N) \in \prod_{j=1}^{N} \mathfrak{X}_j$, for any $i = 1, \dots, N$, any $y_i \in \mathfrak{X}_i$,

$$f(x_1, \dots, x_N) - f(x_1, \dots, x_{i-1}, y_i, x_{i+1}, \dots, x_N) \leq \frac{B_i}{\sqrt{N}}.$$

The reader should understand that we are interested mainly in the case when the constants $B_i$ are of order one. Although all these scaling factors are not really needed for finite sample bounds, we have found them useful to indicate what should be considered as *small* and what should not.

To ensure that $f(X)$ is almost Gaussian, we need to make also an assumption on its second partial differences. This corresponds to conditions on the second partial derivatives of $f$ in the case when the random variables $(X_1, \dots, X_N)$ take their values in some finite dimensional vector space, are bounded, and $f$ is a smooth function.

Let us put for any $(x_1, \dots, x_N) \in \prod_{j=1}^{N} \mathfrak{X}_j$, and any $y_i \in \mathfrak{X}_i$

$$\Delta_i f(x_1^N, y_i) = f(x_1^N) - f(x_1^{i-1}, y_i, x_{i+1}^N).$$

For a fixed value of $y_i$, $\Delta_i f$ may be seen as a function of $x_1^N$, and when we will write $\Delta_j \Delta_i f(x_1^N, y_i, y_j)$, we will mean that we apply $\Delta_j$ to this function and to $y_j$. (A more accurate but lengthy notation would have been $\Delta_j(\Delta_i f(\cdot, y_i))(x_1^N, y_j)$.)

Let us assume that for some non negative exponent $\zeta$, for any $i \neq j$, for some positive constant $C_{i,j}$, and for any $x_1^N \in \prod_{k=1}^N \mathcal{X}_k$, $y_i \in \mathcal{X}_i$, $y_j \in \mathcal{X}_j$,

$$\Delta_i \Delta_j f(x_1^N, y_j, y_i) \leq \frac{C_{i,j}}{N^{3/2-\zeta}}.$$

Note that $\Delta_i \Delta_j f(x_1^N, y_j, y_i) = \Delta_j \Delta_i f(x_1^N, y_i, y_j)$ and therefore that we can assume that $C_{i,j} = C_{j,i}$. We will moreover assume by convention that $C_{i,i} = 0$.

The normalisation is made so that $\zeta = 0$ corresponds to the case of

$$f(X_1, \dots, X_N) = \sqrt{N} g\left(\frac{X_1}{N}, \dots, \frac{X_N}{N}\right),$$

where $g$ is a smooth function with bounded first and second partial derivatives.

Another class of functions satisfying these hypotheses are the functions of the type

$$f(x_1^N) = \frac{1}{N^{3/2}} \sum_{i=1}^{N} \sum_{j=1}^{N} \psi_{i,j}(x_i, x_j),$$

where $\psi_{i,j}$ are bounded measurable functions. Here $\zeta = 0$,

$$B_i = \frac{2}{N} \sum_{j=1}^{N} \|\psi_{i,j}\|_\infty,$$

and

$$C_{i,j} = 4\|\psi_{i,j}\|_\infty.$$

More generally

$$f(x_1^N) = N^{1/2-r} \sum_{(i_1,\dots,i_r)} \psi_{(i_1,\dots,i_r)}(x_{i_1}, \dots, x_{i_r})$$

also satisfies our hypotheses, when the functions $\psi_{(i_1,\dots,i_r)}$ are bounded.

**Theorem 6.1.1.** *Under the previous hypotheses, for any positive* $\lambda$,

$$\left|\log \mathbb{E} \exp(\lambda f(X)) - \lambda \mathbb{E}(f(X)) - \frac{\lambda^2}{2} \mathbb{V}\mathrm{ar}(f(X))\right|$$
$$\leq \frac{\lambda^3}{N^{1/2-\zeta}} \frac{B'CB}{4N^2} + \frac{\lambda^3}{\sqrt{N}} \sum_{i=1}^{N} \frac{B_i^3}{3N},$$

**Corollary 6.1.1.** *Thus* $f(X)$ *satisfies the following deviation inequalities :*

$$\mathbb{P}\Big(f(X) \geq \mathbb{E}(f(X)) + \epsilon\Big) \leq \exp\left(-\frac{\epsilon^2}{2\left(\mathbb{V}\mathrm{ar}(f(X)) + \dfrac{\eta\epsilon}{\mathbb{V}\mathrm{ar}(f(X))}\right)}\right),$$

$$\mathbb{P}\Big(f(X) \leq \mathbb{E}\big(f(X)\big) - \epsilon\Big) \leq \exp\left( - \frac{\epsilon^2}{2\left(\mathbb{V}\mathrm{ar}\big(f(X)\big) + \dfrac{\eta\epsilon}{\mathbb{V}\mathrm{ar}\big(f(X)\big)}\right)} \right),$$

*with*

$$\eta = \frac{1}{2N^{1/2-\zeta}} \frac{B'CB}{N^2} + \frac{2}{3\sqrt{N}} \sum_{i=1}^{N} \frac{B_i^3}{N}.$$

*Remark* 6.1.1. We obtain that for some constant $K$ depending on $\max_{1 \leq i \leq N} B_i$ and $\max_{1 \leq i,j \leq N} C_{i,j}$ but not on $N$,

$$\left| \log \mathbb{E} \exp\big(\lambda f(X)\big) - \lambda \mathbb{E}\big(f(X)\big) - \frac{\lambda^2}{2} \mathbb{V}\mathrm{ar}\big(f(X)\big) \right| \leq \frac{K\lambda^3}{N^{1/2-\zeta}}.$$

Therefore if we consider a sequence of problems indexed by $N$ such that the constants $B_i$ and $C_{i,j}$ stay bounded, and such that $\mathbb{V}\mathrm{ar}\big(f(X)\big)$ converges, we get a central limit theorem as soon as $\zeta < 1/2$ (with the caveat that the limiting distribution may degenerate to a Dirac mass if the asymptotic variance is 0) : $f(X) - \mathbb{E}\big(f(X)\big)$ converges in distribution to a Gaussian measure.

*Remark* 6.1.2. The critical value $\zeta_c = 1/2$ is sharp, since when

$$f(X) = g\left( \frac{1}{\sqrt{N}} \sum_{i=1}^{N} X_i \right),$$

the central limit theorem obviously does not hold in general, and $\zeta = 1/2$.

*Proof.* After decomposing $Z$ into the sum of its martingale differences, we can view the log-Laplace transform of $Z$ as a function of $N$ equal temperatures :

$$\log \mathbb{E}\big(\exp(\lambda Z)\big) = \log \mathbb{E}\left( \exp\left( \sum_{i=1}^{N} \lambda_i F_i(Z) \right) \right), \qquad \lambda_i = \lambda, i = 1, \dots, N.$$

The first step is to take three derivatives with respect to $\lambda_i$, for $i$ ranging from $N$ backward to 1 :

$$\log \mathbb{E} \exp\left(\lambda Z_i\right) = \log \mathbb{E} \exp\left(\lambda Z_{i-1}\right) + \int_0^{\lambda} \mathbb{E}_{\lambda Z_{i-1} + \alpha F_i(Z)} \left(F_i(Z)\right) d\alpha.$$

Therefore

$$\log \mathbb{E} \exp(\lambda Z) = \sum_{i=1}^{N} \int_0^{\lambda} \mathbb{E}_{\lambda Z_{i-1} + \alpha F_i(Z)} \big(F_i(Z)\big) d\alpha$$

$$= \sum_{i=1}^{N} \int_0^{\lambda} (\lambda - \beta) \mathbb{E}_{\lambda Z_{i-1} + \beta F_i(Z)} \Bigl( \bigl( F_i(Z) - \mathbb{E}_{\lambda Z_{i-1} + \beta F_i(Z)} (F_i(Z)) \bigr)^2 \Bigr) d\beta$$

$$= \sum_{i=1}^{N} \frac{\lambda^2}{2} \mathbb{E}_{\lambda Z_{i-1}} (F_i(Z)^2)$$

$$+ \int_0^{\lambda} \frac{(\lambda - \gamma)^2}{2} \mathbb{E}_{\lambda Z_{i-1} + \gamma F_i(Z)} \Bigl( \bigl( F_i(Z) - \mathbb{E}_{\lambda Z_{i-1} + \gamma F_i(Z)} (F_i(Z)) \bigr)^3 \Bigr) d\gamma.$$

Thus

$$\left| \log \mathbb{E} \exp(\lambda Z) - \frac{\lambda^2}{2} \sum_{i=1}^{N} \mathbb{E}_{\lambda \mathbb{E}^{\mathfrak{F}_{i-1}}(Z)} (F_i(Z)^2) \right|$$

$$\leq \sum_{i=1}^{N} \frac{2B_i}{\sqrt{N}} \int_0^{\lambda} \frac{(\lambda - \gamma)^2}{2} \mathbb{E}_{\lambda Z_{i-1} + \gamma F_i(Z)} \Bigl( F_i(Z) - \mathbb{E}_{\lambda Z_{i-1} + \gamma F_i(Z)} (F_i(Z)) \Bigr)^2 d\gamma$$

$$\leq \sum_{i=1}^{N} \frac{\lambda^3 B_i^3}{3N^{3/2}}.$$

We have proved that

**Lemma 6.1.1.**

$$\left| \log \mathbb{E} \exp(\lambda Z) - \frac{\lambda^2}{2} \sum_{i=1}^{N} \mathbb{E}_{\lambda \mathbb{E}^{\mathfrak{F}_{i-1}}(Z)} (F_i(Z)^2) \right| \leq \sum_{i=1}^{N} \frac{\lambda^3 B_i^3}{3N^{3/2}}$$

We have now to approximate $\mathbb{E}_{\lambda Z_{i-1}} \Bigl( F_i(Z)^2 \Bigr)$ by $\mathbb{E} \Bigl( F_i(Z)^2 \Bigr)$, in order to get the variance of $Z$, that can be written as

$$\sum_{i=1}^{N} \mathbb{E} \Bigl( F_i(Z)^2 \Bigr).$$

Let us put for short $V_i = F_i(Z)^2$ and let us introduce its martingale differences:

$$\mathbb{E}_{\lambda Z_{i-1}} (V_i - \mathbb{E}(V_i)) = \sum_{j=1}^{i-1} \mathbb{E}_{\lambda Z_{i-1}} (F_j(V_i)).$$

To deal with the $j$th term of this sum, we introduce the conditional expectation with respect to $\mathfrak{G}_j$ :

$$\mathbb{E}_{\lambda Z_{i-1}}^{\mathfrak{G}_j} \bigl( F_j(V_i) \bigr) = \mathbb{E}_{\lambda G_j(Z_{i-1})}^{\mathfrak{G}_j} \bigl( F_j(V_i) \bigr)$$

$$= \underbrace{\mathbb{E}^{\mathfrak{G}_j} \bigl( F_j(V_i) \bigr)}_{\overset{\text{i}}{=} 0}$$

$$+ \int_0^\lambda \mathbb{E}^{\mathfrak{G}_j}_{\alpha G_j(Z_{i-1})}\Big(F_j(V_i)\big(G_j(Z_{i-1}) - \mathbb{E}^{\mathfrak{G}_j}_{\alpha G_j(Z_{i-1})} G_j(Z_{i-1})\big)\Big) d\alpha.$$

As a consequence

$$\Big|\mathbb{E}^{\mathfrak{G}_j}_{\lambda Z_{i-1}}(F_j(V_i))\Big| \leq \int_0^\lambda \Big\{ \mathbb{E}^{\mathfrak{G}_j}_{\alpha G_j(Z_{i-1})}\left(F_j(V_i)^2\right) \times \mathbb{E}^{\mathfrak{G}_j}_{\alpha G_j(Z_{i-1})}\Big[\Big(G_j(Z_{i-1}) - \mathbb{E}^{\mathfrak{G}_j}_{\alpha G_j(Z_{i-1})}(G_j(Z_{i-1}))\Big)^2\Big]\Big\}^{1/2} d\alpha$$

Now we can observe that

$$G_j(Z_{i-1}) \stackrel{i}{=} \mathbb{E}^{\mathfrak{F}_{i-1}} G_j(Z),$$

and therefore that its conditional range is upper bounded by

$$\operatorname{ess\,sup}\big(G_j(Z_{i-1}) \,|\, \mathfrak{G}_j\big) - \operatorname{ess\,inf}\big(G_j(Z_{i-1}) \,|\, \mathfrak{G}_j\big) \leq \frac{B_j}{\sqrt{N}}.$$

This implies that its variance is bounded by $\dfrac{B_j^2}{4N}$.

Moreover, let us consider on some enlarged probability space two independent random variables $X_i'$ and $X_j'$, such that $(X_1, \ldots, X_N, X_j', X_i')$ is distributed according to $\bigotimes_{k=1}^N \mu_k \otimes \mu_j \otimes \mu_i$. We have that

$$F_j(F_i(Z)^2) = F_j\left(\left(\mathbb{E}^{\mathfrak{F}_i}\left(\Delta_i f(X_1^N, X_i')\right)\right)^2\right).$$

Moreover for any function $h(X_1^N)$,

$$\begin{aligned} \left|F_j(h(X)^2)\right| &= \left|\mathbb{E}^{\mathfrak{F}_j} \Delta_j h^2(X_1^N, X_j')\right| \\ &= \left|\mathbb{E}^{\mathfrak{F}_j}\Big(\big(h(X_1^N) + h(X_1^{j-1}, X_j', X_{j+1}^N)\big)\Delta_j h(X_1^N, X_j')\Big)\right| \\ &\leq 2 \operatorname{ess\,sup}|h(X)|\ \mathbb{E}^{\mathfrak{F}_j}\left|\Delta_j h(X_1^N, X_j')\right|. \end{aligned}$$

Applying this to $h(X) = \mathbb{E}^{\mathfrak{F}_i}\big(\Delta_i f(X_1^N, X_i')\big)$, we get that

$$\begin{aligned} \left|F_j\big(F_i(Z)^2\big)\right| &\leq 2\frac{B_i}{\sqrt{N}} \mathbb{E}^{\mathfrak{F}_j}\left|\mathbb{E}^{\mathfrak{F}_i} \Delta_j \Delta_i f(X_1^N, X_i', X_j')\right| \\ &\leq \frac{2B_i C_{i,j}}{N^{2-\zeta}}. \end{aligned}$$

Therefore

$$\left|\mathbb{E}_{\lambda Z_{i-1}} F_j(V_i)\right| \leq \lambda \frac{B_i C_{i,j} B_j}{N^{5/2-\zeta}}.$$

This, combined with lemma 6.1.1, ends the proof of theorem 6.1.1. The derivation of its corollary is standard : it is obtained by taking

$$\lambda = \frac{\epsilon}{V + \eta\epsilon/V},$$

and by applying successively the theorem to $f$ and $-f$. □

## 6.2 Extension to unbounded ranges

The boundedness assumption of the finite differences of $f$ can be relaxed to exponential moment assumptions.

Indeed, in the proof of lemma 6.1.1 we can bound

$$\left|\int_0^\lambda \frac{(\lambda-\gamma)^2}{2}\mathbb{E}_{\lambda Z_{i-1}+\gamma F_i(Z)}\Big(F_i(Z) - \mathbb{E}_{\lambda Z_{i-1}+\gamma F_i(Z)}(F_i(Z))\Big)^3 d\gamma\right|$$
$$\leq \int_0^\lambda \frac{(\lambda-\gamma)^2}{2}\mathbb{E}_{\lambda Z_{i-1}+\gamma F_i(Z)}\Big(\big|F_i(Z)\big| + \mathbb{E}_{\lambda Z_{i-1}+\gamma F_i(Z)}\big|F_i(Z)\big|\Big)^3 d\gamma$$
$$\leq \int_0^\lambda \frac{(\lambda-\gamma)^2}{2}\mathbb{E}_{\lambda Z_{i-1}+\gamma F_i(Z)}\Big(4\big|F_i(Z)\big|^3\Big) + 4\Big(\mathbb{E}_{\lambda Z_{i-1}+\gamma F_i(Z)}\big|F_i(Z)\big|\Big)^3 d\gamma$$
$$\leq 4\int_0^\lambda (\lambda-\gamma)^2\mathbb{E}_{\lambda Z_{i-1}+\gamma F_i(Z)}\Big(\big|F_i(Z)\big|^3\Big)d\gamma$$
$$\leq \mathbb{E}_{\lambda Z_{i-1}}\left(4\int_0^\lambda(\lambda-\gamma)^2\mathbb{E}^{\mathfrak{F}_{i-1}}\Big(\exp\big(\gamma\big|F_i(Z)\big|\big)\big|F_i(Z)\big|^3\Big)d\gamma\right)$$
$$\leq 4\operatorname{ess\,sup}\mathbb{E}^{\mathfrak{F}_{i-1}}\left(\int_0^\lambda(\lambda-\gamma)^2\big|F_i(Z)\big|^3\exp\big(\gamma\big|F_i(Z)\big|\big)d\gamma\right).$$

Thus, considering the function

$$\Phi(x) = \exp(x) - 1 - x - \frac{x^2}{2} = \int_0^x \frac{(x-y)^2}{2}\exp(y)dy, \tag{6.2.1}$$

we obtain

**Lemma 6.2.1.**

$$\left|\log\mathbb{E}\exp(\lambda Z) - \frac{\lambda^2}{2}\sum_{i=1}^N \mathbb{E}_{\lambda Z_{i-1}}(F_i(Z)^2)\right| \leq 8\sum_{i=1}^N \operatorname{ess\,sup}\mathbb{E}^{\mathfrak{F}_{i-1}}\Phi(\lambda|F_i(Z)|).$$

Moreover

$$\left|\int_0^\lambda \mathbb{E}^{\mathfrak{G}_j}_{\gamma G_j(Z_{i-1})}\Big(\big(G_j(Z_{i-1}) - \mathbb{E}^{\mathfrak{G}_j}_{\gamma G_j(Z_{i-1})}G_j(Z_{i-1})\big)F_j(F_i(Z)^2)\Big)d\gamma\right|$$

$$\leq \int_0^\lambda \mathbb{E}^{\mathfrak{G}_j}_{\gamma G_j(Z_{i-1})}\Big(\big|G_j(Z_{i-1})F_j\big(F_i(Z)^2\big)\big|\Big)d\gamma$$
$$+ \int_0^\lambda \mathbb{E}^{\mathfrak{G}_j}_{\gamma G_j(Z_{i-1})}\Big(\big|G_j(Z_{i-1})\big|\Big)\mathbb{E}^{\mathfrak{G}_j}_{\gamma G_j(Z_{i-1})}\Big(\big|F_j\big(F_i(Z)^2\big)\big|\Big)d\gamma.$$

But

$$\mathbb{E}^{\mathfrak{G}_j}_{\gamma G_j(Z_{i-1})}\Big(\big|G_j(Z_{i-1})F_j\big(F_i(Z)^2\big)\big|\Big) \leq \mathbb{E}^{\mathfrak{G}_j}\Big(\big|G_j(Z_{i-1})\exp\big(\gamma G_j(Z_{i-1})\big)F_j\big(F_i(Z)^2\big)\big|\Big)$$

and

$$\Big|G_j(Z_{i-1})\exp\big(\gamma G_j(Z_{i-1})\big)\Big| \leq \mathbb{E}^{\mathfrak{F}_{i-1}}\big(|G_j(Z)|\big)\exp\Big(\gamma\mathbb{E}^{\mathfrak{F}_{i-1}}\big(|G_j(Z)|\big)\Big)$$
$$\leq \mathbb{E}^{\mathfrak{F}_{i-1}}\Big(|G_j(Z)|\exp\big(\gamma|G_j(Z)|\big)\Big),$$

(as for any integrable random variable $h$,

$$\mathbb{E}(h) \leq \mathbb{E}_{\gamma h}(h),$$

that is

$$\mathbb{E}(h)\mathbb{E}\big(\exp(\gamma h)\big) \leq \mathbb{E}\big(h\exp(\gamma h)\big),$$

because $\frac{\partial}{\partial\gamma}\mathbb{E}_{\gamma h}(h) = \mathbb{E}_{\gamma h}\big(h - \mathbb{E}_{\gamma h}(h)\big)^2 \geq 0$.) Thus

$$\mathbb{E}^{\mathfrak{G}_j}_{\gamma G_j(Z_{i-1})}\Big(\big|G_j(Z_{i-1})F_j\big(F_i(Z)^2\big)\big|\Big)$$
$$\leq \mathbb{E}^{\mathfrak{F}_{i-1}}\mathbb{E}^{\mathfrak{G}_j}\Big(|G_j(Z)|\exp\big(\gamma|G_j(Z)|\big)\Big|F_j\big(F_i(Z)^2\big)\Big|\Big)$$
$$\leq \operatorname{ess\,sup}\sqrt{\mathbb{E}^{\mathfrak{G}_j}\Big(G_j(Z)^2\exp\big(2\gamma|G_j(Z)|\big)\Big)}$$
$$\times\sqrt{\mathbb{E}^{\mathfrak{G}_j}\Big(F_j\big(F_i(Z)^2\big)^2\Big)}.$$

In the same way

$$\mathbb{E}^{\mathfrak{G}_j}_{\gamma G_j(Z_{i-1})}\big(|G_j(Z_{i-1})|\big)\mathbb{E}^{\mathfrak{G}_j}_{\gamma G_j(Z_{i-1})}\Big(\big|F_j\big(F_i(Z)^2\big)\big|\Big)$$
$$\leq \operatorname{ess\,sup}\mathbb{E}^{\mathfrak{G}_j}\Big(|G_j(Z)|\exp\big(\gamma|G_j(Z)|\big)\Big)$$
$$\times \operatorname{ess\,sup}\sqrt{\mathbb{E}^{\mathfrak{G}_j}\Big(\exp\big(2\gamma|G_j(Z)|\big)\Big)}\sqrt{\mathbb{E}^{\mathfrak{G}_j}\Big(F_j\big(F_i(Z)^2\big)^2\Big)}.$$

Thus

$$\left| \int_0^\lambda \mathbb{E}^{\mathfrak{G}_j}_{\gamma G_j(Z_{i-1})} \Big( \big(G_j(Z_{i-1}) - \mathbb{E}^{\mathfrak{G}_j}_{\gamma G_j(Z_{i-1})} G_j(Z_{i-1})\big) F_j\big(F_i(Z)^2\big) \Big) d\gamma \right|$$

$$\leq \lambda \operatorname{ess\,sup} \sqrt{\mathbb{E}^{\mathfrak{G}_j} \Big( F_j\big(F_i(Z)^2\big)^2 \Big)}$$

$$\times \operatorname{ess\,sup} \sqrt{\mathbb{E}^{\mathfrak{G}_j} \Big( G_j(Z)^2 \exp\big(2\lambda |G_j(Z)|\big) \Big)}$$

$$\times \left( 1 + \operatorname{ess\,sup} \sqrt{\mathbb{E}^{\mathfrak{G}_j} \Big( \exp\big(2\lambda |G_j(Z)|\big) \Big)} \right).$$

Moreover, viewing $F_i(Z) = h(X)$ as a function of $X$, and putting

$$\overset{j}{X'} = (X_1, \dots, X_{j-1}, X'_j, X_{j+1}, \dots, X_N),$$

where $X'_j$ is an independent copy of $X_j$, we see that

$$\sqrt{\mathbb{E}^{\mathfrak{G}_j} \Big( F_j\big(F_i(Z)^2\big)^2 \Big)} = \sqrt{\mathbb{E}^{\mathfrak{G}_j} \Big( \Big( \mathbb{E}^{\mathfrak{F}_j} \big( h(X)^2 - h(\overset{j}{X'})^2 \big) \Big)^2 \Big)}$$

$$\leq \sqrt{\mathbb{E}^{\mathfrak{G}_j} \Big( \mathbb{E}^{\mathfrak{F}_j} \Big( \big(h(X) + h(\overset{j}{X'})\big)^2 \Big) \mathbb{E}^{\mathfrak{F}_j} \Big( \big(h(X) - h(\overset{j}{X'})\big)^2 \Big) \Big)}$$

$$\leq \Big( \mathbb{E}^{\mathfrak{F}_{j-1}} \Big( \big(h(X) + h(\overset{j}{X'})\big)^4 \Big) \Big)^{1/4} \Big( \mathbb{E}^{\mathfrak{F}_{j-1}} \Big( \big(h(X) - h(\overset{j}{X'})\big)^4 \Big) \Big)^{1/4}$$

$$\leq 2 \Big( \mathbb{E}^{\mathfrak{F}_{j-1}} \big( h(X)^4 \big) \Big)^{1/4} \Big( \mathbb{E}^{\mathfrak{F}_{j-1}} \Big( \big(h(X) - h(\overset{j}{X'})\big)^4 \Big) \Big)^{1/4}.$$

To proceed, let us consider an independent copy $X'_i$ of $X_i$, and let us remark that

$$h(X) = \mathbb{E}\big( \Delta_i f(X, X'_i) \,|\, X_1^i, X'_j \big)$$

$$h(\overset{j}{X'}) = \mathbb{E}\big( \Delta_i f(\overset{j}{X'}, X'_i) \,|\, X_1^i, X'_j \big).$$

This shows that

$$\sqrt{\mathbb{E}^{\mathfrak{G}_j} \Big( F_j\big(F_i(Z)^2\big)^2 \Big)}$$

$$\leq 2 \Big( \mathbb{E}^{\mathfrak{F}_{j-1}} \Big( \big(\Delta_i f(X, X'_i)\big)^4 \Big) \Big)^{1/4} \Big( \mathbb{E}^{\mathfrak{F}_{j-1}} \Big( \big(\Delta_j \Delta_i f(X, X'_i, X'_j)\big)^4 \Big) \Big)^{1/4}.$$

We have proved that

$$\left| \log\Big( \mathbb{E} \exp\big(\lambda f(X)\big) \Big) - \lambda \mathbb{E}\big(f(X)\big) - \frac{\lambda^2}{2} \operatorname{Var}\big(f(X)\big) \right|$$

$$
\begin{aligned}
&\leq \sum_{i=1}^{N} 8 \operatorname{ess\,sup} \mathbb{E}^{\mathfrak{F}_{i-1}} \Phi(\lambda |F_i(Z)|) \\
&+ \lambda^3 \sum_{i=1}^{N} \sum_{j=1}^{i-1} \Biggl( \operatorname{ess\,sup} \Bigl( \mathbb{E}^{\mathfrak{F}_{j-1}} \bigl( (\Delta_i f(X, X_i'))^4 \bigr) \Bigr)^{1/4} \\
&\times \Bigl( \mathbb{E}^{\mathfrak{F}_{j-1}} \bigl( (\Delta_j \Delta_i f(X, X_i', X_j'))^4 \bigr) \Bigr)^{1/4} \\
&\times \operatorname{ess\,sup} \sqrt{\mathbb{E}^{\mathfrak{G}_j} \bigl( G_j(Z)^2 \exp(2\lambda |G_j(Z)|) \bigr)} \\
&\times \Bigl( 1 + \operatorname{ess\,sup} \sqrt{\mathbb{E}^{\mathfrak{G}_j} \bigl( \exp(2\lambda |G_j(Z)|) \bigr)} \Bigr) \Biggr).
\end{aligned}
$$

Using the fact that

$$
\begin{aligned}
G_j(Z) &= \mathbb{E}^{\mathfrak{F}_N} \bigl( f(X) - f(\overset{j}{X'}) \bigr) \\
F_i(Z) &= \mathbb{E}^{\mathfrak{F}_i} \bigl( f(X) - f(\overset{i}{X'}) \bigr)
\end{aligned}
$$

and remarking that $\Phi$ is convex on the positive real axis and that for any positive random variable $h$, any exponent $\alpha \geq 1$,

$$
\bigl(\mathbb{E}(h)\bigr)^{\alpha} \exp(\lambda \mathbb{E}(h)) \leq \mathbb{E}\Bigl( h^{\alpha} \exp(\lambda h) \Bigr),
$$

— because $\bigl(\mathbb{E}(h)\bigr)^{\alpha} \leq \bigl(\mathbb{E}_{\lambda h}(h)\bigr)^{\alpha} \leq \mathbb{E}_{\lambda h}\bigl(h^{\alpha}\bigr)$ — we can derive from the previous inequality a theorem involving only the finite differences of $f$ :

**Theorem 6.2.1.** *As soon as $(X_1, \dots, X_N)$ are independent random variables and all the expressions below are integrable, we have the following bound on the third term of the expension of the* log-*Laplace transform of $f(X_1, \dots, X_N)$:*

$$
\begin{aligned}
&\left| \log\Bigl( \mathbb{E} \exp(\lambda f(X)) \Bigr) - \lambda \mathbb{E}\bigl(f(X)\bigr) - \frac{\lambda^2}{2} \mathbb{V}\mathrm{ar}\bigl(f(X)\bigr) \right| \\
&\leq \sum_{i=1}^{N} 8 \operatorname{ess\,sup} \mathbb{E}^{\mathfrak{F}_{i-1}} \Phi(\lambda |\Delta_i f(X, X_i')|) \\
&+ \lambda^3 \sum_{i=1}^{N} \sum_{j=1}^{i-1} \Biggl( \operatorname{ess\,sup} \Bigl( \mathbb{E}^{\mathfrak{F}_{j-1}} \bigl( (\Delta_i f(X, X_i'))^4 \bigr) \Bigr)^{1/4} \\
&\times \Bigl( \mathbb{E}^{\mathfrak{F}_{j-1}} \bigl( (\Delta_j \Delta_i f(X, X_i', X_j'))^4 \bigr) \Bigr)^{1/4} \\
&\times \operatorname{ess\,sup} \sqrt{\mathbb{E}^{\mathfrak{G}_j} \Bigl( \bigl(\Delta_j f(X, X_j')\bigr)^2 \exp(2\lambda |\Delta_j f(X, X_j')|) \Bigr)}
\end{aligned}
$$

$$\times \left(1 + \operatorname{ess\,sup} \sqrt{\mathbb{E}^{\mathfrak{G}_j}\left(\exp(2\lambda|\Delta_j f(X, X'_j)|)\right)}\right)\Bigg),$$

*where $\Phi$ is defined by* (6.2.1)*, $X'_i$ and $X'_j$ are independent copies of $X_i$ and $X_j$ (i.e. where $(X_1, \dots, X_N, X'_j, X'_i)$ is distributed according to $\left(\bigotimes_{k=1}^N \mu_k\right) \otimes \mu_j \otimes \mu_i$).*

## 6.3 Generalization to Markov chains

We will study here the case when $(X_1, \dots, X_N)$ is a Markov chain. The assumptions on $f$ will be the same as in the first section. Therefore we will assume throughout this section that

$$\Delta_i f(x_1^N, y_i) \leq \frac{B_i}{\sqrt{N}},$$

and that

$$\Delta_j \Delta_i f(x_1^N, y_i, y_j) \leq \frac{C_{i,j}}{N^{3/2-\zeta}}.$$

When the random variables $(X_1, \dots, X_N)$ are dependent, we have to modify the definition of the operators $G_i(Z)$ and of the sigma algebras $\mathfrak{G}_i$. Indeed to generalise the first part of the proof we would like to have the identity

$$\mathbb{E}^{\mathfrak{F}_i}(G_i(Z)) = F_i,$$

where $G_i$ is "as small as possible". This identity does not hold in the general case with the definition we had for $G_i$, so we will have to change it.

To generalise the second part of the proof, we need to consider a new definition of the sigma algebra $\mathfrak{G}_j$ for which

$$\mathbb{E}^{\mathfrak{G}_j}(F_j(W)) = 0,$$

and $Z - \mathbb{E}^{\mathfrak{G}_j}(Z)$ is as small as possible.

We propose a solution here where the two objects $G_i$ and $\mathfrak{G}_j$ are built with the help of coupled processes.

We will define auxiliary random variables that will be coupled with the process $(X_1, \dots, X_N)$ in a suitable way. For this, we will enlarge the probability space : instead of working on the canonical space $(\prod_{i=1}^N \mathcal{X}_i, \bigotimes_{i=1}^N \mathfrak{B}_i)$, we will work on some enlarged probability space $(\Omega, \mathfrak{B})$, where we will jointly define the process $(X_1, \dots, X_N)$, and $N$ other processes $\{\overset{i}{Y}{}_{j=1}^N\, ; \, i = 1, \dots, N\}$ that will be useful for the construction of the operators $\{G_i\, ; \, i = 1, \dots, N\}$. In the following the symbol $\mathbb{E}$ will stand for the expectation on $\Omega$.

The basic construction of coupled processes we will need is the following : we consider, on some augmented probability space $\Omega$, $N+1$ stochastic processes $(X_1, \dots, X_N)$ and $\{(\overset{i}{Y}_1, \dots, \overset{i}{Y}_N)\, ; \, i = 1, \dots, N\}$ satisfying the following properties :

- The distribution of each $\overset{i}{Y}$ is equal to the distribution of $X$.
- Almost surely $\overset{i}{Y}{}_1^{i-1} = X_1^{i-1}$.
- Given $X$, the $N$ processes $\{\overset{i}{Y}\,;\, i = 1, \dots, N\}$ are independent.
- Given $X_1^{i-1}$, $\overset{i}{Y}{}_i^N$ is independent from $X_i$ (but not from $X_{i+1}^N$, the interesting thing will be on the contrary to have a maximal coupling between $\overset{i}{Y}{}_{i+1}^N$ and $X_{i+1}^N$).

The general method to build such processes is the following :

- **Choice of $\Omega$ :** Take for $\Omega$ the canonical space of $\Bigl(X_1^N, (\overset{i}{Y}{}_{j=1}^N)_{i=1}^N\Bigr)$, that is $\Bigl(\bigotimes_{i=1}^N(\mathfrak{X}_i, \mathfrak{B}_i)\Bigr)^{\otimes(N+1)}$. For any random variable $W$ defined on $\Omega$, we will use the notation $\mathbb{P}(dW)$ to denote the distribution of $W$. We will assume without further notice that all the conditionnal distributions we need exist and have regular versions. This will always be the case when we deal with Polish spaces $(\mathfrak{X}_i, \mathfrak{B}_i)$.
- **Construction of the distribution of the pair $(X, \overset{i}{Y})$ :** The distribution of $\mathbb{P}(dX_1^{i-1})$ is the original one. We define $\mathbb{P}\bigl(d\overset{i}{Y}{}_{j=1}^{i-1} \,|\, X_1^{i-1}\bigr)$ by letting $\overset{i}{Y}{}_1^{i-1} = X_1^{i-1}$, a.s..
  For $j$ ranging from $i$ to $N$, we build the conditional distribution
  $$\mathbb{P}\bigl(dX_j, d\overset{i}{Y}_j \,|\, X_1^{j-1}, \overset{i}{Y}{}_1^{j-1}\bigr)$$
  by putting
  $$\mathbb{P}\bigl(dX_i, d\overset{i}{Y}_i \,|\, X_1^{i-1}, \overset{i}{Y}{}_1^{i-1}\bigr) = \mathbb{P}\bigl(dX_i \,|\, X_1^{i-1}\bigr) \otimes \mathbb{P}\bigl(d\overset{i}{Y}_i \,|\, \overset{i}{Y}{}_1^{i-1}\bigr),$$
  and for any $j > i$, we choose for
  $$\mathbb{P}(dX_j, d\overset{i}{Y}_j \,|\, X_1^{j-1}, \overset{i}{Y}{}_1^{j-1})$$
  some maximally coupled distribution with marginals
  $$\begin{cases} \mathbb{P}\bigl(dX_j \,|\, X_1^{j-1}, \overset{i}{Y}{}_1^{j-1}\bigr) = \mathbb{P}\bigl(dX_j \,|\, X_1^{j-1}\bigr) \\ \mathbb{P}\bigl(d\overset{i}{Y}_j \,|\, X_1^{j-1}, \overset{i}{Y}{}_1^{j-1}\bigr) = \mathbb{P}\bigl(d\overset{i}{Y}_j \,|\, \overset{i}{Y}{}_1^{j-1}\bigr), \end{cases}$$
  where the second marginal is defined by the requirement that $\mathbb{P}(d\overset{i}{Y})$ be the same as $\mathbb{P}(dX)$.

- **Last step of the construction :** Once we have built the distribution of each couple of processes $\mathbb{P}(dX, d\overset{i}{Y})$, separately for each $i$, we build the joint distribution of $(X, \overset{i}{Y}, i = 1, \dots, N)$ on its canonical space. For the time being, we will not really use this joint distribution, but it is more simple to deal with one probability space $\Omega$ than with $N$ probability spaces $\Omega_i$, so let us say that we build $\mathbb{P}(dX)$ first and then let

$$\mathbb{P}(d\overset{i}{Y}{}_{i=1}^{N} \,|\, X) = \bigotimes_{i=1}^{N} \mathbb{P}(d\overset{i}{Y} \,|\, X).$$

We will need to refine this in the last part of the proof.

It is immediate to see from this construction that

$$\begin{aligned}\mathbb{P}\big(dX_i, d\overset{i}{Y}{}_i^N \,|\, X_1^{i-1}\big) &= \mathbb{P}(dX_i \,|\, X_1^{i-1}) \otimes \mathbb{P}(d\overset{i}{Y}{}_i \,|\, X_1^{i-1}) \otimes \mathbb{P}\big(d\overset{i}{Y}{}_{i+1}^N \,|\, X_1^i, \overset{i}{Y}{}_i\big)\\ &= \mathbb{P}(dX_i \,|\, X_1^{i-1}) \otimes \mathbb{P}(d\overset{i}{Y}{}_i \,|\, X_1^{i-1}) \otimes \mathbb{P}\big(d\overset{i}{Y}{}_{i+1}^N \,|\, \overset{i}{Y}{}_1^i\big)\\ &= \mathbb{P}(dX_i \,|\, X_1^{i-1}) \otimes \mathbb{P}\big(d\overset{i}{Y}{}_i^N \,|\, \overset{i}{Y}{}_1^{i-1}\big).\end{aligned}$$

This proves that conditionally to $(X_1, \dots, X_{i-1})$, the random variable $X_i$ is independent from the sigma algebra generated by $(\overset{i}{Y}{}_i, \dots, \overset{i}{Y}{}_N)$.

*Remark* 6.3.1. We have also exactly in the same way

$$\mathbb{P}\big(dX_i^N, d\overset{i}{Y}{}_i \,|\, X_1^{i-1}\big) = \mathbb{P}\big(dX_i^N \,|\, X_1^{i-1}\big) \otimes \mathbb{P}\big(\overset{i}{Y}{}_i \,|\, \overset{i}{Y}{}_1^{i-1}\big).$$

As in the previous sections, $\mathfrak{F}_i$ will be the sigma algebra generated by $(X_1, \dots, X_i)$, and we will put

$$Z(X) = f(X) - \mathbb{E}\big(f(X)\big).$$

For any bounded measurable function $h(X)$ we will define

$$\begin{aligned}G_i(h(X)) &= h(X) - \mathbb{E}^{\mathfrak{F}_N}\Big(h(\overset{i}{Y})\Big),\\ F_i(h(X)) &= \mathbb{E}^{\mathfrak{F}_i}(h(X)) - \mathbb{E}^{\mathfrak{F}_{i-1}}(h(X))\\ &= \mathbb{E}^{\mathfrak{F}_i}\Big(G_i\big(h(X)\big)\Big).\end{aligned}$$

The last line holds because

$$\begin{aligned}\mathbb{E}^{\mathfrak{F}_i}\mathbb{E}^{\mathfrak{F}_N}\big(h(\overset{i}{Y})\big) &= \mathbb{E}\Big(h(\overset{i}{Y}) \,|\, X_1, \dots, X_i\Big)\\ &= \mathbb{E}\Big(h(\overset{i}{Y}) \,|\, X_1, \dots, X_{i-1}\Big)\\ &= \mathbb{E}^{\mathfrak{F}_{i-1}}\big(h(X)\big).\end{aligned}$$

*Remark* 6.3.2. In the case when the random variables $X_1, \ldots, X_N$ are independent, we can take for $\overset{i}{Y}_i$ an independent copy of $X_i$ and we can put $\overset{i}{Y}{}_{i+1}^N = X_{i+1}^N$ a.s.. With this choice, the definition of $G_i$ given here coincides with that given in the first section.

We have

$$|G_i(Z)| = \left| \mathbb{E}^{\mathfrak{F}_N} \left( f(X) - f(\overset{i}{Y}) \right) \right| \leq \mathbb{E}^{\mathfrak{F}_N} \left( \sum_{j=1}^N \mathbf{1}(X_j \neq \overset{i}{Y}_j) \frac{B_j}{\sqrt{N}} \right).$$

Consequently

$$|F_i(Z)| \leq \operatorname{ess\,sup} \ \mathbb{E}^{\mathfrak{F}_i} \left( \sum_{j=1}^N \mathbf{1}(X_j \neq \overset{i}{Y}_j) \frac{B_j}{\sqrt{N}} \right).$$

Let us introduce the notation

$$\tilde{B}_i = \operatorname{ess\,sup} \ \mathbb{E} \left( \sum_{j=1}^N \mathbf{1}(X_j \neq \overset{i}{Y}_j) B_j \,|\, \mathfrak{F}_i, \overset{i}{Y}_i \right). \tag{6.3.1}$$

We have established that

$$|F_i(Z)| \leq \frac{\tilde{B}_i}{\sqrt{N}}.$$

We can now proceed exactly in the same way as in the independent case to prove that

**Lemma 6.3.1.**

$$\left| \log \mathbb{E} \exp(\lambda Z) - \frac{\lambda^2}{2} \sum_{i=1}^N \mathbb{E}_{\lambda \mathbb{E}^{\mathfrak{F}_{i-1}}(Z)} \left( F_i(Z)^2 \right) \right| \leq \sum_{i=1}^N \frac{\lambda^3 \tilde{B}_i^{\,3}}{3 N^{3/2}}.$$

We would like now to bound

$$\mathbb{E}_{\lambda \mathbb{E}^{\mathfrak{F}_{i-1}}(Z)} \left( F_i(Z)^2 \right) - \mathbb{E}\left( F_i(Z)^2 \right),$$

which we will decompose as in the independent case into

$$\sum_{j=1}^{i-1} \mathbb{E}_{\lambda Z_{i-1}} \Big( F_j \big( F_i(Z)^2 \big) \Big).$$

Among other things, we will have to bound $\operatorname{ess\,sup} F_j\big(F_i(Z)^2\big)$. Let us start with this. For any bounded measurable function $h(X)$, we have

$$\begin{aligned}\left|F_j(h(X)^2)\right| &= \left|\mathbb{E}^{\mathfrak{F}_j}(h(X)^2 - h(\overset{j}{Y})^2)\right| \\ &= \left|\mathbb{E}^{\mathfrak{F}_j}\Big((h(X)+h(\overset{j}{Y}))(h(X)-h(\overset{j}{Y}))\Big)\right| \\ &\le 2\,\mathrm{ess\,sup}|h(X)|\;\mathbb{E}^{\mathfrak{F}_j}\big|h(X)-h(\overset{j}{Y})\big|.\end{aligned}$$

We will apply this to $h(X) = F_i(Z)$, and in this case, we will try to express $h(X) - h(\overset{j}{Y})$ as a difference "of order two" of four coupled processes. Let us build these processes right now, since we cannot proceed without them. We will call them $(X, \overset{j}{Y}, \overset{i}{Y}, \overset{i}{U})$. The distribution of $(X, \overset{j}{Y})$ and $(X, \overset{i}{Y})$ on their canonical spaces will be as previously defined. Let us repeat this construction here, to make precise the fact that we can build them in such a way that they satisfy the Markov property, when $X$ does :

We build $\mathbb{P}(dX_1^{i-1}, d\overset{i}{Y}{}_1^{i-1})$ as

$$\mathbb{P}(dX_1^{i-1}) \otimes \delta_{X_1^{i-1}}(d\overset{i}{Y}{}_1^{i-1}),$$

where $\delta_{X_1^{i-1}}$ is the Dirac mass at point $X_1^{i-1}$ in $\prod_{k=1}^{i-1} \mathcal{X}_k$.

We then put

$$\mathbb{P}(dX_i, d\overset{i}{Y}_i \,|\, X_1^{i-1}, \overset{i}{Y}{}_1^{i-1}) = \mathbb{P}(dX_i \,|\, X_{i-1}) \otimes \mathbb{P}(d\overset{i}{Y}_i \,|\, \overset{i}{Y}_{i-1}),$$

and for $k > i$ we build $\mathbb{P}(dX_k, d\overset{i}{Y}_k \,|\, (X, \overset{i}{Y})_1^{k-1})$ as some maximal coupling between $\mathbb{P}(dX_k \,|\, X_{k-1})$ and $\mathbb{P}(d\overset{i}{Y}_k \,|\, \overset{i}{Y}_{k-1})$, which we choose in a fixed way, independent of $(X, \overset{i}{Y})_1^{k-2}$. Thus built, $(X, \overset{i}{Y})$ is a Markov chain. We build $(X, \overset{j}{Y})$ in the same way, with the index $i$ replaced by $j$. Then we define the distribution of $(\overset{j}{Y}, \overset{i}{U})$ on its canonical space to be the same as the distribution of $(X, \overset{i}{Y})$.

These preliminaries being set, we are ready to define the distribution of $(X, \overset{j}{Y}, \overset{i}{Y}, \overset{i}{U})$ on its canonical space. Let us put for convenience $T_k = (X_k, \overset{j}{Y}_k, \overset{i}{Y}_k, \overset{i}{U}_k)$. We set

$$\mathbb{P}(dX_k, d\overset{j}{Y}_k \,|\, T_1^{k-1}) = \mathbb{P}(dX_k, d\overset{j}{Y}_k \,|\, X_{k-1}, \overset{j}{Y}_{k-1}),$$

which we have already defined, and we take for

$$\mathbb{P}(d\overset{i}{Y}_k, d\overset{i}{U}_k \,|\, T_1^{k-1}, X_k, \overset{j}{Y}_k)$$

some maximally coupled distribution depending only on $(T_{k-1}, X_k, \overset{j}{Y}_k)$ with marginals

$$\mathbb{P}(d\overset{i}{Y}_k \,|\, X_{k-1}, \overset{i}{Y}_{k-1}, X_k)$$

and

$$\mathbb{P}(d\overset{i}{U}_k \,|\, \overset{j}{Y}_{k-1}, \overset{i}{U}_{k-1}, \overset{j}{Y}_k)$$

which we have already defined.

*Remark* 6.3.3. The processes $\overset{i}{Y}$ and $\overset{j}{Y}$ are independent knowing $X$, therefore this construction is compatible with the previous one. Indeed

$$\begin{aligned}\mathbb{P}(dX, d\overset{j}{Y}, d\overset{i}{Y}) &= \prod_{k=1}^{N} \mathbb{P}\big(dX_k, d\overset{j}{Y}_k, d\overset{i}{Y}_k \,|\, (X, \overset{j}{Y}, \overset{i}{Y})_1^{k-1}\big) \\ &= \prod_{k=1}^{N} \mathbb{P}(dX_k \,|\, X_{k-1}) \mathbb{P}(d\overset{j}{Y}_k \,|\, X_k, X_{k-1}, \overset{j}{Y}_{k-1}) \\ &\qquad \times \mathbb{P}(d\overset{i}{Y}_k \,|\, X_k, X_{k-1}, \overset{i}{Y}_{k-1}),\end{aligned}$$

thus

$$\begin{aligned}\mathbb{P}(d\overset{j}{Y}, d\overset{i}{Y} \,|\, X) &= \prod_{k=1}^{N} \mathbb{P}(d\overset{j}{Y}_k \,|\, X_k, X_{k-1}, \overset{j}{Y}_{k-1}) \prod_{k=1}^{N} \mathbb{P}(d\overset{i}{Y}_k \,|\, X_k, X_{k-1}, \overset{i}{Y}_{k-1}) \\ &= \mathbb{P}(d\overset{j}{Y} \,|\, X) \otimes \mathbb{P}(d\overset{i}{Y} \,|\, X).\end{aligned}$$

The following lemma will be important to carry the computations :

**Lemma 6.3.2.**

$$\begin{aligned}h(X) &= \mathbb{E}(f(X) - f(\overset{i}{Y}) \,|\, X_1^i) \\ &= \mathbb{E}(f(X) - f(\overset{i}{Y}) \,|\, X_1^i, \overset{j}{Y}_1^i)\end{aligned}$$

*and in the same way*

$$\begin{aligned}h(\overset{j}{Y}) &= \mathbb{E}(f(\overset{j}{Y}) - f(\overset{i}{U}) \,|\, \overset{j}{Y}_1^i) \\ &= \mathbb{E}(f(\overset{j}{Y}) - f(\overset{i}{U}) \,|\, X_1^i, \overset{j}{Y}_1^i).\end{aligned}$$

*Proof.* Let us remark first that

$$\mathbb{E}\big(f(X) \,|\, X_1^i\big) = \mathbb{E}\big(f(X) \,|\, X_1^i, \overset{j}{Y}_1^i\big),$$

because $\big(X_{i+1}^N \perp\!\!\!\perp \overset{j}{Y}_1^i \,|\, X_1^i\big)$.

Moreover, from the construction of the coupled process $T$, we see that

$$\mathbb{P}(d\overset{i}{Y}{}_1^N, dX_1^i, d\overset{j}{Y}{}_1^i) = \prod_{k=1}^{i} \mathbb{P}(dX_k \mid X_{k-1})\mathbb{P}\big(d\overset{j}{Y}{}_k \mid X_k, X_{k-1}, \overset{j}{Y}{}_{k-1}\big) \times \mathbb{P}\big(d\overset{i}{Y}{}_k \mid X_k, X_{k-1}, \overset{i}{Y}{}_{k-1}\big) \times \prod_{k=i+1}^{N} \mathbb{P}\big(d\overset{i}{Y}{}_k \mid \overset{i}{Y}{}_{k-1}\big),$$

and therefore that

$$\mathbb{P}\big(d\overset{i}{Y}{}_1^N \mid X_1^i, \overset{j}{Y}{}_1^i\big) = \prod_{k=1}^{i} \mathbb{P}\big(d\overset{i}{Y}{}_k \mid X_k, X_{k-1}, \overset{i}{Y}{}_{k-1}\big) \prod_{k=i+1}^{N} \mathbb{P}\big(d\overset{i}{Y}{}_k \mid \overset{i}{Y}{}_{k-1}\big) = \mathbb{P}\big(d\overset{i}{Y}{}_1^N \mid X_1^i\big).$$

As the couples of random variables $(X, \overset{i}{Y})$ and $(\overset{j}{Y}, \overset{i}{U})$ play symmetric roles (they can be chosen to be exchangeable by a proper construction of $T$, but even without this refinement, the proof applies mutatis mutandis when the roles of $(X, \overset{i}{Y})$ and $(\overset{j}{Y}, \overset{i}{U})$ are exchanged), we have in the same way

$$\mathbb{E}\big(f(\overset{j}{Y}) \mid X_1^i\big) = \mathbb{E}\big(f(\overset{j}{Y}) \mid X_1^i, \overset{j}{Y}{}_1^i\big)$$
$$\mathbb{E}\big(f(\overset{i}{U}) \mid X_1^i\big) = \mathbb{E}\big(f(\overset{i}{U}) \mid X_1^i, \overset{j}{Y}{}_1^i\big).$$

□

We deduce from the previous lemma that

$$\mathbb{E}^{\mathfrak{F}_j}\Big|h(X) - h(\overset{j}{Y})\Big| = \mathbb{E}^{\mathfrak{F}_j}\Big|\mathbb{E}\Big(f(X) - f(\overset{i}{Y}) - f(\overset{j}{Y}) + f(\overset{i}{U}) \mid (X, \overset{j}{Y})_1^i\Big)\Big| \leq \mathbb{E}^{\mathfrak{F}_j}\Big(\Big|f(X) - f(\overset{i}{Y}) - f(\overset{j}{Y}) + f(\overset{i}{U})\Big|\Big).$$

To write the right-hand side of this last inequality as far as possible as a function of the second differences $\Delta_\ell \Delta_k f$, we need one more lemma : let us introduce the two stopping times

$$\tau_i = \inf\{k \geq i \mid \overset{i}{Y}{}_k = X_k\},$$
$$\tau_j = \inf\{k \geq j \mid \overset{j}{Y}{}_k = X_k\}.$$

**Lemma 6.3.3.** *With the previous construction, we have*

$$\mathbb{P}(\overset{i}{U}{}_i^N = \overset{i}{Y}{}_i^N \mid \tau_j < i) = 1.$$

*In other words, on the event* $(\tau_j < i)$ *it is almost surely true that* $\overset{i}{U}{}_i^N = \overset{i}{Y}{}_i^N$.

*Proof.* We have obviously $\overset{j}{Y}{}_{\tau_j}^N = X_{\tau_j}^N$ almost surely. Now when $\tau_j < i$, then a.s. $\overset{j}{Y}_{i-1} = X_{i-1} = \overset{i}{Y}_{i-1} = \overset{i}{U}_{i-1}$, and so $\overset{i}{U}_i$ and $\overset{i}{Y}_i$ knowing the past are maximally coupled and have the same marginals, therefore they are almost surely equal. Then we can carry on the same reasoning for $k = i+1, \dots, N$ and thus prove by induction that for all these values of $k$, $\overset{i}{Y}_k = \overset{i}{U}_k$ a. s.. □

Resuming the previous chain of inequalities, we can write, as a consequence of this lemma, that

$$\begin{aligned}
&\mathbb{E}^{\mathfrak{F}_j}|h(X) - h(\overset{j}{Y})| \\
&\quad \leq \mathbb{P}^{\mathfrak{F}_j}(\tau_j \geq i)\frac{2\tilde{B}_i}{\sqrt{N}} \\
&\qquad + \mathbb{E}^{\mathfrak{F}_j}\Bigg(\mathbf{1}(\tau_j < i)\Big|f(X) - f(X_1^{i-1}, \overset{i}{Y}{}_i^N) \\
&\qquad\qquad - f(X_1^{j-1}, \overset{j}{Y}{}_j^{\tau_j - 1}, X_{\tau_j}^N) + f(X_1^{j-1}, \overset{j}{Y}{}_j^{\tau_j-1}, X_{\tau_j}^{i-1}, \overset{i}{Y}{}_i^N)\Big|\Bigg) \\
&\quad \leq \mathbb{P}^{\mathfrak{F}_j}(\tau_j \geq i)\frac{2\tilde{B}_i}{\sqrt{N}} \\
&\qquad + \mathbb{E}^{\mathfrak{F}_j}\Bigg(\mathbf{1}(\tau_j < i)\Big|\sum_{k=i}^{\tau_i - 1} \Delta_k f((X_1^k, \overset{i}{Y}{}_{k+1}^N), \overset{i}{Y}_k) \\
&\qquad\qquad - \Delta_k f((X_1^{j-1}, \overset{j}{Y}{}_j^{\tau_j-1}, X_{\tau_j}^k, \overset{i}{Y}{}_{k+1}^N), \overset{i}{Y}_k)\Big|\Bigg) \\
&\quad \leq \mathbb{P}^{\mathfrak{F}_j}(\tau_j \geq i)\frac{2\tilde{B}_i}{\sqrt{N}} \\
&\quad + \mathbb{E}^{\mathfrak{F}_j}\Bigg(\mathbf{1}(\tau_j < i)\Big|\sum_{k=i}^{\tau_i-1}\sum_{\ell=j}^{\tau_j-1} \Delta_\ell \Delta_k f((X_1^\ell, \overset{j}{Y}{}_{\ell+1}^{\tau_j-1}, X_{\tau_j}^k, \overset{i}{Y}{}_{k+1}^N), \overset{i}{Y}_k, \overset{j}{Y}_\ell)\Big|\Bigg) \\
&\qquad\qquad \leq \mathbb{P}^{\mathfrak{F}_j}(\tau_j \geq i)\frac{2\tilde{B}_i}{\sqrt{N}} + \mathbb{E}^{\mathfrak{F}_j}\left(\sum_{k=i}^{\tau_i-1}\sum_{\ell=j}^{\tau_j-1}\frac{C_{\ell,k}}{N^{3/2-\zeta}}\right)
\end{aligned}$$

Let us put

$$\tilde{C}_{i,j} = \operatorname{ess\,sup} \mathbb{E}^{\mathfrak{F}_j}\left(\sum_{k=i}^{\tau_i-1}\sum_{\ell=j}^{\tau_j-1} C_{k,\ell}\right). \tag{6.3.2}$$

We get

$$\left|F_j\left(F_i(Z)^2\right)\right| \leq \frac{2\tilde{B}_i\tilde{C}_{i,j}}{N^{2-\zeta}} + \mathbb{P}^{\mathfrak{F}_j}(\tau_j \geq i)\frac{4\tilde{B}_i^2}{N}.$$

Let us now define $\mathfrak{G}_j$ to be $\sigma(\overset{j}{Y})$, the sigma algebra generated by $(\overset{j}{Y}_1, \dots, \overset{j}{Y}_N)$. We have

$$\mathbb{E}^{\mathfrak{G}_j}\Big(F_j\big(F_i(Z)^2\big)\Big) = 0,$$

because $X_j$ and $\overset{j}{Y}{}_j^N$ are conditionally independent knowing $X_1^{j-1}$. Let moreover

$$\begin{aligned}
\tilde{G}_j &= Z_{i-1}(X) - Z_{i-1}(\overset{j}{Y}) \\
&= \mathbb{E}\big(f(X)\,|\,X_1^{i-1}\big) - \mathbb{E}\big(f(\overset{j}{Y})\,|\,(\overset{j}{Y})_1^{i-1}\big) \\
&= \mathbb{E}\big(f(X) - f(\overset{j}{Y})\,|\,X_1^{i-1}, \overset{j}{Y}{}_1^{i-1}\big)
\end{aligned}$$

We have

$$\begin{aligned}
\mathbb{E}_{\lambda Z_{i-1}}\left(F_j\big(F_i(Z)^2\big)\right) &= \mathbb{E}_{\lambda Z_{i-1}}\Big(\mathbb{E}^{\mathfrak{G}_j}_{\lambda Z_{i-1}}\left(F_j\big(F_i(Z)^2\big)\right)\Big) \\
&= \mathbb{E}_{\lambda Z_{i-1}}\mathbb{E}^{\mathfrak{G}_j}_{\lambda \tilde{G}_j}\left(F_j\big(F_i(Z)^2\big)\right) \\
&= \mathbb{E}_{\lambda Z_{i-1}}\int_0^{\lambda}\mathbb{E}^{\mathfrak{G}_j}_{\alpha\tilde{G}_j}\Big(F_j\big(F_i(Z)^2\big)\big(\tilde{G}_j - \mathbb{E}^{\mathfrak{G}_j}_{\alpha\tilde{G}_j}\tilde{G}_j\big)\Big)\,d\alpha.
\end{aligned}$$

Therefore

$$\begin{aligned}
&\left|\mathbb{E}_{\lambda Z_{i-1}}\Big(F_j\big(F_i(Z)^2\big)\Big)\right| \\
&\qquad\leq \operatorname{ess\,sup}\left|F_j\big(F_i(Z)^2\big)\right|\mathbb{E}_{\lambda Z_{i-1}}\int_0^{\lambda}\mathbb{E}^{\mathfrak{G}_j}_{\alpha\tilde{G}_j}\left|\tilde{G}_j - \mathbb{E}^{\mathfrak{G}_j}_{\alpha\tilde{G}_j}\tilde{G}_j\right|d\alpha \\
&\qquad\leq 2\operatorname{ess\,sup}\left|F_j\big(F_i(Z)^2\big)\right|\mathbb{E}_{\lambda Z_{i-1}}\int_0^{\lambda}\mathbb{E}^{\mathfrak{G}_j}_{\alpha\tilde{G}_j}|\tilde{G}_j|d\alpha \\
&\qquad\leq 2\operatorname{ess\,sup}\left|F_j\big(F_i(Z)^2\big)\right|\ \operatorname{ess\,sup}\mathbb{E}^{\mathfrak{G}_j}\left(\int_0^{\lambda}\exp\left(\alpha\tilde{G}_j\right)|\tilde{G}_j|d\alpha\right) \\
&\qquad\qquad\times\Big(\mathbb{E}^{\mathfrak{G}_j}\big(\exp(-\lambda|\tilde{G}_j|)\big)\Big)^{-1} \\
&\qquad\leq 2\operatorname{ess\,sup}\left|F_j\big(F_i(Z)^2\big)\right|\ \operatorname{ess\,sup}\mathbb{E}^{\mathfrak{G}_j}\Big(\exp(\lambda|\tilde{G}_j|) - 1\Big)
\end{aligned}$$

$$\times \mathbb{E}^{\mathfrak{G}_j}\big(\exp(\lambda|\tilde{G}_j|)\big).$$

Moreover

$$\begin{aligned}
\mathbb{E}^{\mathfrak{G}_j}\big(\exp(\lambda|\tilde{G}_j|)\big) &= \mathbb{E}\Big(\exp\Big(\lambda\big|\mathbb{E}(f(X) - f(\overset{j}{Y}) \mid X_1^{i-1}, \overset{j}{Y}{}_1^{i-1})\big|\Big) \mid \overset{j}{Y}\Big) \\
&\le \mathbb{E}\Big(\mathbb{E}\Big(\exp(\lambda|f(X) - f(\overset{j}{Y})|)\Big) \mid X_1^{i-1}, \overset{j}{Y}{}_1^{i-1}\Big) \mid \overset{j}{Y}\Big) \\
&= \mathbb{E}\Big(\exp\Big(\lambda|f(X) - f(\overset{j}{Y})|\Big) \mid \overset{j}{Y}{}_1^{i-1}\Big)
\end{aligned}$$

because $(X_1^{i-1} \perp\!\!\!\perp \overset{j}{Y}{}_i^N \mid \overset{j}{Y}{}_1^{i-1})$

$$\le \operatorname{ess\,sup} \mathbb{E}^{\mathfrak{G}_j}\left(\exp\left(\lambda \sum_{k=j}^{\tau_j - 1} \frac{B_k}{\sqrt{N}}\right)\right).$$

Let us put

$$\tilde{\tilde{B}}_j(\lambda) = \operatorname{ess\,sup} \frac{\sqrt{N}}{\lambda} \mathbb{E}^{\mathfrak{G}_j}\left(\exp\left(\lambda \sum_{k=j}^{\tau_j - 1} \frac{B_k}{\sqrt{N}}\right) - 1\right) \times \mathbb{E}^{\mathfrak{G}_j}\left(\exp\left(\lambda \sum_{k=j}^{\tau_j - 1} \frac{B_k}{\sqrt{N}}\right)\right). \tag{6.3.3}$$

We have

$$\mathbb{E}_{\lambda Z_{i-1}}\big(F_j(F_i(Z)^2)\big) \le \frac{2\lambda}{\sqrt{N}} \tilde{\tilde{B}}_j(\lambda) \frac{2\tilde{B}_i \tilde{C}_{i,j}}{N^{2-\zeta}} + \operatorname{ess\,sup} \mathbb{P}^{\mathfrak{F}_j}(\tau_j \ge i) \frac{4\tilde{B}_i^2}{N}.$$

Thus

$$\begin{aligned}
\sum_{i=1}^{N} \sum_{j=1}^{i-1} \frac{\lambda^2}{2} \mathbb{E}_{\lambda Z_{i-1}}\Big(F_j\big(F_i(Z)^2\big)\Big) &\le \frac{\lambda^3}{N^{1/2-\zeta}} \sum_{1 \le j < i \le N} \frac{2\tilde{B}_i \tilde{C}_{i,j} \tilde{\tilde{B}}_j(\lambda)}{N^2} \\
&\quad + \frac{\lambda^3}{\sqrt{N}} \sum_{j=1}^{N} \sum_{i=j+1}^{N} \operatorname{ess\,sup} \mathbb{P}^{\mathfrak{F}_j}(\tau_j \ge i) \frac{4\tilde{B}_i^2 \tilde{\tilde{B}}_j(\lambda)}{N}.
\end{aligned}$$

Therefore if we put

$$\check{B}_j = \sqrt{\sum_{i=j+1}^{N} \tilde{B}_i^2 \operatorname{ess\,sup} \mathbb{P}^{\mathfrak{F}_j}(\tau_j \ge i)}, \tag{6.3.4}$$

we obtain the following theorem :

**Theorem 6.3.1.** *When $(X_1, \dots, X_N)$ satisfies the Markov property and the function $f$ satisfies*

$$\sup_{x,y_i} \Delta_i f(x, y_i) \leq \frac{B_i}{\sqrt{N}},$$
$$\sup_{x,y_i,y_j} \Delta_j \Delta_i f(x, y_i, y_j) \leq \frac{C_{i,j}}{N^{3/2-\varsigma}},$$

*then*

$$\left| \log \mathbb{E}\big(\exp(\lambda Z)\big) - \frac{\lambda^2}{2} \mathbb{E}(Z^2) \right| \leq \frac{\lambda^3}{N^{1/2-\varsigma}} \sum_{1 \leq j < i \leq N} \frac{2\tilde{B}_i \tilde{C}_{i,j} \tilde{\tilde{B}}_j(\lambda)}{N^2} + \frac{\lambda^3}{\sqrt{N}} \sum_{i=1}^{N} \left( \frac{\tilde{B}_i^3}{3N} + \frac{4\check{B}_i^2 \tilde{\tilde{B}}_i(\lambda)}{N} \right),$$

*where the constants $\tilde{B}_i$ are defined by (6.3.1), the constants $\tilde{C}_{i,j}$ are defined by (6.3.2), the constants $\tilde{\tilde{B}}_i(\lambda)$ are defined by (6.3.3) and the constants $\check{B}_i$ are defined by (6.3.4).*

**Corollary 6.3.1.** *Let us assume that $(X_1, \dots, X_N)$ is a Markov chain such that for some positive constants $A$ and $\rho$*

$$\mathbb{P}\big(\tau_i > i + k \,|\, \mathfrak{G}_i, X_i\big) \leq A\rho^k, \quad a.s., \tag{6.3.5}$$
$$\mathbb{P}\big(\tau_i > i + k \,|\, \mathfrak{F}_N, \overset{i}{Y}_i\big) \leq A\rho^k, \quad a.s., \tag{6.3.6}$$

*and let us put $B = \max_i B_i$ and $C = \max_{i,j} C_{i,j}$. Then*

$$\left| \log \mathbb{E}\big(\exp(\lambda Z)\big) - \frac{\lambda^2}{2} \mathbb{E}(Z^2) \right| \leq \frac{\lambda^3}{N^{1/2-\varsigma}} \frac{BCA^3}{(1-\rho)^3} \left( \frac{\rho \log(\rho^{-1})}{2AB} - \frac{\lambda}{\sqrt{N}} \right)_+^{-1} + \frac{\lambda^3}{\sqrt{N}} \left( \frac{B^3 A^3}{3(1-\rho)^3} + \frac{4B^2A^3}{(1-\rho)^3} \left( \frac{\rho \log(\rho^{-1})}{2AB} - \frac{\lambda}{\sqrt{N}} \right)_+^{-1} \right).$$

*Consequently*

$$\mathbb{P}\left(f(X) \geq \mathbb{E}\big(f(X)\big) + \epsilon\right) \leq \exp\left( - \frac{\epsilon^2}{2\Big(\mathbf{Var}\big(f(X)\big) + \frac{2\eta\epsilon}{\mathbf{Var}(f(X))}\Big)} \right),$$

$$\mathbb{P}\left(f(X) \leq \mathbb{E}\big(f(X)\big) - \epsilon\right) \leq \exp\left( - \frac{\epsilon^2}{2\Big(\mathbf{Var}\big(f(X)\big) + \frac{2\eta\epsilon}{\mathbf{Var}(f(X))}\Big)} \right),$$

*where*

$$\eta = \frac{1}{N^{1/2-\varsigma}} \frac{BCA^3}{(1-\rho)^3} \left( \frac{\rho \log(\rho^{-1})}{2AB} - \frac{\epsilon}{\mathbb{V}\mathrm{ar}(f(X))\sqrt{N}} \right)_+^{-1}$$
$$+ \frac{1}{\sqrt{N}} \left( \frac{B^3 A^3}{3(1-\rho)^3} + \frac{4B^2A^3}{(1-\rho)^3} \left( \frac{\rho \log(\rho^{-1})}{2AB} - \frac{\epsilon}{\mathbb{V}\mathrm{ar}(f(X))\sqrt{N}} \right)_+^{-1} \right).$$

*Remark* 6.3.4. If we choose the distribution of the pair $(X, \overset{i}{Y})$ to be exchangeable, and this can always be done, then the two conditions (6.3.5) and (6.3.6) are equivalent and one is of course superfluous.

*Remark* 6.3.5. The hypotheses are for example fulfilled by any irreducible aperiodic homogeneous Markov chain on a finite state space.

*Proof of the corollary.* We have

$$\tilde{B}_i \leq B \operatorname{ess\,sup} \mathbb{E}^{\mathfrak{F}_N}(\tau_j - j) = B \sum_{k=0}^{+\infty} \operatorname{ess\,sup} \mathbb{P}^{\mathfrak{F}_N}(\tau_j > j + k)$$
$$\leq AB \sum_{k=0}^{+\infty} \rho^k = \frac{BA}{1-\rho}.$$

In the same way

$$\tilde{C}_{i,j} \leq C \operatorname{ess\,sup} \mathbb{E}^{\mathfrak{F}_N}\big((\tau_i - i)(\tau_j - j)\big)$$
$$= C \operatorname{ess\,sup} \mathbb{E}^{\mathfrak{F}_N}(\tau_i - i) \mathbb{E}^{\mathfrak{F}_N}(\tau_j - j) \leq \frac{CA^2}{(1-\rho)^2},$$

where we have used the fact that $(\overset{i}{Y} \perp\!\!\!\perp \overset{j}{Y} \,|\, X)$. We also have

$$\mathbb{E}^{\mathfrak{G}_j}\left( \exp\left( \lambda \sum_{k=j}^{\tau_j - 1} \frac{B_k}{\sqrt{N}} \right) - 1 \right) = \int_0^{+\infty} \mathbb{P}^{\mathfrak{G}_j}\left( \exp\left( \lambda \sum_{k=j}^{\tau_j - 1} \frac{B_k}{\sqrt{N}} \right) - 1 \geq \xi \right) d\xi$$
$$\leq \int_0^{+\infty} \mathbb{P}^{\mathfrak{G}_j}\left( (\tau_j - j) \geq \frac{\sqrt{N}}{\lambda B} \log(1+\xi) \right) d\xi$$
$$\leq \int_0^{+\infty} \frac{A}{\rho} \exp\left( \frac{\sqrt{N}}{\lambda B} \log(\rho) \log(1+\xi) \right) d\xi$$
$$\leq \frac{A}{\rho} \left( \frac{\sqrt{N} \log(\rho^{-1})}{\lambda B} - 1 \right)_+^{-1}.$$

Thus

$$\begin{aligned}
\frac{\lambda}{\sqrt{N}}\tilde{\tilde{B}}_j(\lambda) &\leq \frac{A}{\rho}\left(\frac{\sqrt{N}\log(\rho^{-1})}{\lambda B} - 1\right)_+^{-1} \\
&\quad \times \left(1 + \frac{A}{\rho}\left(\frac{\sqrt{N}\log(\rho^{-1})}{\lambda B} - 1\right)_+^{-1}\right) \\
&\leq \frac{2A}{\rho}\left(\frac{\sqrt{N}\log(\rho^{-1})}{\lambda B} - \frac{2A}{\rho}\right)_+^{-1} \\
&= \left(\frac{\sqrt{N}\rho\log(\rho^{-1})}{2\lambda AB} - 1\right)_+^{-1},
\end{aligned}$$

and

$$\tilde{\tilde{B}}_j(\lambda) \leq \left(\frac{\rho\log(\rho^{-1})}{2AB} - \frac{\lambda}{\sqrt{N}}\right)_+^{-1}.$$

On the other hand

$$\begin{aligned}
\breve{B}_j &\leq \frac{BA}{1-\rho}\sqrt{\sum_{i=j+1}^{N} A\rho^{i-j-1}} \\
&\leq \frac{BA^{3/2}}{(1-\rho)^{3/2}}.
\end{aligned}$$

Substituting all these upper bounds in the theorem proves its corollary. □

## Conclusion

We have shown that under quite natural boundedness or exponential moment assumptions, it is possible to get non asymptotic bounds for the distance between the log Laplace transform of a function of $N$ random variables and the transform of the corresponding Gaussian random variable. In particular, no convexity assumption is required and we can deal not only with independent random variables, but also with a large class of Markov chains.

# 7

# Markov chains with exponential transitions

## 7.1 Model definition

The aim of this chapter is to study the behaviour of any given family of Markov chains

$$\mathcal{F} = \left(E^{\mathbb{N}}, (X_n)_{n\in\mathbb{N}}, \mathcal{B}, \mathbb{P}_\beta\right)_{\beta\in\mathbb{R}_+},$$

where

- $E$ is a finite set,
- The canonical process on $E^{\mathbb{N}}$ is considered. Let us remind that it is defined as

$$\begin{array}{rcl} X_n : E^{\mathbb{N}} & \longrightarrow & E \\ (\omega_k)_{k\in\mathbb{N}} & \longmapsto & \omega_n. \end{array}$$

- $\mathcal{B}$ is the $\sigma$-algebra generated by the coordinate functions $X_n$, $n \in \mathbb{N}$.
- $\mathbb{P}_\beta$ is the probability distribution on $\mathbb{E}^{\mathbb{N}}$ of a homogeneous Markov chain with transition matrix $p_\beta : E \times E \longrightarrow [0,1]$ : namely for any $n \in \mathbb{N}$ and any $n$-tuple $x_1^n \in E^n$

$$\begin{aligned} \mathbb{P}_\beta\big(X_n = x_n \,|\, X_1^{n-1} = x_1^{n-1}\big) &= \mathbb{P}_\beta\big(X_n = x_n \,|\, X_{n-1} = x_{n-1}\big) \\ &= p_\beta(x_{n-1}, x_n), \quad \mathbb{P}_\beta \text{ a.s.}, \end{aligned}$$

  where we used the notation $X_k^\ell \overset{\text{def}}{=} (X_k, \dots, X_\ell)$.
- The transition matrices $(p_\beta)_{\beta\in\mathbb{R}_+}$ satisfy a large deviation principle with rate function $V$ and speed $\beta$ :

  **Definition 7.1.1 (Hypothesis $LD(V)$).** We will say that the family of Markov matrices $(p_\beta)_{\beta\in\mathbb{R}_+}$ satisfies hypothesis $LD(V)$, where the large deviation rate function

$$V : E \times E \longrightarrow \mathbb{R}_+ \cup \{+\infty\}$$

  is irreducible in the sense that the matrix

$$\left[\exp(-V(x,y))\right]_{(x,y)\in E^2}$$

is irreducible [1], when for any $x \neq y \in E$

$$\lim_{\beta\to+\infty} \frac{1}{\beta} \log p_\beta(x,y) = -V(x,y).$$

The family of processes $\mathcal{F}$ is sometimes called a "generalised Metropolis algorithm".

To each generalised Metropolis algorithm $\mathcal{F}$, corresponds by a diagonalisation scheme a *generalised simulated annealing algorithm*: it is defined as the *nonhomogeneous* family of Markov chains

$$\mathcal{R} = \left(E^{\mathbb{N}}, \left(X_n\right)_{n\in\mathbb{N}}, \mathcal{B}, \mathbb{P}_{(\beta)}\right)_{(\beta)\in\mathcal{C}}$$

indexed by the set $\mathcal{C}$ of increasing sequences in $\mathbb{R}_+^{\mathbb{N}}$, defined by the following properties:

- $E$, $\left(X_n\right)_{n\in\mathbb{N}}$ and $\mathcal{B}$ are defined as previously,
- $\mathbb{P}_{(\beta)}$ satisfies the Markov property and the transition at time $n$ follows the same distribution as for the Metropolis algorithm at inverse temperature $\beta_n$:

$$\begin{aligned}\mathbb{P}_{(\beta)}\left(X_n = y \,|\, X_1^{n-1} = x_1^{n-1}\right) &= \mathbb{P}_{\beta_n}\left(X_n = y \,|\, X_{n-1} = x_{n-1}\right)\\ &= p_{\beta_n}(x_{n-1}, y).\end{aligned}$$

The link with Gibbs measures is the following:

**Proposition 7.1.1 (The "classical" Metropolis algorithm).** *Let us assume that*

- $\mathbb{P}_\beta(X_n = y \,|\, X_{n-1} = x) = p_\beta(x,y)$,
- $p_\beta(x,y) = q(x,y)\exp(-\beta(U(y) - U(x))_+)$, $\quad x \neq y \in E$, [2]
- *the Markov matrix $q$ is irreducible and reversible with respect to the distribution $\mu \in \mathcal{M}_+^1$. Namely for any $x, y \in E$*

$$\mu(x)q(x,y) = \mu(y)q(y,x).$$

- *The energy function $U : E \longrightarrow \mathbb{R}$ is not constant.*

*Then the Metropolis algorithm described by the family of distributions $\mathbb{P}_\beta$ is a family of irreducible aperiodic Markov chains, whose invariant probability distribution is the Gibbs measure $\mu_{\beta U}$* [3].

---

[1] i.e. for any couple of states $(x,y) \in E^2$, there exists a path

$$x_0 = x, x_1, \ldots, x_{r-1}, x_r = y$$

in $E^{r+1}$ joining $x$ to $y$ such that $V(x_{i-1}, x_i) < +\infty$, $i = 1, \ldots, r$.

[2] Where for any real number $r$, we used the notation $r_+ = \max\{r, 0\}$.

[3] defined by $\mu_{\beta U}(x) \propto \exp\left[-\beta U(x)\right]\mu(x)$

*Proof.* It is sufficient to check that $p_\beta$ is reversible with respect to $\mu_{\beta U}$: for any $x \neq y \in E$,

$$\mu(x)\exp(-\beta U(x))q(x,y)\exp\Bigl(-\beta\bigl(U(y)-U(x)\bigr)_+\Bigr) = \mu(y)\exp(-\beta U(y))q(y,x)\exp\Bigl(-\beta\bigl(U(x)-U(y)\bigr)_+\Bigr),$$

because

$$U(x)+\bigl(U(y)-U(x)\bigr)_+ = U(y)+\bigl(U(x)-U(y)\bigr)_+.$$

□

We are going to show that the general case is similar to the classical setting: indeed the invariant probability distribution of any generalised Metropolis algorithm has a logarithmic equivalent, which we call its "virtual energy".

To establish this fact, we are going to describe a construction which will also be used to study the behavious of the trajectories of the process at low temperatures.

## 7.2 The reduction principle

Let us consider a generalised Metropolis algorithm

$$\mathcal{F} = \Bigl(E^{\mathbb{N}}, (X_n)_{n\in\mathbb{N}}, \mathcal{B}, \mathbb{P}_\beta\Bigr)_{\beta\in\mathbb{R}_+}$$

with transition matrix $(p_\beta)_{\beta\in\mathbb{R}_+}$. Let us assume that for any $\beta \in \mathbb{R}_+$, any $(x,y) \in E^2$ such that $V(x,y) \in \mathbb{R}$, the transition probability from $x$ to $y$ is positive: $p_\beta(x,y) > 0$. This hypothesis is always satisfied then $\beta$ is large enough, thus it does not restrict the generality of the following discussion, and allows to avoid repeating everywhere the condition "for large enough values of $\beta$".

We will also systematically use for any event $\mathcal{A} \in \mathcal{B}$ and any initial point $x \in E$ the concise notation

$$\mathbb{P}_\beta^x(\mathcal{A}) = \mathbb{P}_\beta(\mathcal{A}\,|\,X_0 = x).$$

Moreover $\theta_k$ will denote the $k$th iterate of the shift operator on $E^{\mathbb{N}}$, defined by

$$\theta_k(\omega) = (\omega_{k+n})_{n\in\mathbb{N}} \in E^{\mathbb{N}}, \qquad \omega \in E^{\mathbb{N}}.$$

Let us recall the strong Markov property: for any stopping time $\tau$, let $\mathcal{B}_\tau$ be the $\sigma$-algebra of events prior to $\tau$, defined by

$$\mathcal{B}_\tau = \bigl\{\mathcal{A} \in \mathcal{B} : \mathcal{A} \cap \{\tau = n\} \in \sigma(X_0,\dots,X_n)\bigr\}.$$

Then for any random variable $Y$ which is measurable with respect to $\mathcal{B}$, for any initial point $x \in E$

$$\mathbb{E}_\beta^x\bigl(Y\circ\theta_\tau\,\mathbb{1}(\tau < +\infty)\,|\,\mathcal{B}_\tau\bigr) = \mathbb{1}(\tau < +\infty)\,\mathbb{E}_\beta^{X_\tau}(Y).$$

**Definition 7.2.1.** For any non empty sub-domain $A \subset E$, let us define the stopping times

$$\tau_0(A) = \inf\{k \geq 0 \,:\, X_k \in A\}$$
$$\tau_n(A) = \inf\{k > \tau_{n-1}(A) \,:\, X_k \in A\}, \qquad n > 0.$$

Let us then define the process $X^A$ reduced to the sub-domain $A$ as

$$X_n^A = X_{\tau_n(A)}, \qquad n \in \mathbb{N}.$$

**Proposition 7.2.1.** *The process $(X_n^A)_{n\in\mathbb{N}}$ is an irreducible homogeneous Markov chain and its invariant probability distribution is $\mu_\beta^A = \mu_\beta(\cdot \,|\, A)$, where $\mu_\beta$ is the invariant probability distribution of $(X_n)_{n\in\mathbb{N}}$ under $\mathbb{P}_\beta$.*

*Proof.* The stopping times $\tau_n(A)$ are almost surely finite. This comes from the fact that $(X_n)$ is irreducible. Let us show first that $\tau_0(A) < +\infty$ almost surely. For any $x \in E$, there is a time $k(x)$ such that $a(x) \overset{\text{def}}{=} \mathbb{P}_\beta\big(X_{k(x)} \in A \,|\, X_0 = x\big) > 0$. Let

$$a = \inf_{x\in E} a(x)$$
$$k = \sup_{x\in E} k(x).$$

As $E$ *is finite*, $a > 0$ and $k < +\infty$. For any integer $m$, we can apply the Markov property at time $(m-1)k$ to prove that

$$\begin{aligned}\mathbb{P}_\beta^x(\tau_0(A) > mk) &= \mathbb{E}_\beta^x\Big(\mathbb{1}\big(\tau_0(A) \circ \theta_{(m-1)k} > k\big)\,\mathbb{1}\big(\tau_0(A) > (m-1)k\big)\Big)\\ &= \mathbb{E}_\beta^x\Big(\mathbb{P}_\beta^{X_{(m-1)k}}\big(\tau_0(A) > k\big)\,\mathbb{1}\big(\tau_0(A) > (m-1)k\big)\Big).\end{aligned}$$

But

$$\begin{aligned}\mathbb{P}_\beta^y\big(\tau_0(A) > k\big) &\leq \mathbb{P}_\beta^y\big(\tau_0(A) > k(y)\big)\\ &\leq 1 - a(y)\\ &\leq 1 - a,\end{aligned}$$

hence

$$\begin{aligned}\mathbb{P}_\beta^x\big(\tau_0(A) > mk\big) &\leq (1-a)\mathbb{P}_\beta^x\big(\tau_0(A) > (m-1)k\big)\\ &\leq (1-a)^m,\end{aligned}$$

d'où

$$\mathbb{P}_\beta^x\big(\tau_0(A) = +\infty\big) = \lim_{m\to+\infty} \mathbb{P}_\beta^x\big(\tau_0(A) > mk\big) = 0.$$

Let us then notice that $\tau_1(A)$ is such that

$$\mathbb{1}(X_0 \in A)\tau_1(A) = \mathbb{1}(X_0 \in A)(\tau_0(A) \circ \theta_1).$$

In the case when the starting point $x$ belongs to the sub-domain $A$, we can apply the Markov property at time 1 and get

$$\begin{aligned} \mathbb{P}_\beta^x\big(\tau_1(A) = +\infty\big) &= \mathbb{P}_\beta^x\big(\tau_0 \circ \theta_1 = +\infty\big) \\ &= \mathbb{E}_\beta^x\Big(\mathbb{P}_\beta^{X_1}\big(\tau_0 = +\infty\big)\Big) = 0. \end{aligned}$$

We can then apply the strong Markov property to the stopping time $\tau_{n-1}(A)$ to prove that

$$\begin{aligned} \mathbb{P}_\beta^x\big(\tau_n(A) = +\infty\big) &= \mathbb{P}_\beta^x\big(\tau_1 \circ \theta_{\tau_{n-1}(A)} = +\infty \text{ and } \tau_{n-1}(A) < +\infty\big) \\ &\quad + \mathbb{P}_\beta^x\big(\tau_{n-1}(A) = +\infty\big) \\ &= \mathbb{E}_\beta^x\Big(\mathbb{1}\big(\tau_{n-1}(A) < +\infty\big)\mathbb{P}_\beta^{X_{\tau_{n-1}(A)}}\big(\tau_1(A) = +\infty\big)\Big) \\ &\quad + \mathbb{P}_\beta^x\big(\tau_{n-1}(A) = +\infty\big) \\ &= \mathbb{P}_\beta^x\big(\tau_{n-1}(A) = +\infty\big). \end{aligned}$$

Thus by induction on $n$

$$\mathbb{P}_\beta^x\big(\tau_n(A) = +\infty\big) = \mathbb{P}_\beta^x\big(\tau_0(A) = +\infty\big) = 0.$$

Thus the stopping times $\tau_n(A)$ are almost surely finite, which justifies the construction of $X^A$. The fact that $X^A$ is a homogeneous Markov chain is a straightforward consequence of the strong Markov property.

This Markov chain is indeed irreducible, because for any couple $(x, y)^2 \in A^2$, there is a path $x_0 = x, x_1, \ldots, x_m = y$ leading from $x$ to $y$ which is followed by the chain $X$ with a positive probability. The trace of this path on $A$ (i.e. the subsequence of points of this path belonging to $A$) is followed by the reduced chain $X^A$ with at least the same (positive !) probability. Therefore the invariant probability $\mu_\beta^A$ of $X^A$ is unique[4]

Let us then consider some point $x \in A$. Let $\tau_x \overset{\text{def}}{=} \tau_1(\{x\})$ the the first return time of the chain $X$ in $x$. Let $\tau_x^A$ be the first return time in $x$ of

[4] The invariant probability measure of an irreducible homogeneous Markov chain on a finite state space $E$ is unique. Hints for a proof : it is enough to show that any invariant signed measure $\rho$ assigning a null measure $\rho(E) = 0$ to the whole space is the null measure. Consider the inward and outward flows of the iterates of the transition matrix $p$ of the chain from the support of the positive part $S$ of $\rho$. Show that

$$\sum_{x \in S} \sum_{y \notin S} \rho(x) p^k(x, y) = \sum_{x \notin S} \sum_{y \in S} \rho(x) p^k(x, y).$$

Deduce from this that $\rho(x)p^k(x, y) = 0$, for any $(x, y) \in S \times (E \setminus S) \cup (E \setminus S) \times S$, and any $k \in \mathbb{N}$, and therefore that $\rho(x) = 0$ for any $x \in E$.

the reduced chain $X^A$. These two stopping times are linked by the relation $\tau_{\tau_x^A}(A) = \tau_x$. Let $\mu_\beta^A$ be the invariant probability measure of $X^A$. It is well known that for any point $y$ of $A$, the invariant probability measure of the chain $X$ can be expressed as

$$\frac{\mu_\beta(y)}{\mu_\beta(x)} = \mathbb{E}_\beta^x \left( \sum_{k=0}^{\tau_x - 1} \mathbb{1}(X_k = y) \right). \tag{7.2.1}$$

In the same way

$$\frac{\mu_\beta^A(y)}{\mu_\beta^A(x)} = \mathbb{E}_\beta^x \left( \sum_{n=0}^{\tau_x^A - 1} \mathbb{1}(X_n^A = y) \right).$$

As the number of visits to $y$ before the first return to $x$ of the chains $X$ and $X^A$ are the same, these two expressions are equal, this shows that $\mu_\beta^A$ is indeed equal to $\mu_\beta$ conditioned by the event $A$. □

The reduction method consists in ordering the state space in some way, putting $E = \{g_1, \dots, g_{|E|}\}$, and in considering the reduction sets

$$A_k = \{g_1, \dots, g_k\}.$$

A backward induction schemes then allows to compute equivalents for the transitions of the reduced chains $X^{A_k}$, letting $k$ range from $|E|$ to 2. Indeed for any $x, y \in A_{k-1}$,

$$\begin{aligned} \mathbb{P}_\beta^x(X_1^{A_{k-1}} = y) &= \mathbb{P}_\beta^x(X_1^{A_k} = y) \\ &\quad + \sum_{n=0}^{+\infty} \mathbb{P}_\beta^x(X_1^{A_k} = g_k) \left( \mathbb{P}_\beta^{g_k}(X_1^{A_k} = g_k) \right)^n \mathbb{P}_\beta^{g_k}(X_1^{A_k} = y) \\ &= \mathbb{P}_\beta^x(X_1^{A_k} = y) + \frac{\mathbb{P}_\beta^x(X_1^{A_k} = g_k)\, \mathbb{P}_\beta^{g_k}(X_1^{A_k} = y)}{\mathbb{P}_\beta^{g_k}(X_1^{A_k} \neq g_k)}. \end{aligned}$$

Choosing in a suitable way the ordering $(g_k)_{k=1}^{|E|}$ of the state space leads to the following theorem.

**Theorem 7.2.1.** *For any subdomain $A \subset E$ containing at least two points, there exists a rate function*

$$V^A : E \times E \longrightarrow \mathbb{R}_+ \cup \{+\infty\}$$

*such that the transitions of the reduced chain $X^A$ satisfy a large deviation principle with rate function $V^A$ and speed $\beta$ : for any $x, y \in A$*

$$\lim_{\beta \to +\infty} \frac{1}{\beta} \log \mathbb{P}_\beta^x(X_1^A = y) = -V^A(x, y).$$

*Moreover, for any given reduction sequence* $(g_k)_{k=1}^{|E|}$[5], *letting* $A_k = \{g_1, \dots, g_k\}$ *denote the corresponding reduction sets, the rate functions* $V^{A_k}$ *can be explicitly built through the following backward inductive computation :*

$$V^{A_{k-1}}(x,y) = \min\Big\{V^{A_k}(x,y),\\ V^{A_k}(x,g_k) + V^{A_k}(g_k,y) - \min\{V^{A_k}(g_k,z)\ :\ z \in A_{k-1}\}\Big\}.$$

*This formula also shows that the rate functions of the reduced chains are* irreducible.

Let us now chose two distinct states $x$ and $y$ in $E$, and let us consider the reduction set $\{x,y\}$. It is clear from the previous discussion that

$$\frac{\mu_\beta(y)}{\mu_\beta(x)} = \frac{\mu_\beta^{\{x,y\}}(y)}{\mu_\beta^{\{x,y\}}(x)} = \frac{\mathbb{P}_\beta^x(X_1^{\{x,y\}} = y)}{\mathbb{P}_\beta^y(X_1^{\{x,y\}} = x)}.$$

Letting $x$ be fixed and letting $y$ range through $E$, we see that

$$\mu_\beta(y) = \frac{\mathbb{P}_\beta^x(X_1^{\{x,y\}} = y)}{\mathbb{P}_\beta^y(X_1^{\{x,y\}} = x)} \left(\sum_{z\in E} \frac{\mathbb{P}_\beta^x(X_1^{\{x,z\}} = z)}{\mathbb{P}_\beta^z(X_1^{\{x,z\}} = x)}\right)^{-1} \tag{7.2.2}$$

satisfies a large deviation principle stated in the following theorem:

**Theorem 7.2.2.** *The invariant probability measure* $\mu_\beta$ *of* $p_\beta$ *satisfies a large deviation principle: There exists a function* $U : E \to \mathbb{R}_+$, *called the* virtual energy function *of the Metropolis algorithm, such that for any state* $x \in E$,

$$\lim_{\beta\to+\infty} \frac{1}{\beta} \log \mu_\beta(x) = -U(x).$$

*Moreover, if* $(g_k)_{k=1}^{|E|}$ *is a reduction sequence, and* $A_k$ *the corresponding reduction sets, the relative energy levels* $U(g_k) - U(g_1)$ *can be computed through the following forward inductive computation:*

$$U(g_k) = \inf\{U(y) + V^{A_k}(y,g_k)\ :\ y \in A_{k-1}\} - \inf\{V^{A_k}(g_k,y)\ :\ y \in A_{k-1}\}.$$

*The value of* $U(g_1)$ *can then be deduced from the normalizing condition*

$$\min\{U(x)\ :\ x \in E\} = 0.$$

*Proof.* The existence of $U(x) \in \mathbb{R}_+ \cup \{+\infty\}$ is a consequence of (7.2.2). We need also to show that $U(x)$ cannot take the value $+\infty$. We saw on the occasion of the previous theorem that $V^{\{x,y\}}$ was irreducible. This implies that for any reduction set reduced to two states $x$ and $y$, both $V^{\{x,y\}}(x,y)$ and $V^{\{x,y\}}(y,x)$

[5] i.e. some ordering of the state space $E$.

are finite. Formula (7.2.2) then shows that $U(x)$ is always finite. The inductive computation of the virtual energy function can be deduced from the fact that $\mu_\beta(\ |A_k)$ is invariant with respect to $X^{A_k}$ :

$$\mu_\beta(g_k)\mathbb{P}_\beta^{g_k}(X_1^{A_k} \neq g_k) = \sum_{y\in A_{k-1}} \mu_\beta(y)\mathbb{P}_\beta^{y}(X_1^{A_k} = g_k).$$

□

## 7.3 Larve deviation estimates for the excursions from a domain

The framework will be the same as previously. We are going to state a large deviation principle for the trajectories

$$k \longmapsto X_{\tau_{n-1}(A)+k}, \qquad k = 0, \dots, \tau_n(A) - \tau_{n-1}(A).$$

of the excursions of the chain $X$ from a non empty domain $A$ of $E$. To this purpose, some notion of loop erased trajectories will be useful.

**Definition 7.3.1.** Let $\gamma = (\gamma_i)_{i=0}^r \in E^{r+1}$ be any given path. The definition of hitting times $\tau_n(A)$, which was given for infinite sequences of states $\omega \in E^{\mathbb{N}}$, can be generalized to the finite path $\gamma$. For this, it is enough to put

$$\tau_0(A)(\gamma) = \inf\{k \geq 0 \,:\, \gamma_k \in A\},$$
$$\tau_n(A)(\gamma) = \inf\{k > \tau_{n-1}(A)(\gamma) \,:\, \gamma_k \in A\},$$

considering that, in the case when the set of indices on the righthand side is empty, either because $\tau_{n-1} \geq r$, or because the condition $\gamma_k \in A$ is not fulfilled for any valid value of $k$, we set by convention $\inf \varnothing = +\infty$. For any domain $A$ of $E$, the reduced path $(\gamma_j^A)_{j=1}^{r^A(\gamma)}$ can then be defined by the formula

$$r^A(\gamma) = \inf\{n \in \mathbb{N} \,|\, \tau_{n+1}(A)(\gamma) = +\infty\},$$
$$\gamma_n^A = \gamma_{\tau_n(A)(\gamma)}, \qquad 0 \leq n \leq r^A(\gamma).$$

**Definition 7.3.2.** Let $(g_k)_{k=1}^{|E|}$ be a reduction sequence of the state space $E$ and let $A_k$ be the corresponding sequence of reduction sets. For any $1 \leq k \leq j \leq |E|$, let us define the set $\mathcal{V}_k^j$ of reduction paths from $A_j$ to $A_k$ by the formula:

$$\mathcal{V}_k^j = \{(\gamma_i)_{i=0}^n \in E^{r+1};\, r \geq 1, \{\gamma_0, \gamma_r\} \subset A_k,$$
$$(\gamma_i)_{i=1}^{r-1} \in (A_j \setminus A_k)^{r-1}, \gamma_{i-1}^{A_m\setminus A_k} \neq \gamma_i^{A_m\setminus A_k}, k < m \leq j, 0 < i \leq r^{A_m\setminus A_k}(\gamma)\}.$$

Let us stress the fact that a reduction path $\gamma^{A_m\setminus A_k}$ should not contain two times the same state in a row, but that $\gamma_0 = \gamma_r$ is allowed. Let us also notice that $\mathcal{V}_k^k = A_k^2$.

**Lemma 7.3.1.** *For any $1 \leq k < j \leq |E|$, the set $\mathcal{V}_k^j$ is finite.*

*Proof.* Indeed, in any path $\gamma \in \mathcal{V}_k^j$, the state $g_{k+1}$ can appear only zero or one time. In each of these two cases, the state $g_{k+2}$ can be present or not in two positions at most, on the lefthand and righthand side of $g_{k+1}$, when this state is present. More generally, the state $g_{k+\ell}$ can appear or not at $2^{\ell-1}$ positions at most, so that $|\mathcal{V}_j^k| \leq k^2 2^{2^{j-k}-1}$. □

**Definition 7.3.3.** For any $1 \leq k < |E|$, let us define the support of the trajectories of the excursions from $A_k$ by

$$\mathcal{E}_k = \Big\{ (\gamma_i)_{i=0}^r \in E^{r+1} \;:\; r \in \mathbb{N}, \{\gamma_0, \gamma_r\} \subset A_k, (\gamma_i)_{i=1}^{r-1} \in \big(E \setminus A_k\big)^{r-1} \Big\}.$$

For any excursion path $\gamma \in \mathcal{E}_k$, let us consider the times of first and last visit to the state $g_{k+1}$:

$$\begin{aligned} \tau_{k+1}(\gamma) &= \inf\{i \in \mathbb{N} \;:\; \gamma_i = g_{k+1}\} \in \mathbb{N} \cup \{+\infty\} \\ \sigma_{k+1}(\gamma) &= \sup\{i \in \mathbb{N} \;:\; \gamma_i = g_{k+1}\} \in \mathbb{N} \cup \{-\infty\}, \end{aligned}$$

where the values $+\infty$ and $-\infty$ respectively are assigned when the index sets are empty.

Let us define the concatenation $\big((\gamma \odot \gamma')_i\big)_{i=0}^{r+r'}$ of two given paths $(\gamma_i)_{i=0}^r$ and $(\gamma'_j)_{j=0}^{r'}$ such that $\gamma_r = \gamma'_0$ by

$$(\gamma \odot \gamma')_i = \begin{cases} \gamma_i & \text{when } 0 \leq i \leq r, \\ \gamma'_{i-r} & \text{when } r < i \leq r + r'. \end{cases}$$

(Let us stress the fact that the concatenation point is not repeated in the concatenated path.)

For any $1 \leq k \leq j \leq |E|$, let us then define the *erasor function*

$$\Gamma_k^j : \mathcal{E}_k \longrightarrow \mathcal{V}_k^j,$$

by the following backward induction on $k$:

$$\Gamma_j^j\big((\gamma_i)_{i=0}^r\big) = (\gamma_0, \gamma_r),$$

$$\Gamma_k^j\big((\gamma_i)_{i=0}^r\big) = \begin{cases} \Gamma_{k+1}^j\Big((\gamma_i)_{i=0}^{\tau_{k+1}(\gamma)}\Big) \odot \Gamma_{k+1}^j\Big((\gamma_i)_{i=\sigma_{k+1}(\gamma)}^r\Big), & \tau_{k+1}(\gamma) < +\infty, \\ \Gamma_{k+1}^j\big((\gamma_i)_{i=0}^r\big), & \tau_{k+1}(\gamma) = +\infty. \end{cases}$$

**Definition 7.3.4.** Some reduction sequence $(g_k)_{k=1}^{|E|}$ of the state space $E$ being fixed, a *depth function* $H : E \to \mathbb{R}_+$ can be defined by the following formula:

$$H(g_k) = \min\big\{V^{A_k}(g_k, y) : y \in A_{k-1}\big\}.$$

Let us stress the fact that *this definition depends on the choice of the reduction sequence*, and is therefore not intrinsic.

**Theorem 7.3.1 (Large deviation principle for excursions).** *Let us define the rate functions*

$$R_j(\gamma) = V^{A_j}(\gamma_0, \gamma_1) + \sum_{i=2}^{r} \Bigl( V^{A_j}(\gamma_{i-1}, \gamma_i) - H(\gamma_{i-1}) \Bigr),$$

$$\gamma \in \mathcal{V}_k^j, 1 \leq k \leq j \leq |E|.$$

*For any $\gamma \in \mathcal{V}_k^j$ the excursions satisfy*

$$\lim_{\beta \to +\infty} \frac{1}{\beta} \log \mathbb{P}_\beta^{\gamma_0} \left( \Gamma_k^j \left( (X_i)_{i=0}^{\tau_1(A_k)} \right) = \gamma \right) = -R_j(\gamma). \tag{7.3.1}$$

*Consequently, for any $(x, y) \in A_k^2$, any $k < j$,*

$$V^{A_k}(x, y) = \inf\{ R_j(\gamma) \,:\, (\gamma_i)_{i=0}^r \in \mathcal{V}_k^j,\ \gamma_0 = x,\ \gamma_r = y \}.$$

*Remark* 7.3.1. The fact that $\mathcal{V}_k^i$ is finite is crucial to get a large deviation principle. This is why it is legitimate to write

$$\lim_{\beta \to +\infty} \frac{1}{\beta} \log \mathbb{P}_\beta^x \Bigl( \Gamma_k^j \Bigl( X_0^{\tau_1(A_k)} \Bigr) \in \Lambda \Bigr) = - \inf_{\gamma \in \Lambda} R_j(\gamma),$$

and to deduce from this that

$$V^{A_k}(x, y) = \inf\bigl\{ R_j(\gamma) \,:\, (\gamma_i)_{i=0}^r \in \mathcal{V}_k^j,\ \gamma_0 = x,\ \gamma_r = y \bigr\}.$$

This is to be compared with the fact that, although

$$\lim_{\beta \to +\infty} \frac{1}{\beta} \log \mathbb{P}_\beta^x \bigl( X_0^{\tau_1(A_k)} = (\gamma_i)_{i=0}^r \bigr) = \sum_{i=1}^{r} V(\gamma_{i-1}, \gamma_i),$$

we *cannot* deduce from this that

$$V^{A_k}(x,\ y) \overset{\text{false!}}{=} \inf \left\{ \sum_{i=1}^{r} V(\gamma_{i-1}, \gamma_i) \,:\, \gamma_0 = x,\ \gamma_r = y,\ (\gamma_i)_{i=1}^{r-1} \in (E \setminus A_k)^{r-1} \right\}$$

*Proof.* In the case when $k = j$, the stated result is nothing but the large deviation principle for the transitions of reduced Markov chains, as it is stated in theorem 7.2.1.

The proof then proceeds by backward induction on $k$.

In the case when $\tau_{k+1}(\gamma) = +\infty$, (this is the case when the path $\gamma$ does not visit $g_{k+1}$,) then $\Gamma_{k+1}^j(\gamma) = \Gamma_k^j(\gamma)$ and

$$\Bigl\{ \Gamma_k^j \bigl( X_0^{\tau_1(A_k)} = \gamma \bigr) \Bigr\} = \Bigl\{ \Gamma_{k+1}^j \bigl( X_0^{\tau_1(A_{k+1})} = \gamma \bigr) \Bigr\}.$$

Therefore we can assume that equation (7.3.1) is satisfied in this case, according to our induction hypothesis.

In the case when $\tau_{k+1}(\gamma) < +\infty$, the paths $\gamma_0^{\tau_{k+1}(\gamma)}$ and $\gamma^r_{\tau_{k+1}(\gamma)}$ are excursion paths belonging to $\mathcal{E}_{k+1}$, since state $g_{k+1}$ is not repeated in $\gamma$. The definition of the erasor function then shows that

$$\left\{\Gamma_k^j(X_0^{\tau_1(A_k)}) = \gamma\right\} = \mathcal{A}_1 \cap \mathcal{A}_2,$$

where

$$\mathcal{A}_1 = \left\{\Gamma_{k+1}^j(X_0^{\tau_1(A_{k+1})}) = \gamma_0^{\tau_{k+1}(\gamma)}\right\},$$
$$\mathcal{A}_2 = \left\{\Gamma_{k+1}^j(X_{\sigma_1(A_k)}^{\tau_1(A_k)}) = \gamma^r_{\tau_{k+1}(\gamma)}\right\},$$

the last hitting time $\sigma_1(A_k)$ before $\tau_1(A_k)$ being defined as

$$\sigma_1(A_k) = \sup\{n < \tau_1(A_k) \,:\, X_n = g_{k+1}\}.$$

The strong Markov property can then be applied at time $\tau_1(A_{k+1})$, noting that under the event $\mathcal{A}_1$ it is always true that $X_{\tau_1(A_{k+1})} = \gamma_{\tau_{k+1}(\gamma)} = g_{k+1}$ and that moreover $\mathbb{1}_{\mathcal{A}_2} = \mathbb{1}_{\mathcal{A}_3} \circ \theta_{\tau_1(A_{k+1})}$, where

$$\mathcal{A}_3 = \left\{\Gamma_{k+1}^j\left(X_{\sigma_0(A_k)}^{\tau_0(A_k)}\right) = \gamma^r_{\tau_{k+1}(\gamma)}\right\},$$

with $\sigma_0(A_k) = \sup\{n < \tau_0(A_k) \,:\, X_n = g_{k+1}\} \in \mathbb{N} \cup \{-\infty\}$. We get

$$\mathbb{P}_\beta^{\gamma_0}(\mathcal{A}_1 \cap \mathcal{A}_2) = \mathbb{P}_\beta^{\gamma_0}(\mathcal{A}_1)\mathbb{P}_\beta^{g_{k+1}}(\mathcal{A}_3).$$

The probability $\mathbb{P}_\beta^{g_{k+1}}(\mathcal{A}_3)$ can then be decomposed into

$$\mathbb{P}_\beta^{g_{k+1}}(\mathcal{A}_3) = \sum_{n=0}^{+\infty} \mathbb{P}_\beta^{g_{k+1}}\left(\sigma_0(A_k) = n,\, \Gamma_{k+1}^j\left(X_{\sigma_0(A_k)}^{\tau_0(A_k)}\right) = \gamma^r_{\tau_{k+1}(\gamma)}\right),$$

and the two following remarks can be made:

- As $\sigma_0(A_k)$ is the last hitting time of the trajectory $X_0^{\tau_0(A_k)}$ at state $g_{k+1}$, $\tau_0(A_k)$ is also the first time $X$ hits $A_{k+1}$ after $\sigma_0(A_k)$. This leads to

$$\left(X_{\sigma_0(A_k)}^{\tau_0(A_k)}\right)(\omega) = \left(X_0^{\tau_1(A_{k+1})}\right) \circ \theta_{\sigma_0(A_k)}(\omega);$$

- The last time when the trajectory $X_0^{\tau_0(A_k)}$ hits $g_{k+1}$ is equal to $n$ if and only if $X_n = g_{k+1}$, $X$ did not hit back $A_k$ before and the next visit to $A_{k+1}$ is made at some state in $A_k$, that is not at state $g_{k+1}$. This can be written as

$$\{\sigma_0(A_k) = n\} = \left\{\tau_0(A_k) > n,\, X_n = g_{k+1} \text{ and } \left(X_{\tau_1(A_{k+1})}\right) \circ \theta_n \neq g_{k+1}\right\}.$$

These two remarks show that

$$\left\{\sigma_0(A_k) = n \text{ and } \Gamma^j_{k+1}\left(X^{\tau_0(A_k)}_{\sigma_0(A_k)}\right) = \gamma^r_{\tau_{k+1}(\gamma)}\right\}$$
$$= \left\{\tau_0(A_k) > n,\ X_n = g_{k+1} \text{ and } \Gamma^j_{k+1}\left(\left(X_0^{\tau_1(A_{k+1})}\right)\circ\theta_n\right) = \gamma^r_{\tau_{k+1}(\gamma)}\right\}.$$

Applying the Markov property at time $n$, we deduce that

$$\mathbb{P}^{g_{k+1}}_\beta(\mathcal{A}_3) = \sum_{n=0}^{+\infty}\mathbb{P}^{g_{k+1}}_\beta\big(\tau_0(A_k) > n,\ X_n = g_{k+1}\big)$$
$$\times\, \mathbb{P}^{g_{k+1}}_\beta\left(\Gamma^j_{k+1}\left(X_0^{\tau_1(A_{k+1})}\right) = \gamma^r_{\tau_{k+1}(\gamma)}\right).$$

Moreover

$$\sum_{n=0}^{+\infty}\mathbb{P}^{g_{k+1}}_\beta\big(\tau_0(A_k) > n,\ X_n = g_{k+1}\big)$$
$$= \mathbb{E}^{g_{k+1}}_\beta\left(\sum_{n=0}^{+\infty}\mathbb{1}(X_n = g_{k+1})\mathbb{1}\big(\tau_0(A_k) > n\big)\right)$$
$$= \mathbb{E}^{g_{k+1}}_\beta\left(\sum_{n=0}^{+\infty}\mathbb{1}(X_{\tau_n(A_{k+1})} = g_{k+1})\mathbb{1}\big(\tau_0(A_k) > \tau_n(A_{k+1})\big)\right)$$
$$= \sum_{n=0}^{+\infty}\mathbb{P}^{g_{k+1}}_\beta\big(X_m^{A_{k+1}} = g_{k+1},\ 0 \le m \le n\big)$$
$$= \sum_{n=0}^{+\infty}\mathbb{P}^{g_{k+1}}_\beta\big(X_1^{A_{k+1}} = g_{k+1}\big)^n$$
$$= \left(\mathbb{P}^{g_{k+1}}_\beta\big(X_1^{A_{k+1}} \neq g_{k+1}\big)\right)^{-1}.$$

This ends to prove that, when $\gamma \in \mathcal{V}^j_k$ and $\tau_{k+1}(\gamma) < +\infty$,

$$\mathbb{P}^{\gamma_0}_\beta\left(\Gamma^j_k\left(X_0^{\tau_1(A_k)}\right) = \gamma\right) = \mathbb{P}^{\gamma_0}_\beta\left(\Gamma^j_{k+1}\left(X_0^{\tau_1(A_{k+1})}\right) = \gamma_0^{\tau_{k+1}(\gamma)}\right)$$
$$\times\, \mathbb{P}^{g_{k+1}}_\beta\left(\Gamma^j_{k+1}\left(X_0^{\tau_1(A_{k+1})}\right) = \gamma^r_{\tau_{k+1}(\gamma)}\right)$$
$$\times \left(\mathbb{P}^{g_{k+1}}_\beta\big(X_1^{A_{k+1}} \neq g_{k+1}\big)\right)^{-1}.$$

Applying the induction hypothesis to the two first factors of this product and remembering the definition of $H(g_{k+1})$, we see that

$$-\lim_{\beta\to+\infty}\frac{1}{\beta}\log\mathbb{P}^{\gamma_0}_\beta\left(\Gamma^j_k\left(X_0^{\tau_1(A_k)}\right) = \gamma\right)$$

$$= V^{A_j}(\gamma_0, \gamma_1) + \sum_{i=2}^{\tau_{k+1}(\gamma)} \Big( V^{A_j}(\gamma_{i-1}, \gamma_i) - H(\gamma_{i-1}) \Big)$$
$$+ V^{A_j}\big(\gamma_{\tau_{k+1}(\gamma)}, \gamma_{\tau_{k+1}(\gamma)+1}\big)$$
$$+ \sum_{i=\tau_{k+1}(\gamma)+2}^{r} \Big( V^{A_j}(\gamma_{i-1}, \gamma_i) - H(\gamma_{i-1}) \Big) - H(\gamma_{\tau_{k+1}(\gamma)})$$
$$= R_j(\gamma).$$

□

## 7.4 Fast reduction algorithm

We will show in this section that it is sometimes possible to simplify the reduction algorithm described above by performing several steps at a time.

Let us first make some addition to the large deviation principle for the excursions from $A_k$ showing that the energy level of any state $y$ in $A_j \setminus A_k$ can be directly expressed as a function of the rate function $V^{A_j}$ and the energy level of the states of the domain $A_k$. Let us introduce on this occasion the set $\mathcal{U}_k^j$ of "truncated reduction paths":

**Definition 7.4.1.** Let us consider a reduction sequence $(g_k)_{k=1}^{|E|}$ and the corresponding reduction sets $A_k$. For any $k < j$, let us define the set of truncated reduction paths $\mathcal{U}_k^j$ as

$$\mathcal{U}_k^j = \Big\{ (\gamma_i)_{i=0}^r \in E^{r+1} \ : r \geq 1,\ \gamma_0 \in A_k,\ (\gamma_i)_{i=1}^r \in \big(A_j \setminus A_k\big)^r,$$
$$\gamma_{i-1}^{A_m} \neq \gamma_i^{A_m},\ k < m \leq j,\ 1 \leq i \leq r^{A_m}(\gamma) \Big\}.$$

The name given to $\mathcal{U}_k^j$ is meant to recall that for any reduction path $(\gamma_i)_{i=0}^r \in \mathcal{V}_k^j$, the truncated paths $(\gamma_i)_{i=0}^\ell$ (where $0 < \ell < r$) belong to $\mathcal{U}_k^j$.

The rate function $R_j$ can be extended to $\mathcal{U}_k^j$, putting for any $(\gamma_i)_{i=1}^r \in \mathcal{U}_k^j$

$$R_j(\gamma) = V^{A_j}(\gamma_0,\ \gamma_1) + \sum_{i=1}^{r} V^{A_j}(\gamma_{i-1},\ \gamma_i) - H(\gamma_{i-1}).$$

**Theorem 7.4.1.** *In the situation described in the previous definition, for any $k < j$, any state $y \in A_j \setminus A_k$,*

$$U(y) = \inf\{U(\gamma_0) + R_j(\gamma) - H(y)\ :\ r > 0,\ (\gamma_i)_{i=0}^r \in \mathcal{U}_k^j,\ \gamma_r = y\}.$$

*Proof.* We are going to make the proof by induction for $\ell = k+1, \ldots, j$ and $y = g_\ell$. Let us notice first that

$$\begin{aligned} U(g_\ell) &= \inf\{U(x) + V^{A_\ell}(x, y) - H(g_\ell) : x \in A_{\ell-1}\} \\ &= \inf\{U(\gamma_0) + R_j(\gamma) - H(g_\ell) : r > 0, \\ &\qquad (\gamma_i)_{i=0}^r \in \mathcal{V}_\ell^j, \gamma_0 \in A_{\ell-1}, \gamma_r = g_\ell\} \\ &= \inf\{U(\gamma_0) + R_j(\gamma) - H(g_\ell) : r > 0, (\gamma_i)_{i=0}^r \in \mathcal{U}_{\ell-1}^j, \gamma_r = y\}. \end{aligned}$$

This proves the theorem when $\ell = k+1$. In the case when $\ell > k+1$, the same equation, combined with the induction hypothesis, shows that

$$\begin{aligned} U(g_\ell) = \inf\{&U(\zeta_0) + R_j(\zeta) - H(\gamma_0) + R_j(\gamma) - H(g_\ell) : (\zeta_i)_{i=0}^s \in \mathcal{U}_k^j ; r > 0, \\ &(\gamma_i)_{i=0}^r \in \mathcal{U}_{\ell-1}^j, r > 0, \gamma_0 = \zeta_s\} \\ &\wedge \inf\{U(\gamma_0) + R_j(\gamma) - H(g_\ell) : (\gamma_i)_{i=0}^r \in \mathcal{U}_{\ell-1}^j, \gamma_0 \in A_k\}. \end{aligned}$$

But we can now make the following decomposition, based on the fact that the last time some path of $\mathcal{U}_k^j$ hits $A_{\ell-1}$ can be 0 or positive,

$$\begin{aligned} \mathcal{U}_k^j = \{\zeta \odot \gamma : (\zeta_i)_{i=0}^s \in \mathcal{U}_k^j, (\gamma_i)_{i=0}^r \in \mathcal{U}_{\ell-1}^j, \zeta_s = \gamma_0\} \\ \cup \{(\gamma_i)_{i=0}^r \in \mathcal{U}_{\ell-1}^j : \gamma_0 \in A_k\}. \end{aligned}$$

This ends the proof. □

Theorems 7.3.1 and 7.4.1 can be used to skip some steps in the recursive computation of the rate functions $V^{A_k}$ and the of the energy function $U$, as soon as it is possible to compute in one shot the depths $H(y)$ for any $y \in A_j \setminus A_k$. This is what the following proposition is about:

**Proposition 7.4.1.** *Let $(g_k)_{k=1}^{|E|}$ be any given reduction sequence and $A_k$ the corresponding reduction sets. Let us assume that for any $y \in A_j \setminus A_k$, there is a path $(\gamma_i)_{i=0}^r \in E^{r+1}$ such that $\gamma_0 = y$, $\gamma_r \in A_k$, $(\gamma_i)_{i=0}^{r-1} \in (A_j \setminus A_k)^r$ and such that moreover*

$$V^{A_j}(\gamma_{i-1}, \gamma_i) = \inf\{V(\gamma_{i-1}, z) : z \in A_j \setminus \{\gamma_{i-1}\}\}.$$

*In this case for any state $y \in A_j \setminus A_k$*

$$H(y) = \inf\{V^{A_j}(y, z) : z \in A_j \setminus \{z\}\}.$$

*Using this expression, it is possible to compute directly $V^{A_k}$ from $V^{A_j}$ in the backward phase of the reduction algorithm, with the help of the last part of theorem 7.3.1. Then in the forward phase when energy levels are computed, $U(y)$ can be directly computed for any $y \in A_j \setminus A_k$ from the values of $U$ in $A_k$ and of $V^{A_j}$, with the help of theorem 7.4.1.*

*Proof.* For any couple of imbedded domains $B \subset A$, let us consider the hitting times

$$\alpha^A(B) \overset{\text{def}}{=} \inf\{n > 0 : X_n^A \in B\}.$$

Let us notice first that for any $x \in A_j$

$$H(x) \geq \inf\{V^{A_j}(x,z) \,:\, z \in A_j \setminus \{x\}\}.$$

Indeed, in the case when $\ell$ is such that $x = g_\ell$,

$$\begin{aligned} H(x) &= \lim_{\beta\to+\infty} \frac{1}{\beta} \log \mathbb{E}_\beta^{g_\ell}\left( \sum_{n=0}^{\alpha^{A_\ell}(A_{\ell-1})} \mathbb{1}(X_n^{A_\ell} = g_\ell) \right) \\ &= \lim_{\beta\to+\infty} \frac{1}{\beta} \log \mathbb{E}_\beta^{g_\ell}\left( \sum_{n=0}^{\alpha^{A_j}(A_{\ell-1})} \mathbb{1}(X_n^{A_j} = g_\ell) \right) \\ &\geq \lim_{\beta\to+\infty} \frac{1}{\beta} \log \mathbb{E}_\beta^{g_\ell}\left( \sum_{n=0}^{\alpha^{A_j}(A_j\setminus\{g_\ell\})} \mathbb{1}(X_n^{A_j} = g_\ell) \right) \\ &= \inf\{V^{A_j}(g_\ell, z) \,:\, z \in A_j \setminus \{g_\ell\}\}, \end{aligned}$$

because $A_{\ell-1} \subset A_j \setminus \{g_\ell\}$, and therefore $\alpha^{A_j}(A_{\ell-1}) \geq \alpha(A_j \setminus \{g_\ell\})$.

The hypotheses we made show that there exists a path $(\gamma_i)_{i=0}^r \in E^{r+1}$ such that $\gamma_0 = g_\ell$, $\gamma_r \in A_{\ell-1}$, $(\gamma_i)_{i=0}^{r-1} \in (A_j \setminus A_{\ell-1})^r$ and

$$V^{A_j}(\gamma_{i-1}, \gamma_i) = \inf\{V(\gamma_{i-1}, z) \,:\, z \in A_j \setminus \{\gamma_{i-1}\}\}.$$

It is obtained by truncating the path appearing in the hypotheses at the first time it hits $A_{\ell-1}$. It is moreover possible to assume that $\gamma \in \mathcal{V}_\ell^j$, replacing it if necessary with $\Gamma_\ell^j(\gamma)$, which has the same properties. Theorem 7.3.1 then shows that

$$H(g_\ell) \leq R_j(\gamma).$$

In the same time, for any $i = 2, \ldots, r$,

$$V^{A_j}(\gamma_{i-1},\gamma_i) - H(\gamma_{i-1}) \leq V^{A_j}(\gamma_{i-1},\gamma_i) - \inf\{V^{A_j}(\gamma_{i-1}, z),\ z \in A_j \setminus \{z\}\} = 0,$$

implying that

$$R_j(\gamma) \leq V^{A_j}(\gamma_0, \gamma_1) = \inf\{V^{A_j}(g_\ell, z) \,:\, z \in A_j \setminus \{g_\ell\}\},$$

which ends the proof. □

## 7.5 Elevation function and cycle decomposition

In this section we are going to describe the energy landscape in a more intuitive way, generalizing the notion of attractor sets to decompose the state space into a hierarchy of cycles.

**Proposition 7.5.1.** *For any energy level $\lambda \in \mathbb{R}_+$, let us define the communication relation at level $\lambda$ as*

$$\mathcal{R}_\lambda = \{(x, y) \in E^2 : U(x) + V^{\{x,\, y\}}(x,\, y) < \lambda\} \cup \{(x, x) : x \in E\}.$$

*As it is defined, $\mathcal{R}_\lambda$ is an equivalence relation.*

*Proof.* The relation $\mathcal{R}_\lambda$ is symmetric. Indeed, we have already seen that

$$U(x) + V^{\{x,\, y\}}(x, y) = U(y) + V^{\{x,\, y\}}(y, x).$$

Let us show that it is also transitive. To this purpose, let us consider three points $x,\ y,\ z \in E^3$. Let us assume first that

$$V^{\{x,y,z\}}(y, z) \leq V^{\{x,y,z\}}(y, x).$$

In this case,

$$\begin{aligned} V^{\{x,z\}}(x, z) &= \min\Big\{V^{\{x,y,z\}}(x, z), \\ &\qquad V^{\{x,y,z\}}(x, y) + V^{\{x,y,z\}}(y, z) \\ &\qquad\qquad - \min\{V^{\{x,y,z\}}(y, x),\ V^{\{x,y,z\}}(y, z)\}\Big\} \\ &= \min\{V^{\{x,y,z\}}(x, z),\ V^{\{x,y,z\}}(x, y)\} \\ &\leq \min\Big\{V^{\{x,y,z\}}(x,\ z) + V^{\{x,y,z\}}(z,\ y) \\ &\qquad - \min\{V^{\{x,y,z\}}(z, x), V^{\{x,y,z\}}(z, y)\}, \\ &\qquad\quad V^{\{x,y,z\}}(x,\ y)\Big\} \\ &= V^{\{x,\, y\}}(x,\ y). \end{aligned}$$

In the same way, in the case when

$$V^{\{x,y,z\}}(y, z) > V^{\{x,y,z\}}(y, x),$$

it is seen by exchanging $x$ and $z$ that

$$V^{\{x,z\}}(z, x) \leq V^{\{y,z\}}(z, y),$$

hence

$$\begin{aligned} V^{\{x,z\}}(x,\ z) &= V^{\{x,\, z\}}(z, x) + U(z) - U(x) \\ &\leq V^{\{z,\, y\}}(z, y) + U(z) - U(x) \\ &= U(y) + V^{\{y,\, z\}}(y,\ z) - U(x). \end{aligned}$$

In any case we are thus sure that

$$U(x) + V^{\{x,z\}}(x, z) \leq \max\{U(x) + V^{\{x,y\}}(x, y),\ U(y) + V^{\{y,z\}}(y, z)\}.$$

The fact that $\mathcal{R}_\lambda$ is transitive for any value of $\lambda$ immediatly follows. □

**Definition 7.5.1.** The set of cycles $\mathcal{C}$ is defined as the set of all the components of $E$ for all the equivalence relations $\mathcal{R}_\lambda$ :

$$\mathcal{C} \stackrel{\text{def}}{=} \bigcup_{\lambda \in \mathbb{R}_+} E/\mathcal{R}_\lambda,$$

**Proposition 7.5.2.** *Any cycle $C \in \mathcal{C}$ containing at least two states is such that:*

$$V^{(E\setminus C)\cup\{x,\, y\}}(x,\, y) = V^{\{x,\, y\}}(x,\, y), \qquad x \neq y \in C$$
$$V^{(E\setminus C)\cup\{x,\, y\}}(x,\, y) < V^{(E\setminus C)\cup\{x,\, y\}}(x, z), \qquad x \neq y \in C,\ z \notin C.$$

*(The differences of energy levels between two states of $C$ therefore only depend on the restriction of $V$ to $C \times E$. Moreover, at low temperatures, an exponential number of round trips between any two points $x$ and $y$ of $C$ are performed before leaving* [6]*.)*

*Proof.* The probability to reach $y$ from $x$ before coming back to $x$ is not greater than the probability to reach $(E \setminus C) \cup \{y\}$ before coming back to $x$. Thus

$$\begin{aligned} V^{\{x,\, y\}}(x,\, y) &\geq \min\bigl\{V^{(E\setminus C)\cup\{x,y\}}(x,\, w) \,:\, w \in (E \setminus C) \cup \{y\}\bigr\} \\ &\geq \min\bigl\{V^{(E\setminus C)\cup\{x,y\}}(x,\, y),\ V^{\{x,w\}}(x,\, w) \,:\, w \in (E \setminus C)\bigr\}. \end{aligned}$$

As on the other hand, by definition of $C$,

$$V^{\{x,y\}}(x,\, y) < \min\bigl\{V^{\{x,w\}}(x,\, w) \,:\, w \in (E \setminus C)\bigr\},$$

it follows that necessarily

$$V^{\{x,y\}}(x,\, y) \geq V^{(E\setminus C)\cup\{x,y\}}(x, y).$$

Therefore these two quantities are equal. This shows the first part of the proposition. The second part follows from

$$\begin{aligned} V^{(E\setminus C)\cup\{x,\, y\}}(x,\, z) &\geq V^{\{x,z\}}(x,\, z) \\ &> V^{\{x,\, y\}}(x,\, y). \end{aligned}$$

□

**Lemma 7.5.1.** *For any path $(\gamma_i)_{i=0}^r \in \mathcal{V}_k^j$, any $i = 1, \ldots, r$, the two following quantities are non negative:*

$$V(\gamma_0,\, \gamma_1) + \sum_{j=2}^{i} V(\gamma_{j-1},\, \gamma_j) - H(\gamma_{j-1}) \geq 0 \tag{7.5.1}$$

$$\sum_{j=i+1}^{r} V(\gamma_{j-1},\, \gamma_j) - H(\gamma_{j-1}) \geq 0. \tag{7.5.2}$$

[6] It is an interesting exercise to show that this property is characteristic of cycles.

*Proof.* Let us prove it by backward induction on $k$. When $k = j$, the lemma just claims that $V^{A_j}$ is non negative. Let us assume now that the lemma is proved for any path of $\mathcal{V}_{k+1}^j$. Let us consider any given path $(\gamma_i)_{i=0}^r$ in $\mathcal{V}_k^j \setminus \mathcal{V}_{k+1}^j$. This path necessarily goes (only once) through the state $g_{k+1}$, let $m$ be the index satisfying $\gamma_m = g_{k+1}$. As $\gamma_0^m$ and $\gamma_m^r$ are both in $\mathcal{V}_{k+1}^j$, this shows equation (7.5.1) when $i \leq r$ and (7.5.2) when $i \geq m$. In the two other cases, after dividing the sums appearing in equations (7.5.1) and (7.5.2) at time $m$, one sees from the induction hypothesis that they are the sum of two non negative terms. □

**Definition 7.5.2.** Let us define the elevation of the path $(\gamma_i)_{i=0}^r$ as

$$\mathfrak{L}(\gamma) = \max_{i=1}^{r} U(\gamma_{i-1}) + V(\gamma_{i-1}, \gamma_i).$$

**Proposition 7.5.3.** *For any path* $(\gamma_i)_{i=0}^r \in \mathcal{V}_k^{|E|}$

$$R_j(\gamma) \geq \mathfrak{L}(\gamma) - U(\gamma_0).$$

*Consequently, for any couple* $(x, y) \in A_k^2$

$$V^{A_k}(x, y) \geq \min\bigl\{\mathfrak{L}((\gamma_i)_{i=0}^r) - U(\gamma_0) : (\gamma_i)_{i=1}^{r-1} \in (E \setminus A_k)^{r-1}, \gamma_0 = x, \gamma_r = y\bigr\}.$$

*Proof.* From the preceding lemma and theorem 7.4.1

$$R_j(\gamma) \geq V(\gamma_0, \gamma_1) + \sum_{j=2}^{i} V(\gamma_{i-1}, \gamma_i) - H(\gamma_{i-1})$$
$$\geq U(\gamma_{i-1}) + V(\gamma_{i-1}, \gamma_i) - U(\gamma_0).$$

□

**Proposition 7.5.4.** *For any couple of states* $(x, y) \in E^2$*, it is true that*

$$V^{\{x,y\}}(x, y) = \min\bigl\{\mathfrak{L}(\gamma) - U(\gamma_0) : (\gamma_i)_{i=0}^r \in E^{r+1}, r > 0, \gamma_0 = x, \gamma_r = y\bigr\}.$$

*Proof.* From the preceding proposition, it is enough to show that

$$V^{\{x,y\}}(x, y) \leq \min\bigl\{\mathfrak{L}(\gamma) - U(\gamma_0) : (\gamma_i)_{i=0}^r \in E^{r+1}, r > 0, \gamma_0 = x, \gamma_r = y\bigr\}.$$

Let therefore $(\gamma_i)_{i=0}^r \in E^{r+1}$ be a path leading from $x$ to $y$. We want to prove that $\mathfrak{L}(\gamma) \geq U(x) + V^{\{x,y\}}(x, y)$. The elevation of $\gamma$ is not increased when all the loops are removed (without exceptions, following some arbitrary scheme to do this). It is therefore enough to assume that $\gamma$ does not visit the same state twice. Let $S(\gamma)$ be the support of $\gamma$:

$$S(\gamma) = \bigl\{\gamma_i : i = 0, \ldots, r\bigr\}.$$

Let $\mathfrak{L}^{S(\gamma)}(\gamma)$ be the elevation of $\gamma$ for the reduced rate function $V^{S(\gamma)}$:

$$\mathfrak{L}^{S(\gamma)}(\gamma) \stackrel{\text{def}}{=} \max_{i=1}^{r} U(\gamma_{i-1}) + V^{S(\gamma)}(\gamma_{i-1}, \gamma_i).$$

As the reduction process cannot but decrease the rate function, $\mathfrak{L}^{S(\gamma)}(\gamma) \leq \mathfrak{L}(\gamma)$. It is therefore enough to show that $\mathfrak{L}^{S(\gamma)}(\gamma) \geq U(x) + V^{\{x,y\}}(x, y)$. This formula is true by definition when $\gamma$ is of length one (i.e. is reduced to two points forming a single edge). We are going to show that as long as $\gamma$ is of length larger than one, it can be replaced without increasing $\mathfrak{L}^{S(\gamma)}(\gamma)$ by a shorter path joining $x$ to $y$ or $y$ to $x$. This will show the proposition, due to the previous remark, because $U(x) + V^{\{x,y\}}(x, y) = U(y) + V^{\{x,y\}}(y, x)$.

For any domain $A \subsetneq E$, let

$$H^A(x) = \min\{V^A(x, y) \,:\, y \in A \setminus \{x\}\}, \qquad x \in A.$$

If for two indices $i$ and $j$ such that $0 < i < j \leq r$ it is true that

$$V^{S(\gamma)}(\gamma_i, \gamma_j) = H^{S(\gamma)}(\gamma_i),$$

then as a first step $\gamma_{i+1}, \dots, \gamma_{j-1}$ can be removed, and therefore it may be assumed that $j = i + 1$. It may then be noticed that

$$V^{S(\gamma)\setminus\{\gamma_i\}}(\gamma_{i-1}, \gamma_{i+1}) \leq V^{S(\gamma)}(\gamma_{i-1}, \gamma_i),$$

and $\gamma_i$ can be removed from $\gamma$. Let us assume now that for any $i > 0$, $H^{S(\gamma)}(\gamma_i)$ is reached at some point $\gamma_{j(i)}$ such that $j < i$. In this case, let us consider the path $(\gamma'_i)_{i=0}^{r'}$ joining $y$ to $x$ and defined by $\gamma'_i = \gamma_{\ell_i}$, with $\ell_i = j(\ell_{i-1})$. When $i > 0$,

$$\begin{aligned} U(\gamma'_i) + V^{S(\gamma')}(\gamma'_i, \gamma'_{i+1}) &\leq U(\gamma'_i) + V^{S(\gamma)}(\gamma'_i, \gamma'_{i+1}) \\ &= U(\gamma_{\ell_i}) + H^{S(\gamma)}(\gamma_{\ell_i}) \\ &\leq U(\gamma_{\ell_i}) + V^{S(\gamma)}(\gamma_{\ell_i}, \gamma_{\ell_i+1}) \\ &\leq \mathfrak{L}^{S(\gamma)}(\gamma). \end{aligned}$$

Moreover

$$\begin{aligned} U(y) + V^{S(\gamma')}(\gamma'_0, \gamma'_1) &\leq U(y) + V^{S(\gamma)}(\gamma'_0, \gamma'_1) \\ &= U(y) + H^{S(\gamma)} \\ &\leq U(\gamma_{r-1}) + V^{S(\gamma)}(\gamma_{r-1}, \gamma_r) \\ &\leq \mathfrak{L}^{S(\gamma)}(\gamma). \end{aligned}$$

From these two last equations, it follows that $\mathfrak{L}^{S(\gamma')}(\gamma') \leq \mathfrak{L}^{S(\gamma)}(\gamma)$. By construction of $\gamma'$,

$$V^{S(\gamma')}(\gamma'_1, \gamma'_2) \leq V^{S(\gamma)}(\gamma'_1, \gamma'_2) = H^{S(\gamma)}(\gamma'_1) \leq H^{S(\gamma')}(\gamma'_1).$$

Thus $V^{S(\gamma')}(\gamma_1', \gamma_2') = H^{S(\gamma')}(\gamma_1')$. It is seen from this discussion that $\gamma_1'$ can be removed from the path $\gamma'$, to form a shorter path. All the possible cases have been considered, and it has been shown that in each of them a shorter path of lower reduced elevation joining either $x$ to $y$ or $y$ to $x$ could be found. This ends the proof. □

*Remark* 7.5.1. We thus have shown that relation $\mathcal{R}_\lambda$ could also be defined from the minimal elevation of paths joining $x$ to $y$. Cycles can therefore also be considered as "the level sets of some virtual energy landscape".

**Proposition 7.5.5.** *For any cycle $C \in \mathcal{C}$, any $x \in C$, and any $z \notin C$,*

$$U(x) + V^{(E\setminus C)\cup\{x\}}(x, z) = \min\{U(y) + V(y, z) \ : \ y \in C\}$$

*Proof.* The elevation of any path joining $x$ to $z$ while remaining in $C \setminus \{x\}$ is not lower than $\min\{U(y) + V(y, z) \ : \ y \in C\}$. Therefore it is seen from proposition 7.5.3 that

$$U(x) + V^{(E\setminus C)\cup\{x\}}(x, z) \geq \min\{U(y) + V(y, z) \ : \ y \in C\}.$$

On the other hand, for any $y \in C$,

$$V^{(E\setminus C)\cup\{x\}}(x,\ z) \leq V^{(E\setminus C)\cup\{x,y\}}(x,\ y) + V^{(E\setminus C)\cup\{x,y\}}(y,\ z) - H^{(E\setminus C)\cup\{x,y\}}(y).$$

But from proposition 7.5.2,

$$V^{(E\setminus C)\cup\{x,y\}}(x,\ y) = V^{\{x,y\}}(x,\ y)$$

and

$$H^{(E\setminus C)\cup\{x,y\}}(y) = V^{\{x,y\}}(y,\ x),$$

thus

$$V^{(E\setminus C)\cup\{x,y\}}(x,\ y) - H^{(E\setminus C)\cup\{x,y\}}(y) = U(y) - U(x).$$

In the same way $V^{(E\setminus C)\cup\{x,y\}}(y,\ z) \leq V(y, z)$, hence

$$V^{(E\setminus C)\cup\{x\}}(x,\ z) \leq U(y) + V(y, z) - U(x).$$

□

**Proposition 7.5.6.** *Let us as previously define for any domain $A$ and any $x \in A$,*

$$H^A(x) = \min\{V^A(x, y) \ : \ y \in A \setminus \{x\}\}.$$

*Let us also define the first hitting time of $A$ :*

$$\alpha(A) = \inf\{n > 0 : X_n \in A\}.$$

*For any domain $A$ of $E$, any $x \in A$, the mean time spent in $x$ before leaving $A$ satisfies:*

$$\lim_{\beta \to +\infty} \frac{1}{\beta} \log \left[ \mathbb{E}_\beta^x \left( \sum_{n=0}^{\alpha(E \setminus A)} \mathbb{1}(X_n = x) \right) \right] = H^{(E\setminus A)\cup\{x\}}(x)$$
$$= \min\{V^{\{x,y\}}(x,y) \,:\, y \notin A\}$$
$$= \min\{\mathfrak{L}(\gamma) - U(x) \,:\, (\gamma_i)_{i=0}^r \in E^{r+1},\ \gamma_0 = x,\ \gamma_r \notin A\}.$$

*Proof.* Let us define more generally the reduced hitting times for any nested domains $A \subset B$ :

$$\alpha^B(A) = \inf\{n > 0 : X_n^B \in A\}.$$

The first equality stated in the proposition is a straightforward consequence of the following identities:

$$\mathbb{E}_\beta^x \left( \sum_{n=0}^{\alpha(E\setminus A)} \mathbb{1}(X_n = x) \right) = \mathbb{E}_\beta^x \Big[ \alpha^{(E\setminus A)\cup\{x\}}(E \setminus A) \Big]$$
$$= \sum_{n=0}^{+\infty} \mathbb{P}_\beta^x \Big[ \alpha^{(E\setminus A)\cup\{x\}}(E \setminus A) > n \Big]$$
$$= \sum_{n=0}^{+\infty} \mathbb{P}_\beta^x \big[ X_1^{(E\setminus A)\cup\{x\}} = x \big]^n$$
$$= \mathbb{P}_\beta^x \big[ X_1^{(E\setminus A)\cup\{x\}} \neq x \big]^{-1}.$$

Now by definition

$$H^{(E\setminus A)\cup\{x\}}(x) = \min\{V^{(E\setminus A)\cup\{x\}}(x,z) \,:\, z \in E \setminus A\}$$
$$= \min\{V^{\{x,y\}}(x,y) \,:\, y \in E \setminus A\}.$$

Indeed, as $\{x,y\} \subset (E \setminus A) \cup \{x\}$,

$$V^{\{x,y\}}(x,y) \leq V^{(E\setminus A)\cup\{x\}}(x,y).$$

On the other hand, if the chain reaches $y$ before coming back to $x$, then it hits $E \setminus A$ before coming back to $x$, and this shows that

$$\mathbb{P}_\beta^x (X_1^{\{x,y\}} = y) \leq \sum_{z \in E\setminus A} \mathbb{P}_\beta^x (X_1^{(E\setminus A)\cup\{x\}} = z),$$

and therefore that

$$V^{\{x,y\}}(x,y) \geq \min\{V^{(E\setminus A)\cup\{x\}}(x,z) \,:\, z \in E \setminus A\}.$$

This proves the second equality of the proposition. The third one is then a consequence of proposition 7.5.4. □

## 7.6 Mean hitting times and ordered reduction

We are going to study in this section the mean hitting times of the subsets of $E$. We will see how to use the reduction algorithm to compute some of them, when the reduction sequence is chosen in an appropriate way. As in the previous proof, we will use the following notation, where $A \subset B$ are two imbeded domains.

$$\alpha^B(A) \stackrel{\text{def}}{=} \inf\{n > 0 \,:\, X_n^B \in A\}.$$

We will also put $\alpha(A) \stackrel{\text{def}}{=} \alpha^E(A)$.

**Proposition 7.6.1 (Local potential).** *For any strict subdomain $A$ of $E$, any couple of states $(x,y) \in (E \setminus A)^2$ not belonging to $A$, the mean number of visits to $y$ starting from $x$ before reaching $A$ satisfies a large deviation principle whose rate function $W^A(x,y)$ will be called the local potential of $y$ in $E \setminus A$ starting from $x$ :*

$$\lim_{\beta \to +\infty} \frac{1}{\beta} \log \mathbb{E}_\beta^x \left( \sum_{n=0}^{\alpha(A)} \mathbb{1}(X_n = y) \right) = -W^A(x,y).$$

*Proof.* Let us apply first the strong Markov property to the first hitting time of $y$ before reaching $A$, which is equal to $\alpha(A \cup \{y\})$ when it exists:

$$\mathbb{E}_\beta^x \left( \sum_{n=0}^{\alpha(A)} \mathbb{1}(X_n = y) \right) = \mathbb{P}_\beta^x(X_{\alpha(A\cup\{y\})} = y) \mathbb{E}_\beta^y \left( \sum_{n=0}^{\alpha(A)} \mathbb{1}(X_n = y) \right). \quad (7.6.1)$$

The first factor on the righthand side can still be decomposed into

$$\begin{aligned}
\mathbb{P}_\beta^x(X_{\alpha(A\cup\{y\})} = y) &= \mathbb{P}_\beta^x(X^{A\cup\{x,y\}}_{\alpha^{A\cup\{x,y\}}(A\cup\{y\})} = y) \\
&= \sum_{n=0}^{+\infty} \left( \mathbb{P}_\beta^x(X_1^{A\cup\{x,y\}} = x) \right)^n \mathbb{P}_\beta^x(X_1^{A\cup\{x,y\}} = y) \\
&= \frac{\mathbb{P}_\beta^x(X_1^{A\cup\{x,y\}} = y)}{\mathbb{P}_\beta^x(X_1^{A\cup\{x,y\}} \neq x)}.
\end{aligned}$$

Moreover the second factor is equal to

$$\begin{aligned}
\mathbb{E}_\beta^y \left( \sum_{n=0}^{\alpha(A)} \mathbb{1}(X_n = y) \right) &= \mathbb{E}_\beta^y \left( \sum_{n=0}^{\alpha^{A\cup\{y\}}(A)} \mathbb{1}(X_n^{A\cup\{y\}} = y) \right) \\
&= \sum_{n=0}^{+\infty} \left( \mathbb{P}_\beta^y(X_1^{A\cup\{y\}} = y) \right)^n
\end{aligned}$$

$$= \Big( \mathbb{P}^y_\beta (X_1^{A \cup \{y\}} \neq y) \Big)^{-1}.$$

Applying theorem 7.2.1 ends the proof. □

**Proposition 7.6.2.** *For any imbeded domains $B \subset A$ the limit*

$$\lim_{\beta \to +\infty} \frac{1}{\beta} \log \mathbb{E}^x_\beta \left( \alpha^A(B) \right)$$

*exists and is equal to*

$$- \min_{y \in A \setminus B} W^B(x, y).$$

*Proof.* It comes from the following observation:

$$\mathbb{E}^x_\beta \left( \alpha^A(B) \right) = \sum_{y \in A \setminus B} \mathbb{E}^x_\beta \left( \sum_{n=0}^{\alpha(B)} \mathbb{1}(X_n = y) \right).$$

□

*Remark* 7.6.1. For any reduction sequence $(g_k)_{k=1}^{|E|}$, the correponding depth function $H$ is such that

$$W^{A_{k-1}}(g_k, g_k) = -H(g_k).$$

This is a straightforward consequence of the definitions.

*Remark* 7.6.2. Up to a constant, the virtual energy is equal to the local potential before reaching a point: for any couple of states $(x, y) \in E^2$

$$W^{\{x\}}(x, y) = U(y) - U(x).$$

This is an immediate consequence of equation (7.2.1).

*Remark* 7.6.3. The loacal potential before reaching a point also satisfies the following obvious relations:

$$W^{\{x\}}(y, y) = -V^{\{x,y\}}(y, x) = U(y) - U(x) - V^{\{x,y\}}(x, y).$$

**Proposition 7.6.3.** *For any cycle $C \in \mathcal{C}$,*

$$W^{E \setminus C}(x, y) = -H^{(E \setminus C) \cup \{y\}}(y), \qquad x, y \in C.$$

*Consequently*

$$\lim_{\beta \to +\infty} \frac{1}{\beta} \log \mathbb{E}^x_\beta \big( \sigma(E \setminus C) \big) = \min \big\{ U(y) + V(y, z) \, : \, y \in C, \; z \notin C \big\} - \min \big\{ U(w) \, : \, w \in C \big\}, \qquad x \in C.$$

*This quantity, which is independent from the initial point $x$, will be called the* depth of cycle $C$, *and will be denoted by $H(C)$.*

*Proof.* We saw that whithin a cycle

$$\lim_{\beta\to+\infty} \mathbb{P}^x_\beta\big(X_{\alpha((E\setminus C)\cup\{y\})} = y\big) = 1.$$

The conclusion is then easily deduced from equation (7.6.1). □

**Proposition 7.6.4.** *For any domain $A \subsetneq E$,*

$$\begin{aligned}\max_{x\in A}\lim_{\beta\to+\infty}\frac{1}{\beta}\log\mathbb{E}^x_\beta\big(\alpha(E\setminus A)\big)\\ =\max_{x\in A}\min\big\{\mathfrak{L}(\gamma)-U(x)\,:\,(\gamma_i)_{i=0}^r\in E^{r+1},\ \gamma_0=x,\ \gamma_r\notin A\big\}\\ =\max\big\{H(C)\,:\,C\in\mathcal{C},\ C\subset A\big\}.\end{aligned}$$

*Let $H(A)$ denote this quantity, which will be called the (maximal) depth of $A$.*

*Proof.* From proposition 7.6.2

$$\begin{aligned}\max_{x\in A}\lim_{\beta\to+\infty}\frac{1}{\beta}\log\mathbb{E}^x_\beta\big(\alpha(E\setminus A)\big) &= \max_{x\in A}\max_{y\in A} -W^A(x,y)\\ &= \max_{y\in A} -W^A(y,y)\\ &= \max_{x\in A} H^{(E\setminus A)\cup\{x\}}(x).\end{aligned}$$

Indeed $-W^A(x,y) \leq -W^A(y,y)$. The conclusion is brought by proposition 7.5.6. □

**Definition 7.6.1.** A reduction sequence $(g_k)_{k=1}^{|E|}$ will be said to be *ordered* if the corresponding reduction sets $A_k$ are such that

$$\min\big\{V^{A_k}(g_k,y)\,:\,y\in A_{k-1}\big\} = \min\big\{V^{A_k}(x,y)\,:\,x\in A_k,\ y\in A_k\setminus\{x\}\big\}.$$

This can equivalently be written as:

$$W^{A_{k-1}}(g_k,g_k) = \max\big\{W^{A_k\setminus\{x\}}(x,x)\,:\,x\in A_k\big\}.$$

(In other words, $A_{k-1}$ is deduced from $A_k$ by removing some state of "minimal depth".)

**Lemma 7.6.1.** *The depth function corresponding to any ordered reduction sequence $(g_k)_{k=1}^{|E|}$ is nonincreasing.*

*Proof.* Indeed for any $k = 1,\dots,|E|$,

$$H(g_k) = H^{A_k}(g_k) \leq H^{A_k}(g_{k-1}) \leq H^{A_{k-1}}(g_{k-1}) = H(g_{k-1}).$$

□

**Proposition 7.6.5.** *If $(g_k)_{k=1}^{|E|}$ is an ordered reduction sequence, then for any corresponding reduction set $A_k$, for any subset $B \subsetneq A_k$*

$$\lim_{\beta\to+\infty}\frac{1}{\beta}\log \mathbb{E}_\beta^x\big(\alpha^{A_k}(B)\big) = \lim_{\beta\to+\infty}\frac{1}{\beta}\log \mathbb{E}_\beta^x\big(\alpha(B)\big), \qquad x\in A_k\setminus B.$$

*In other words, the mean hitting times of the subdomains of $A_k$ can in first approximation be computed on the process reduced to $A_k$.*

*Proof.* By backward induction on $k$, it is enough to show that

$$\lim_{\beta\to+\infty}\frac{1}{\beta}\log \mathbb{E}_\beta^x\big(\alpha^{A_k}(B)\big) = \lim_{\beta\to+\infty}\frac{1}{\beta}\log \mathbb{E}_\beta^x\big(\alpha^{A_{k+1}}(B)\big).$$

Let us introduce the jump times of $X^{A_{k+1}}$ :

$$\begin{aligned}\xi_0^{k+1} &= 0,\\ \xi_n^{k+1} &= \inf\{i > \xi_{n-1}^{k+1} \,:\, X_i^{A_{k+1}} \neq X_{i-1}^{A_{k+1}}\}.\end{aligned}$$

We see that

$$\begin{aligned}\alpha^{A_{k+1}}(B) &= \sum_{n=0}^{+\infty}(\xi_{n+1}^{k+1}-\xi_n^{k+1})\mathbb{1}(\xi_n^{k+1} < \alpha^{A_{k+1}}(B)),\\ \alpha^{A_k}(B) &= \sum_{n=0}^{+\infty}(\xi_{n+1}^{k+1}-\xi_n^{k+1})\\ &\qquad \times \mathbb{1}(\xi_n^{k+1} < \alpha^{A_{k+1}}(B))\mathbb{1}(X_{\xi_n^{k+1}}^{A_{k+1}} \neq g_{k+1}).\end{aligned}$$

Applying the strong Markov property to times $\xi_n^{k+1}$, we get

$$\begin{aligned}&\mathbb{E}_\beta^x\big(\alpha^{A_{k+1}}(B)\big) - \mathbb{E}_\beta^x\big(\alpha^{A_k}(B)\big)\\ &\quad = \sum_{n=0}^{+\infty}\mathbb{E}_\beta^x\Big(\mathbb{1}(\xi_n^{k+1} < \alpha^{A_{k+1}}(B))\mathbb{1}(X_{\xi_n^{k+1}}^{A_{k+1}} = g_{k+1})\Big)\mathbb{E}_\beta^{g_{k+1}}(\xi_1^{k+1}).\end{aligned}$$

On the other hand

$$\begin{aligned}\mathbb{E}_\beta^x\big(\alpha^{A_k}(B)\big) &= \sum_{n=0}^{+\infty}\mathbb{E}_\beta^x\Big(\mathbb{1}(\xi_n^{k+1} < \alpha^{A_{k+1}}(B))\mathbb{1}(X_{\xi_n^{k+1}}^{A_{k+1}} \neq g_{k+1})\\ &\qquad\qquad \times \mathbb{E}_\beta^{X_{\xi_n^{k+1}}^{A_{k+1}}}(\xi_1^{k+1})\Big)\\ &\geq \sum_{n=1}^{+\infty}\mathbb{E}_\beta^x\Big(\mathbb{1}(\xi_n^{k+1} < \alpha^{A_{k+1}}(B))\mathbb{1}(X_{\xi_n^{k+1}}^{A_{k+1}} = g_{k+1})\\ &\qquad\qquad \times \mathbb{E}_\beta^{X_{\xi_{n-1}^{k+1}}^{A_{k+1}}}(\xi_1^{k+1})\Big)\end{aligned}$$

$$\geq \sum_{n=0}^{+\infty} \mathbf{E}_\beta^x\Big(\mathbb{1}(\xi_n^{k+1} < \alpha^{A_{k+1}}(B))\mathbb{1}(X^{A_{k+1}}_{\xi_n^{k+1}} = g_{k+1})\Big)$$
$$\times \min_{y \in A_k \setminus B} \mathbf{E}_\beta^y(\xi_1^{k+1}).$$

This shows that

$$\frac{\mathbf{E}_\beta^x\Big(\alpha^{A_{k+1}}(B) - \alpha^{A_k}(B)\Big)}{\mathbf{E}_\beta^x\Big(\alpha^{A_k}(B)\Big)} \leq \frac{\mathbf{E}_\beta^{g_{k+1}}(\xi_1^{k+1})}{\min_{y \in A_k \setminus B} \mathbf{E}_\beta^y(\xi_1^{k+1})} = \frac{\max_{y \in A_k \setminus B} \mathbf{P}_\beta^y(X_1^{A_{k+1}} \neq y)}{\mathbf{P}_\beta^{g_{k+1}}(X_1^{A_{k+1}} \neq g_{k+1})}.$$

From the definition of an ordered reduction sequence:

$$\lim_{\beta \to +\infty} \frac{1}{\beta} \log \left( \frac{\max_{y \in A_k \setminus B} \mathbf{P}_\beta^y(X_1^{A_{k+1}} \neq y)}{\mathbf{P}_\beta^{g_{k+1}}(X_1^{A_{k+1}} \neq g_{k+1})} \right) \leq 0.$$

This proves that

$$\lim_{\beta \to +\infty} \frac{1}{\beta} \log \mathbf{E}_\beta^x(\alpha^{A_{k+1}}(B)) \leq \lim_{\beta \to +\infty} \frac{1}{\beta} \log \mathbf{E}_\beta^x(\alpha^{A_k}(B)).$$

As moreover $\alpha^{A_k}(B) \leq \alpha^{A_{k+1}}(B)$, this shows that these two limits are equal. □

**Proposition 7.6.6.** (Characterization of the depth functions corresponding to ordered reduction sequences) *Let $(g_k)_{k=1}^{|E|}$ be some ordered reduction sequence. The corresponding depth function $H$ is such that*

$$\begin{aligned} H(g_k) &= \min\Big\{V^{\{g_k, y\}}(g_k,\, y) \;:\; U(y) < U(g_k) \\ &\qquad\qquad \textit{where } \big(U(y) = U(g_k) \textit{ and } y \in A_{k-1}\big)\Big\} \\ &= \min\Big\{V^{\{g_k, y\}}(g_k,\, y) \;:\; U(y) \leq U(g_k),\; y \in A_{k-1}\Big\}, \\ &= \min\Big\{V^{\{g_k, y\}}(g_k,\, y) \;:\; y \in A_{k-1}\Big\}, \qquad k > 1. \end{aligned}$$

*Thus in the non degenerate case when the virtual energy function $U$ is one to one, the depth function has an intrinsic characterization: the depth of a given state is the minimum of the rate function of the time spent in this state before reaching a state of lower energy.*

**Corollary 7.6.1.** *The state $g_1$ is a fundamental state: $U(g_1) = 0$.*

*Proof.* Let $\ell < k$ be such that

$$H(g_k) = V^{A_k}(g_k, g_\ell).$$

As $\mu_\beta(\cdot \mid A_k)$ is invariant under $X^{A_k}$,

$$U(g_\ell) + H^{A_k}(g_\ell) = \min\{U(x) + V^{A_k}(x, g_\ell) \,:\, x \in A_k \setminus \{g_\ell\}\}.$$

As a consequence,

$$U(g_\ell) \le U(g_k) + H(g_k) - H^{A_k}(g_\ell) \le U(g_k).$$

On the other hand, $H(g_k) \le V^{\{g_k, g_\ell\}}(g_k, g_\ell)$ and $V^{\{g_k, g_\ell\}}(g_k, g_\ell) \le V^{A_k}(g_k, g_\ell)$. This shows that $H(g_k) = V^{\{g_k, g_\ell\}}(g_k, g_\ell)$. Moreover, for any $y \in A_{k-1}$, $\{y, g_k\} \subset A_k$, and therefore

$$V^{\{g_k, y\}}(g_k, y) \ge H^{A_k}(g_k) = H(g_k).$$

This proves that

$$\begin{aligned} H(g_k) &= \min\{V^{\{g_k, y\}}(g_k, y) \,:\, y \in A_{k-1}\} \\ &= \min\{V^{\{g_k, y\}}(g_k, y) \,:\, y \in A_{k-1},\, U(y) \le U(g_k)\}. \end{aligned}$$

Let us consider now some index $\ell > k$ such that $U(g_\ell) < U(g_k)$ (if such an index exists). Applying the same reasoning as previously to $g_\ell$, we can build $m < \ell$ such that $H(g_\ell) = V^{\{g_\ell, g_m\}}(g_\ell, g_m)$ and $U(g_m) \le U(g_\ell)$. The symmetry of the elevation function then shows that

$$\begin{aligned} U(g_\ell) + V^{\{g_\ell, g_m\}}(g_\ell, g_m) &= U(g_\ell) + H(g_\ell) \\ &\le U(g_\ell) + V^{\{g_\ell, g_k\}}(g_\ell, g_k) \\ &= U(g_k) + V^{\{g_k, g_\ell\}}(g_k, g_\ell), \end{aligned}$$

consequently, the elevation function being transitive,

$$\begin{aligned} U(g_k) + V^{\{g_k, g_m\}}(g_k, g_m) &\le \max\{U(g_k) + V^{\{g_k, g_\ell\}}(g_k, g_\ell), \\ &\qquad\qquad U(g_\ell) + V^{\{g_\ell, g_m\}}(g_\ell, g_m)\} \\ &= U(g_k) + V^{\{g_k, g_\ell\}}(g_k, g_\ell). \end{aligned}$$

Thus we have built from $g_\ell$ a state $g_m$ such that $m < \ell$, $U(g_m) \le U(g_\ell) < U(g_k)$ and $V^{\{g_k, g_m\}}(g_k, g_m) \le V^{\{g_k, g_\ell\}}(g_k, g_\ell)$. Repeating this operation a finite number of times, we can make the value of $m$ not greater than $k$. The possibility that $g_k = g_m$ is precluded by the inequality $U(g_m) < U(g_k)$, implying that indeed $m < k$. □

## 7.7 Convergence speeds

**Proposition 7.7.1.** *For any domain $A \subsetneq E$, any $x \in A$, any $\epsilon > 0$,*

$$\lim_{\beta \to +\infty} \frac{1}{\beta} \log \mathbb{P}_\beta^x\Big(\alpha(E \setminus A) > e^{\beta\left(H(A)+\epsilon\right)}\Big) = -\infty.$$

*Proof.* Applying the Markov property, we see that for any large enough value of $\beta$

$$\begin{aligned}\mathbb{P}_\beta^x\big(\alpha(E\setminus A) > e^{\beta(H(A)+\epsilon}\big) &\leq \Big(\max_{y\in A}\mathbb{P}_\beta^y\big(\alpha(E\setminus A) > e^{\beta(H(A)+\epsilon/2)}\big)\Big)^{\lfloor\exp(\beta\epsilon/2)\rfloor}\\ &\leq \Big(\max_{y\in A}\mathbb{E}_\beta^y\big(\alpha(E\setminus A)\big)e^{-\beta\left(H(A)+\epsilon/2\right)}\Big)^{\lfloor\exp(\beta\epsilon/2)\rfloor}\\ &\leq \exp\Big(-\beta\big(\exp(\beta\epsilon/2)-1\big)\epsilon/4\Big).\end{aligned}$$

□

**Lemma 7.7.1.** *For any domain $A \subsetneq E$, any $\epsilon > 0$, any $x \in A$ such that $H^{(E\setminus A)\cup\{x\}}(x) = H(A)$,*

$$\limsup_{\beta\to+\infty}\frac{1}{\beta}\log\bigg(\mathbb{P}_\beta^x\Big(\alpha(E\setminus A) < e^{\beta\left(H(A)-\epsilon\right)}\Big)\bigg) \leq -\epsilon.$$

*Proof.* We can notice that

$$\begin{aligned}&\mathbb{P}_\beta^x\Big(\alpha(E\setminus A) < \exp\big(\beta(H(A)-\epsilon)\big)\Big)\\ &\qquad\leq \mathbb{P}_\beta^x\left(\sum_{n=0}^{\alpha(E\setminus A)}\mathbb{1}(X_n = x) < \exp\big(\beta(H(A)-\epsilon\big)\right)\\ &\qquad\qquad = 1-\Big(\mathbb{P}_\beta^x\big(X_1^{(E\setminus A)\cup\{x\}} = x\big)\Big)^{\left\lceil\exp\left(\beta(H(A)-\epsilon)\right)\right\rceil}.\end{aligned}$$

Moreover for any $\eta > 0$ there exists $\beta_0$, such that for any $\beta \geq \beta_0$,

$$\mathbb{P}_\beta^x\big(X_1^{(E\setminus A)\cup\{x\}} \neq x\big) \leq \exp\big(-\beta(H(A)-\eta)\big).$$

Therefore, for any large enough value of $\beta$,

$$\begin{aligned}&\mathbb{P}_\beta^x\Big(\alpha(E\setminus A) < \exp\big(\beta(H(A)-\epsilon)\big)\Big)\\ &\qquad\leq 1-\Big(1-\exp\big(-\beta(H(A)-\eta)\big)\Big)^{\left\lceil\exp\left(\beta(H(A)-\epsilon)\right)\right\rceil}\\ &\qquad\leq 1-\exp\Big(-2\exp\big(-\beta(\epsilon-\eta)\big)\Big)\\ &\qquad\leq 2(1-e^{-1})\exp\big(-\beta(\epsilon-\eta)\big).\end{aligned}$$

Letting $\eta$ tend to 0 then ends the proof. □

**Proposition 7.7.2.** *For any cycle $C \in \mathcal{C}$, any small enough $\epsilon > 0$, any $x \in C$,*

$$\limsup_{\beta\to+\infty}\frac{1}{\beta}\log\mathbb{P}_\beta^x\Big(\alpha(E\setminus C) < e^{\beta\left(H(C)-\epsilon\right)}\Big) \leq -\epsilon.$$

*Proof.* Let $x \in C$ and let $y \in C$ be such that $U(y) = \min_{z \in C} U(z)$. This implies that $H^{(E \setminus C) \cup \{y\}}(y) = H(C)$. Applying the strong Markov property to the first hitting time of $y$ starting from $x$, we see that for any $\chi > 0$,

$$\mathbb{P}_\beta^x\Big(\alpha(E \setminus C) < e^{\beta\chi}\Big) \leq \mathbb{P}_\beta^y\Big(\alpha(E \setminus C) < e^{\beta\chi}\Big)\mathbb{P}_\beta^x(X_{\alpha\left((E\setminus C)\cup\{y\}\right)} = y) + \mathbb{P}_\beta^x\Big(X_{\alpha\left((E\setminus C)\cup\{y\}\right)} \neq y\Big).$$

Let us then put

$$\begin{aligned}\epsilon_0 &= \min_{x,\, x\neq y \in C} \lim_{\beta \to +\infty} -\frac{1}{\beta} \log \mathbb{P}_\beta^x\Big(X_{\alpha\left((E\setminus C)\cup\{y\}\right)} \neq y\Big) \\ &= \min\big\{V^{(E\setminus C)\cup\{x,y\}}(x,z) - V^{(E\setminus C)\cup\{x,y\}}(x,y) \,:\, x \neq y \in C,\ z \in (E \setminus C)\big\} \\ &> 0.\end{aligned}$$

For any $\epsilon < \epsilon_0$, for any large enough $\beta$,

$$\mathbb{P}_\beta^x(\alpha(E \setminus C) < \exp(\beta\chi)) \leq \mathbb{P}_\beta^y(\alpha(E \setminus C) < \exp(\beta\chi)) + \exp(-\beta\epsilon).$$

The proof can now be completed considering the case when $\chi = H(C) - \epsilon$ and applying the preceding lemma. □

**Proposition 7.7.3.** *Let $(g_k)_{k=1}^{|E|}$ be some ordered reduction sequence. The three critical depths $H_1$, $H_2$ and $H_3$ of the rate function $V$ may be defined as*

$$\begin{aligned}H_1 &= \max\{H(x),\, x \in E,\, U(x) > 0\} \\ &= H(U^{-1}(]0, +\infty[) \\ &= \max_{x \in E, U(x) > 0} \min\{V^{\{x,y\}}(x,\, y) \,:\, y \in U^{-1}(\{0\})\}, \\ H_2 &= H(g_2) \\ &= H(E \setminus \{g_1\}) \\ &= \max\{V^{\{x,y\}}(x,\, y) \,:\, U(y) = 0\}, \\ H_3 &= H\big((E \times E) \setminus \{(x,x) \,:\, x \in E\}\big) \\ &= \max_{(x,y)\in E^2} \min_{z \in E} V_2^{\{(x,y),(z,z)\}}\big((x,y),(z,z)\big),\end{aligned}$$

*where we have considered on the product state space $E \times E$ the product chain distributed according to $\mathbb{P}_\beta \otimes \mathbb{P}_\beta$, whose rate function $V_2 : E^2 \times E^2 \longrightarrow \mathbb{R}_+ \cup \{+\infty\}$ is defined by:*

$$V_2\big((x,y),(z,t)\big) = V(x,z) + V(y,t).$$

*These definitions are independent of the choice of the ordered reduction sequence. In general $H_1 \leq H_2 \leq H_3$ may be distinct. However, if the graph* *of null cost jumps on the fundamental states* $\{(x,y) : U(x) = U(y) = 0,\ V(x,y) = 0\}$ *has an aperiodic component, then* $H_2 = H_3$. *Moreover, if there is a unique fundamental state (i.e. a unique state of minimal virtual energy) then* $H_1 = H_2 = H_3$.

*Proof.* The properties of $H_3$ are the only ones which still require a proof. The inequality $H_3 \geq H_2$ can be established by considering in the definition of $H_3$ the case when $x = g_1$ and $y = g_2$. The states $g_1$ and $g_2$ are two fundamental states of two disjoint cycles of $E$ with depths not lower than $H_2$. The desired conclusion comes from the obvious inequality

$$V_2^{\{(x,y),\,(z,z)\}}\big((x,y),\,(z,z)\big) \leq \max\Big\{V^{\{x,z\}}(x,z),\ V^{\{y,z\}}(y,z)\Big\}.$$

Then it remains to show that $H_2 = H_3$ when the graph of null cost jumps on the fundamental states has an aperiodic component. Let $(x,y)$ be some given point in $E \times E$. Let $z$ be a state of null virtual energy and let $n_0$ be some integer such that for any integer $n \geq n_0$ there is a path $(\psi_i^n)_{i=0}^n \in E^{n+1}$ of length $n$ such that $V(\psi_{i-1}^n,\ \psi_i^n) = 0$, $i = 1, \ldots, n$ and $\psi_0 = \psi_n = z$. Let $(\gamma_i^1)_{i=0}^{r_1}$ be some path in $E$ of elevation not greater than $U(x) + H_2$ joining $x$ to $\gamma_{r_1}^1 = z$. Let $(\gamma_i^2)_{i=0}^{r_1}$ be some path of the same length starting from $\gamma_0^2 = y$, such that $V(\gamma_{i-1}^2,\ \gamma_i^2) = 0$, $i = 1, \ldots, r_1$. The elevation of $\gamma^2$ is therefore equal to $U(y)$. Moreover, as the elevation function is symmetric, $U(\gamma_{r_1}^2) \leq U(y)$. Let $(\gamma_i^3)_{i=0}^{r_3}$ be some path of elevation not greater than $U(y) + H_2$ joining $\gamma_{r_1}^2$ to $z$. Let $(\gamma_i^4)_{i=0}^{r_4}$ and $(\gamma_i^5)_{i=0}^{r_5}$ be some paths joining $z$ to itself of null elevation such that $r_4 = r_3 + r_5$. Then in the product state space $E \times E$ the path $(\gamma^1, \gamma^2) \odot (\gamma^4, \gamma^3 \odot \gamma^5)$ goes from $(x,y)$ to $(z,z)$ and has an elevation not greater than $U(x) + U(y) + H_2$. This proves that $H_3 \leq H_2$, and therefore that $H_3 = H_2$, because the reverse inequality is always satisfied. □

**Theorem 7.7.1.** *For any* $\chi > H_1$,

$$\limsup_{\beta \to +\infty} \frac{1}{\beta} \log \mathbb{P}_\beta^x\big(X_{\lfloor \exp(\beta\chi) \rfloor} = y\big) \leq -U(y), \quad (x,y) \in E^2.$$

*For any* $\chi > H_2$,

$$\lim_{\beta \to +\infty} \frac{1}{\beta} \log \mathbb{P}_\beta^x\big(\alpha(\{y\}) > \exp(\beta\chi)\big) = -\infty, \qquad (x,y) \in E^2,\ U(y) = 0.$$

*For any* $\chi > H_3$,

$$\lim_{\beta \to +\infty} \frac{1}{\beta} \mathbb{P}_\beta^x\big(X_{\lfloor \exp(\beta\chi) \rfloor} = y\big) = -U(y), \quad (x,y) \in E^2.$$

*Conversely, for any* $\chi < H_1$, *there is* $x \in E$ *such that*

$$\limsup_{\beta \to +\infty} \log \mathbb{P}_\beta^x \big( U(X_{\lfloor \exp(\beta\chi) \rfloor}) = 0 \big) < 0.$$

*For any $\chi < H_2$, and any $y \in E$ such that $U(y) = 0$, there exists $x \in E$ such that*

$$\limsup_{\beta \to +\infty} \frac{1}{\beta} \log \mathbb{P}_\beta^x \Big( \alpha(\{y\}) \leq \exp(\beta\chi) \Big) < 0.$$

*For any $\chi < H_3$, any $y \in E$ such that $U(y) = 0$, there exists $x \in E$ such that*

$$\liminf_{\beta \to +\infty} \frac{1}{\beta} \log \mathbb{P}_\beta^x \big( X_{\lfloor \exp(\beta\chi) \rfloor} = y \big) < 0.$$

*Remark* 7.7.1. The first assertion of this convergence theorem is the most useful for applications to optimization. It shows that it is necessary to choose a number of iterations $N = \lfloor \exp(\beta\chi) \rfloor$ slightly larger than $\exp(\beta H_1)$. Alternatively, we can choose $\beta = \frac{\log N}{\chi}$ as a function of $N$, to get a probability of error bounded by

$$\limsup_{N \to +\infty} \frac{1}{\log N} \log \max_{x \in E} \mathbb{P}_{\beta(N)}^x \big( U(X_N) \geq \eta \big) \leq -\frac{\eta}{\chi},$$

and therefore decreasing at least as a power of $1/N$ arbitrarily close to the critical constant $\eta/H_1$. In the case when the event of interest is $\big\{ U(X_N) > 0 \big\}$, we get as a critical exponent for the probability of error $\min\big(U^{-1}(]0, +\infty[)\big)/H_1$. We will see how to use an inhomogeneous Markov chain (i.e. a simulated annealing algorithm) to get a larger critical exponent, independent of the discretization step of small energy levels.

*Proof.* Let us put $N = \lfloor \exp(\beta\chi) \rfloor$ and $G = U^{-1}(0)$. For any couple of states $(x, y) \in E^2$,

$$\mathbb{P}_\beta^x (X_n = y) \leq \mathbb{P}_\beta \big( \alpha(G) > \exp(\beta\chi) \big) + \sup_{n \in \mathbb{N}, z \in G} \mathbb{P}_\beta^z (X_n = y).$$

Let us put, for some fixed $z \in G$,

$$f_n(x) = \frac{\mathbb{P}_\beta^z (X_n = x)}{\mu_\beta(x)}.$$

This function follows the following evolution equation:

$$f_{n+1}(y) = \sum_{x \in E} f_n(x) p_\beta(x, y) \frac{\mu_\beta(x)}{\mu_\beta(y)}.$$

As

$$\sum_{x \in E} p_\beta(x, y) \frac{\mu_\beta(x)}{\mu_\beta(y)} = 1,$$

we deduce that

$$\max_{y\in E} f_n(x) \le \max_{y\in E} f_0(x) = \frac{1}{\mu_\beta(z)},$$

end therefore that

$$\sup_{n\in\mathbb{N}} \mathbb{P}_\beta\big(X_n = y \,|\, X_0 = z\big) \le \frac{\mu_\beta(y)}{\mu_\beta(z)}.$$

Thus

$$\mathbb{P}_\beta^x\big(X_n = y\big) \le \mathbb{P}_\beta\big(\alpha(G) > \exp(\beta\chi)\big) + \max_{z\in G} \frac{\mu_\beta(y)}{\mu_\beta(z)},$$

whence the first inequality of the theorem is easily deduced. The second claim is a direct consequence of proposition 7.6.4.

The third claim can be proved by a coupling argument. Let us consider on the product space $\big(E^{\mathbb{N}} \times E^{\mathbb{N}}\big)$ equiped with its canonical process $(X_n, Y_n)_{n\in\mathbb{N}}$ the product measure $\mathbb{P}_\beta^x \otimes \mathbb{P}_\beta^{\mu_\beta}$, where we have written $\mathbb{P}_\beta^{\mu_\beta}$ to denote the distribution of the stationary Markov chain with initial distribution $\mu_\beta$. Let us also define the hitting time of the diagonal

$$\xi = \inf\{n \ge 0 \,:\, X_n = Y_n\}.$$

We see that

$$\begin{aligned}\mathbb{P}_\beta^x\big(X_n = y\big) &= \mathbb{P}_\beta^x \otimes \mathbb{P}_\beta^{\mu_\beta}\big(X_n = y \text{ et } \xi \le n\big) + \mathbb{P}_\beta^x \otimes \mathbb{P}_\beta^{\mu_\beta}\big(X_n = y \text{ et } \xi > n\big)\\ &= \mathbb{P}_\beta^x \otimes \mathbb{P}_\beta^{\mu_\beta}\big(Y_n = y \text{ et } \xi \le n\big) + \mathbb{P}_\beta^x \otimes \mathbb{P}_\beta^{\mu_\beta}\big(X_n = y \text{ et } \xi > n\big)\end{aligned}$$

and therefore that

$$\mu_\beta(y) - \mathbb{P}_\beta^x \otimes \mathbb{P}_\beta^{\mu_\beta}\big(\xi > n\big) \le \mathbb{P}_\beta^x\big(X_n = y\big) \le \mu_\beta(y) + \mathbb{P}_\beta^x \otimes \mathbb{P}_\beta^{\mu_\beta}\big(\xi > n\big).$$

We can now choose $n(\beta) = \lfloor \exp(\beta\chi) \rfloor$ with $\chi > H_3$, to be sure, according to proposition 7.6.4, that

$$\lim_{\beta\to+\infty} \frac{1}{\beta} \log \mathbb{P}_\beta^x \otimes \mathbb{P}_\beta^{\mu_\beta}\big(\xi > n(\beta)\big) = -\infty.$$

The third claim of the theorem follows immediately.

To show the first of the three converse claims, it is enough to consider some initial point $x \in U^{-1}(]0, +\infty[)$ belonging to some cycle $C \subset U^{-1}(]0, +\infty[)$ of maximal depth $H(C) = H(U^{-1}(]0, +\infty[)$ (from proposition 7.6.4) and to apply proposition 7.7.2.

The second converse claim is shown as the first, considering some cycle of maximal depth included in $E \setminus \{y\}$.

As for the third converse claim, let us consider some fundamental state $z$ of null virtual energy $U(z) = 0$, some couple of states $(x, y) \in E^2$ belonging to some cycle of maximal depth of $E^2 \setminus \{(z,z)\}$. According to the definition of $H_3$, for any $\chi < H_3$ the following inequality is then satisfied:

$$\limsup_{\beta \to +\infty} \frac{1}{\beta} \log \mathbb{P}^x_\beta \otimes \mathbb{P}^y_\beta \big(\alpha(\{(z,z)\}) \leq \exp(\beta\chi)\big) < 0.$$

A fortiori it follows that

$$\limsup_{\beta \to +\infty} \frac{1}{\beta} \log \mathbb{P}^x_\beta \otimes \mathbb{P}^y_\beta \big((X,Y)_{\lfloor \exp(\beta\chi) \rfloor} = (z,z)\big) < 0,$$

where $(X,Y)$ denotes the canonical process on $(E \times E)^{\mathbb{N}}$. The desired conclusion now comes from noticing that

$$\min\Big\{\mathbb{P}^x_\beta\big(X_{\lfloor \exp(\beta\chi) \rfloor} = z\big),\ \mathbb{P}^y_\beta\big(X_{\lfloor \exp(\beta\chi) \rfloor} = z\big)\Big\} \\ \leq \sqrt{\mathbb{P}^x_\beta \otimes \mathbb{P}^y_\beta\big((X,Y)_{\lfloor \exp(\beta\chi) \rfloor} = (z,z)\big)},$$

and therefore that one of the two following claims has to be true: either

$$\liminf_{\beta \to +\infty} \frac{1}{\beta} \log \mathbb{P}^x_\beta\big(X_{\lfloor \exp(\beta\chi) \rfloor} = z\big) < 0,$$

or

$$\liminf_{\beta \to +\infty} \frac{1}{\beta} \log \mathbb{P}^y_\beta\big(X_{\lfloor \exp(\beta\chi) \rfloor} = z\big) < 0.$$

□

## 7.8 Generalized simulated annealing algorithm

In this section, we consider the simulated annealing algorithm corresponding to some generalized Metropolis algorithm, that is the family of inhomogeneous Markov chains

$$\mathcal{R} = \Big(E^{\mathbb{N}}, (X_n)_{n \in \mathbb{N}}, \mathcal{B}, \mathbb{P}_{(\beta)}\Big)_{(\beta) \in \mathcal{C}}$$

defined in the beginning of this chapter.

We are going to study its convergence speed as an optimization algorithm. To this purpose, we will characterize the evolution of the probability of error

$$\mathbb{P}_{(\beta)}\big(U(X_N) \geq \eta\big)$$

as a function of the number of iterations $N$, for a suitable choice of the inverse temperature sequence $(\beta_n)_{n=1}^N$ (also known as the cooling schedule).

In addition to the first critical depth $H_1$, a second critical exponent will play a prominent role in this discussion, which we will call *the difficulty* $D$ of the (virtual) energy landscape $(E, V)$. It is defined as

$$D = \max\Bigl\{\frac{H(C)}{U(C)} : C \in \mathcal{C}, \min_{x \in C} U(x) > 0\Bigr\}$$
$$= \max\Bigl\{\frac{\min\bigl\{V^{\{x,y\}}(x,y) : U(y) = 0\bigr\}}{U(x)} : U(x) > 0\Bigr\}.$$

**Theorem 7.8.1.** *Let us consider two bounds $H^*$ and $D_*$ such that $H^* > H_1$ and $0 < D_* < D$. For any threshold $\eta$ such that $0 < \eta < H^*/D_*$, for any integer $r$, let us consider the triangular inverse temperature sequence*

$$\beta_n^N = \frac{1}{H^*} \log\left(\frac{N}{r}\right)\left(\frac{H^*}{D_*\eta}\right)^{\frac{1}{r}\left\lfloor\frac{(n-1)r}{N}\right\rfloor}, \qquad 1 \le n \le N.$$

*Thus parametrized, the simulated annealing algorithm has a probability of error for the worst possible initial state $x \in E$ asymptotically bounded by*

$$\limsup_{N \to +\infty} \frac{1}{\log N} \log \max_{x \in E} \mathbb{P}^x_{(\beta^N_\cdot)}\bigl(U(X_N) \ge \eta\bigr) \le -\frac{1}{D}\left(\frac{D_*\eta}{H^*}\right)^{1/r}.$$

*Remark* 7.8.1. It is thus possible to make the probability of error decrease at least at some power of $1/N$ arbitrarily close to the value $1/D$. It is possible to show (under slightly stronger hypotheses) that no increasing inverse temperature triangular sequence can yield a decrease of the probability of error faster than $1/N^{1/D}$, thus proving that $1/D$ is indeed the critical exponent for the convergence speed of simulated annealing.

*Remark* 7.8.2. Resorting to a *triangular* temperature sequence is crucial. It is possible to show that it is impossible to get close to the critical convergence speed with a non-triangular temperature sequence, in other words a sequence $(\beta_n)_{n=1}^N$ where the inverse temperatures $\beta_n$ are chosen independently of the total number of iterations $N$ to be performed.

*Proof.* Let us put to simplify notations

$$\zeta_k^N = \beta_n^N = \frac{1}{H^*}\log\left(\frac{N}{r}\right)\left(\frac{H^*}{D_*\eta}\right)^{k/r}, \qquad \frac{kN}{r} < n \le \frac{(k+1)N}{r}.$$

Let us also write $\mathbb{P}_{(\beta^N_\cdot)} = \mathbb{P}_N$ and let us assume to make things simpler that $N/r \in \mathbb{N}$. Given some parameter $\xi > 0$ to be fixed later, let us put

$$\eta_k = \frac{H^*}{(1+\xi)D}\left(\frac{H^*}{D_*\eta}\right)^{-(k+1)/r}, \qquad k = 0, \ldots, r-1,$$

as well as

$$\lambda_0 = +\infty,$$

$$\lambda_k = \frac{(1+1/D)H^*}{(1+\xi)} \left( \frac{H^*}{D_* \eta} \right)^{-k/r}, \qquad k = 1, \ldots, r-1.$$

Let us consider the events

$$B_k = \{U(X_n) + V(X_n, X_{n+1}) \le \lambda_k, \ \frac{kN}{r} \le n < \frac{(k+1)N}{r}\},$$
$$A_k = B_k \cap \{U(X_{(k+1)N/r}) < \eta_k\}.$$

Let us notice that

$$\exp\left( H^* \zeta_0^N \right) = \frac{N}{r},$$
$$\exp\left( (1+\xi)\,(1+1/D)^{-1}\,\lambda_k \zeta_k^N \right) = \frac{N}{r}, \qquad k > 0,$$

and that

$$\eta_{r-1} = \frac{D_* \eta}{(\xi+1) D} \le \eta.$$

It follows that

$$\begin{aligned} \mathbb{P}_N^x\big(U(X_N) \ge \eta\big) &\le \mathbb{P}_N^x\big(U(X_N) \ge \eta_{r-1}\big) \\ &\le 1 - \mathbb{P}_N^x \left( \bigcup_{k=0}^{r-1} A_k \right) \\ &\le \sum_{k=0}^{r-1} \mathbb{P}_N^x \left( \overline{A}_k \cap \bigcap_{\ell=0}^{k-1} A_\ell \right). \end{aligned}$$

Moreover

$$\begin{aligned} \mathbb{P}_N^x\big(\overline{A}_k \cap \bigcap_{\ell=0}^{k-1} A_\ell\big) &\le \mathbb{P}_N^x\big(\overline{B}_k \cap \bigcap_{\ell=0}^{k-1} A_\ell\big) \\ &\quad + \mathbb{P}_N^x\big(\{U(X_{(k+1)N/r}) \ge \eta_k\} \cap \bigcap_{\ell=0}^{k-1} A_\ell \cap B_k\big). \end{aligned}$$

For any cycle $C$ of positive fundamental energy $U(C) = \min_{x \in C} U(x) > 0$ whose exit level is such that $H(C) + U(C) \le \lambda_k$,

$$H(C) \le (1+1/D)^{-1}\,\lambda_k.$$

For any state $z \in E$ such that $U(z) < \eta_{k-1}$, let us consider the smallest cycle $C_z \in \mathcal{C}$ containing $z$ such that $U(C_z) + H(C_z) > \lambda_k$. Necessarily $U(C_z) = 0$. Indeed, if it were not the case, that is if $U(C_z) > 0$, then it would follow that

$$U(C_z) + H(C_z) \le U(z)(1+D) \le \eta_{k-1}(1+D) = \lambda_k,$$

in contradiction with the definition of $C_z$.

Let us consider on the space $C_z^{\mathbb{N}}$ of paths staying in $C_z$ the canonical process $(Y_n)_{n\in\mathbb{N}}$, and the family of distributions of the homogeneous Markov chains with respective transition matrices

$$q_\beta(x,y) = \begin{cases} p_\beta(x,y) & \text{when } x \neq y \in C_z, \\ 1 - \sum_{w\in(C_z\setminus\{x\})} q_\beta(x,w) & \text{otherwise.} \end{cases}$$

(The process $Y$ can be viewed as the "reflection" of $X$ on the boundary of $C_z$.) The processes we just defined form a generalized Metropolis algorithm with transition rate function $V_{|C_z\times C_z}$ and first critical depth

$$H_1\big(C_z,\, V_{|C_z\times C_z}\big) \leq (1+1/D)^{-1}\,\lambda_k.$$

Indeed any strict subcycle $C \subsetneq C_z$ such that $U(C) > 0$ is also such that $H(C) \leq (1+1/D)^{-1}\,\lambda_k$.

Let us now notice that

$$\begin{aligned}\mathbb{P}_N\big(X_{(k+1)N/r} = y,\ B_k \,|\, X_{kN/r} = z\big)& \\ \leq \mathbb{P}_N\big(X_{(k+1)N/r} = y,\ X_n \in C_z,\ \frac{kN}{r} < n \leq \frac{(k+1)N}{r} \,|\, X_{kN/r} = z\big).&\end{aligned}$$

Moreover, for any $x,\ y \in C_z$

$$p_{\zeta_k^N}(x,y) \leq q_{\zeta_k^N}(x,y).$$

It follows that

$$\begin{aligned}\mathbb{P}_N\big(X_{(k+1)}N/r = y,\ X_n \in C_z,\ \frac{kN}{r} < n \leq \frac{(k+1)N}{r} \,|\, X_{kN/r} = z\big)& \\ \leq \mathbb{P}^z_{\zeta_k^N}\big(Y_{N/r} = y\big).&\end{aligned}$$

The first claim of theorem 7.7.1 about the convergence speed of Metropolis algorithms can then be applied to $Y$. It shows that for any $\epsilon > 0$, there is some integer $N_0$ such that for any $N \geq N_0$,

$$\mathbb{P}^x_N\big(\{X_{(k+1)N/r} = y\} \cap \bigcap_{\ell=0}^{k-1} A_\ell \cap B_k\big) \leq \exp\Big(-\zeta_k^N\,(U(y)-\epsilon)\Big).$$

It remains now to bound

$$\mathbb{P}^x_N\left(\overline{B}_k \cap \bigcap_{\ell=0}^{k-1} A_\ell\right).$$

For any $z \in E$ such that $U(z) < \eta_{k-1}$, for any $\epsilon > 0$ and for any large enough value of $N$,

$$
\begin{aligned}
&\mathbb{P}_N\big(\overline{B}_k \,|\, X_{kN/r} = z\big)\\
&\quad\le \sum_{n=kN/r+1}^{(k+1)N/r} \mathbb{P}_N\big(U(X_{n-1}) + V(X_{n-1}, X_n) > \lambda_k \,|\, X_{kN/r} = z\big)\\
&\quad= \sum_{\substack{kN/r<n\le(k+1)N/r,\\ (u,v)\in E^2,\\ U(u)+V(u,v)>\lambda_k}} \mathbb{P}_N\big(X_{n-1} = u \,|\, X_{kN/r} = z\big) p_{\zeta_k^N}(u,v)\\
&\quad\le \sum_{\substack{kN/r<n\le(k+1)N/r,\\ (u,v)\in E^2,\\ U(u)+V(u,v)>\lambda_k}} \frac{\mu_{\zeta_k^N}(u)}{\mu_{\zeta_k^N}(z)} p_{\zeta_k^N}(u,\, v)\\
&\quad\le \frac{N}{r} \exp\Big(-\zeta_k^N\big(\lambda_k - U(z) - \epsilon\big)\Big).
\end{aligned}
$$

Thus for any $\epsilon > 0$ and for $N$ large enough,

$$
\begin{aligned}
\mathbb{P}_N^x\left(\overline{B}_k \cap \bigcap_{\ell=0}^{k-1} A_\ell\right) &= \sum_{z,U(z)<\eta_{k-1}} \mathbb{P}_N\big(\overline{B}_k \,|\, X_{kN/r} = z\big)\\
&\qquad\times \mathbb{P}_N^x\Big(\big\{X_{kN/r} = z\big\} \cap \bigcap_{\ell=0}^{k-2} A_\ell \cap B_{k-1}\Big)\\
&\le \sum_{z,\, U(z)<\eta_{k-1}} \frac{N}{r}\exp\Big(-\zeta_k^N\big(\lambda_k - U(z) - \epsilon\big)\Big)\\
&\qquad\times \exp\Big(-\zeta_{k-1}^N\big(U(z) - \epsilon\big)\Big)\\
&\le \frac{N}{r}\exp\Big(-\zeta_k^N\big(\lambda_k - \eta_{k-1} - \epsilon\big)\Big)\\
&\qquad\times \exp\Big(-\zeta_{k-1}^N\big(\eta_{k-1} - 2\epsilon\big)\Big).
\end{aligned}
$$

Consequently, for any $\epsilon > 0$, there is some integer $N_0$ such that for any $N \ge N_0$,

$$
\begin{aligned}
&\mathbb{P}_N^x\big(\overline{A}_k \cap \bigcap_{\ell=0}^{k-1} A_\ell\big)\\
&\quad\le \frac{N}{r}\exp\Big(-(\lambda_k - \eta_{k-1})\zeta_k^N - \eta_{k-1}\zeta_{k-1}^N + \epsilon\big(\zeta_k^N + \zeta_{k-1}^N\big)\Big)\\
&\qquad\qquad + \exp\Big(-(\eta_k - \epsilon)\zeta_k^N\Big).
\end{aligned}
$$

Coming back to the definitions, we see moreover that

$$
\eta_k \zeta_k^N = \log\left(\frac{N}{r}\right)\frac{1}{(1+\xi)D}\left(\frac{D_* \eta}{H^*}\right)^{1/r},
$$

$$\lambda_k \zeta_k^N + \eta_{k-1}(\zeta_{k-1}^N - \zeta_k^N) = \left(1+\frac{1}{D}\right)(1+\xi)^{-1}\log\left(\frac{N}{r}\right)$$
$$+\frac{1}{(1+\xi)D}\left(\left(\frac{\eta D_*}{H^*}\right)^{1/r}-1\right)\log\left(\frac{N}{r}\right)$$
$$=\log\left(\frac{N}{r}\right)\left((1+\xi)^{-1}\left(1+\frac{1}{D}\left(\frac{\eta D_*}{H^*}\right)^{1/r}\right)\right),$$
$$\epsilon\,\zeta_k^N \le \frac{\epsilon}{\eta D_*}\log\left(\frac{N}{r}\right).$$

It follows that

$$\mathbb{P}_N^x\left(\overline{A}_k \cap \bigcap_{\ell=0}^{k-1} A_\ell\right)$$
$$\le \left(\frac{N}{r}\right)^{-(1+\xi)^{-1}\left(\frac{1}{D}\left(\frac{\eta D_*}{H^*}\right)^{1/r}-\xi\right)+\frac{2\epsilon}{\eta D_*}}$$
$$+\left(\frac{N}{r}\right)^{-(1+\xi)^{-1}\frac{1}{D}\left(\frac{D_*\eta}{H^*}\right)^{1/r}+\frac{\epsilon}{\eta D_*}}.$$

Letting $\epsilon$ and $\xi$ tend to 0 eventually yields the desired inequality:

$$\limsup_{N\to+\infty}\frac{1}{\log(N)}\log\max_{x\in E}\mathbb{P}_N^x\big(U(X_N)\ge\eta\big)\le-\frac{1}{D}\left(\frac{D_*\eta}{H^*}\right)^{1/r}.$$

□

# References


1. L. Alonso and R. Cerf, The three-dimensional polyominoes of minimal area. *Electron. J. Combin.* **3** (1996), no. 1, Research Paper 27, approx. 39 pp. (electronic).
2. Y. Baraud *Sélection de modèles et estimation adaptative dans différents cadres de régression,* PHD (1998), Université Paris 11.
3. Y. Baraud, F. Comte and G. Viennet, Adaptive estimation in autoregression or $\beta$-mixing regression via model selection. *Ann. Statist.* **29** (2001), no. 3, 839–875;
4. Y. Baraud, F. Comte and G. Viennet (1999) Model selection for (auto-)regression with dependent data. *ESAIM Probab. Statist.* **5** (2001), 33–49 (electronic).
5. A. Barron, Are Bayes Rules Consistent in Information ? *Open Problems in Communication and Computation, T. M. Cover and B. Gopinath Ed.*, Springer Verlag (1987).
6. A. Barron, L. Birgé and P. Massart, Risk bounds for model selection via penalization. *Probab. Theory Related Fields* **113** (1999), no. 3, 301–413.
7. G. Ben Arous and R. Cerf, Metastability of the three-dimensional Ising model on a torus at very low temperatures. *Electron. J. Probab.* **1** (1996), no. 10, approx. 55 pp. (electronic).
8. L. Birgé, Approximation dans les espaces métriques et théorie de l'estimation. *Z. Wahrsch. Verw. Gebiete* **65** (1983), no. 2, 181–237
9. L. Birgé and P. Massart (1995) Minimum contrast estimators on sieves. *Bernoulli* **4** (1998), no. 3, 329–375.
10. L. Birgé and P. Massart From model selection to adaptive estimation. *Festschrift for Lucien Le Cam* (1997), 55–87, Springer, New York.
11. L. Birgé and P. Massart, An adaptive compression algorithm in Besov spaces. *Constr. Approx.* **16** (2000), no. 1, 1–36.
12. L. Birgé and P. Massart, A generalized $C_p$ criterion for Gaussian model selection, *preprint*, (2001).
13. L. Birgé and P. Massart, Gaussian model selection, *J. Eur. Math. Soc.*, (2001). `http://www.springer.de/link`
14. G. Blanchard, *Méthodes de mélange et d'agrégation d'estimateurs en reconnaissance de formes. Application aux arbres de décision,* thèse de l'Université Paris XIII, janvieur 2001, `http://www.dma.ens.fr/~gblancha/`.

15. G. Blanchard, The "progressive mixture" estimator for regression trees. *Annales de l'I.H.P.*, **35**(6):793-820, (1999).
16. G. Blanchard, A new algorithm for Bayesian MCMC CART sampling. *preprint*, 2000.
17. S. Boucheron, G. Lugosi and P. Massart, A sharp concentration inequality with applications. *Random Structures Algorithms* **16** (2000), no. 3, 277–292.
18. O. Catoni, Rough Large Deviation Estimates for Simulated Annealing: Application to Exponential Schedules, *The Annals of Probability*, **20** (1992) , nb. 3, pp. 1109 - 1146.
19. O. Catoni, Algorithmes de recuit simulé et chaînes de Markov à transitions rares. *Notes de cours de DEA* (1995), université Paris XI, Orsay. English translation: "Simulated Annealing Algorithms and Markov chains with Rare Transitions", *preprint, LMENS-97-09, available on the web at* `http://www.dmi.ens.fr/preprints`.
20. O. Catoni, Simulated annealing algorithms and Markov chains with rare transitions. *Séminaire de Probabilités* **XXXIII** (1999), 69–119, in French 1995, English revised translation at http://www.dmi.ens.fr/preprints 1997, published augmented revision 1999.
21. O. Catoni, A mixture approach to universal model selection, *preprint LMENS - 97 - 30* (1997), http://www.dmi.ens.fr/preprints.
22. O. Catoni, Gibbs estimators, *preprint*, first draft 1998, last revision 2000 (included in these lecture notes).
23. O. Catoni, Data compression and adaptive histograms. In F. Cucker and J.M. Rojas, editors, *Foundations of Computational Mathematics, Proceedings of the Smalefest 2000,* 35-60, World Scientific, 2002.
24. O. Catoni, Laplace transform estimates and deviation inequalities. *Ann. Inst. H. Poincar Probab. Statist.* **39** (2003), no. 1, 1–26.
25. O. Catoni, D. Chen and Jun Xie, The loop erased exit path and the metastability of a biased vote process. *Stochastic Process. Appl.* **86** (2000), 231–261.
26. A. Cohen, I. Daubechies and P. Vial, Wavelets and fast wavelet transform on an interval, *Applied Computational and Harmonic Analysis.* **1** (1993), pp. 54 - 81.
27. T. M. Cover and J. A. Thomas, *Elements of information theory*, John Wiley & Sons Inc., New York, 1991, A Wiley-Interscience Publication.
28. C. Cot and O. Catoni, Piecewise constant triangular cooling schedules for generalized simulated annealing algorithms. *Ann. Appl. Probab.* **8** (1998), no. 2, 375–396.
29. I. Csiszár and J. Körner *Information theory : coding theorems for discrete memoryless systems,* Academic Press, New York (1981).
30. S. Delattre and M. Hoffmann, The Pinsker bound in mixed Gaussian white noise. *Meeting on Mathematical Statistics (Marseille, 2000). Math. Methods Statist.* **10** (2001), no. 3, 283–315.
31. A. Dembo and O. Zeitouni, *Large deviations techniques and applications*, Second edition, Springer, New York, 1998.
32. A. Dembo, O. Zeitouni, Transportation approach to some concentration inequalities in product spaces. *Electron. Comm. Probab.* **1** (1996), no. 9, 83–90 (electronic).
33. D. L. Donoho and I. M. Johnstone, Ideal spatial adaptation by wavelet shrinkage. *Biometrika* **81** (1994), no. 3, 425–455.

34. D. L. Donoho and I. M. Johnstone, Adapting to unknown smoothness via wavelet shrinkage. *J. Amer. Statist. Assoc.* **90** (1995), no. 432, 1200–1224.
35. D. L. Donoho and I. M. Johnstone, Asymptotic minimaxity of wavelet estimators with sampled data. *Statist. Sinica* **9** (1999), no. 1, 1–32.
36. D. L. Donoho, I. M. Johnstone, G. Kerkyacharian and D. Picard, Density estimation by wavelet thresholding. Ann. Statist. **24** (1996), no. 2, 508–539.
37. D. L. Donoho, I. M. Johnstone, G. Kerkyacharian and D. Picard, Universal near minimaxity of wavelet shrinkage. *Festschrift for Lucien Le Cam* (1997), 183–218, Springer, New York.
38. M. Feder and N. Merhav, Hierarchical universal coding. *IEEE Trans. Inform. Theory.* **42** (1996), no. 5, 1354–1364.
39. W. Feller, *An introduction to probability theory and its applications. Vol. I*, Third edition, Wiley, New York, 1968.
40. M. I. Freidlin and A. D. Wentzell, *Random perturbations of dynamical systems*, second ed., Springer-Verlag, New York, 1998, Translated from the 1979 Russian original by Joseph Szücs.
41. A. Goldenshluger and A. Nemirovski, On spatially adaptive estimation of nonparametric regression. *Math. Methods Statist.* **6** (1997), no. 2, 135–170.
42. W. Härdle, G. Kerkyacharian, D. Picard and A. Tsybakov, Wavelets, approximation and statistical applications. *Lecture Notes in Statistics,* **129** (1998). Springer-Verlag, New York. xviii+265 pp.
43. W. Hoeffding, Probability inequalities for sums of bounded random variables. J. Amer. Statist. Assoc. **58** (1963), 13–30;
44. I. Johnstone, Function estimation and wavelets, *Lecture Notes, ENS Paris* (1998).
45. I. Johnstone, Wavelets and the theory of non-parametric function estimation. *R. Soc. Lond. Philos. Trans. Ser. A Math. Phys. Eng. Sci.* **357** (1999), no. 1760, 2475–2493.
46. A. Juditsky and A. Nemirovski. Functional aggregation for nonparametric regression, *in 4th World Congress of the Bernouilli Society, Vienna, Austria.* August 26-31 (1996).
47. U. Krengel, *Ergodic theorems*, Walter de Gruyter & Co., Berlin, 1985, With a supplement by Antoine Brunel.
48. M. Ledoux and M. Talagrand, *Probability in Banach spaces,* Springer, Berlin (1991) .
49. M. Ledoux, Isoperimetry and Gaussian analysis. *Lectures on probability theory and statistics (Saint-Flour, 1994),* 165–294, Lecture Notes in Math., 1648, Springer, Berlin, 1996.
50. M. Ledoux, (1995) On Talagrand's deviation inequalities for product measures. *ESAIM Probab. Statist.* **1** (1995/97), 63–87
51. M. Ledoux, Concentration of measure and logarithmic Sobolev inequalities, Notes (Berlin, 1997).
52. G. Lugosi and A. Nobel, Consistency of data-driven histogram methods for density estimation and classification. *Ann. Statist.* **24** (1996), no. 2, 687–706.
53. D. A. McAllester, Some PAC-Bayesian Theorems, *Proceedings of the Eleventh Annual Conference on Computational Learning Theory (Madison, WI, 1998),* 230–234 (electronic), ACM, New York, 1998;
54. D. A. McAllester, PAC-Bayesian Model Averaging, *Proceedings of the Twelfth Annual Conference on Computational Learning Theory (Santa Cruz, CA, 1999),* 164–170 (electronic), ACM, New York, 1999;

55. C. McDiarmid, Concentration *Probabilistic Methods for Algorithmic Discrete Mathematics*, Habib M., McDiarmid C. and Reed B. Eds., Springer, 1998.
56. K. Marton, Bounding $\bar{d}$-distance by informational divergence: a method to prove measure concentration. *Ann. Probab.* **24** (1996), no. 2, 857–866.
57. K. Marton, A measure concentration inequality for contracting Markov chains. *Geom. Funct. Anal.* **6** (1996), no. 3, 556–571.
58. K. Marton, Measure concentration for a class of random processes. *Probab. Theory Related Fields.* **110** (1998), no. 3, 427–439.
59. P. Massart, Optimal constants for Hoeffding type inequalities. *Prépublication d'Orsay.* **86** (18/12/1998), `http://www.math.u-psud.fr/~biblio/html/ppo.html`
60. P. Massart, About the constants in Talagrand's concentration inequalities for empirical processes. *Ann. Probab.* **28** (2000), no. 2, 863–884.
61. N. Merhav and M. Feder, *A strong version of the redundancy-capacity theorem of universal coding*, IEEE Trans. Inform. Theory **41** (1995), no. 3, 714–722.
62. Meyer, Yves (1990) *Wavelets and operators*, Translated from the 1990 French original by D. H. Salinger, Cambridge Studies in Advanced Mathematics, 37, Cambridge University Press, Cambridge, 1992. xvi+224 pp.
63. A. Nemirovski, Topics in non-parametric statistics. *Lectures on probability theory and statistics (Saint-Flour, 1998),* 85–277, Lecture Notes in Math., 1738, Springer, Berlin, 2000.
64. A. Nobel, Histogram regression estimation using data-dependent partitions. *Ann. Statist.* **24** (1996), no. 3, 1084–1105.
65. M. Nussbaum, Asymptotic equivalence of density estimation and Gaussian white noise. *Ann. Statist.* **24** (1996), no. 6, 2399–2430.
66. M. S. Pinsker, *Information and information stability of random variables and processes*, Holden-Day, San Fransisco (1964).
67. B. Y. Ryabko, Twice-universal coding, *Probl. Inform. Transm.*, **20** (1984), no 3, pp. 24-28, July-Sept.
68. W. Rudin, *Real and Complex Analysis*, McGraw-Hill, (1966).
69. L. Saloff-Coste Lectures on finite Markov chains, *Ecole d'été de Probabilités de Saint-Flour XXVI, 1996, Lectures on probability theory and statistics,* Lecture Notes in Mathematics 1665, Springer (1997).
70. P.-M. Samson, Inégalités de concentration de la mesure pour des chaînes de Markov et des processus $\Phi$- mélangeants. PhD, Université Paul Sabatier, Toulouse, France, (June 1998).
71. P.-M. Samson, Concentration of measure inequalities for Markov chains and $\Phi$-mixing processes. *Ann. Probab.* **28** (2000), no. 1, 416–461.
72. M. Talagrand Concentration of measure and isoperimetric inequalities in product spaces, *Publications Mathématiques de l'I.H.E.S.*, **81** (1995), 73-205.
73. A. Trouvé, Cycle decompositions and simulated annealing, *SIAM J. Control Optim.* **34** (1996), no. 3, 966–986.
74. A. Trouvé, Rough large deviation estimates for the optimal convergence speed exponent of generalized simulated annealing algorithms, *Ann. Inst. H. Poincaré Probab. Statist.* **32** (1996), no. 3, 299–348.
75. A. Trouvé, (1995) Rough Large Deviation Estimates for the Optimal Convergence Speed Exponent of Generalized Simulated Annealing Algorithms, *Ann. Inst. H. Poincaré, Probab. Statist.*, 32(2), 1996.
76. V. N. Vapnik, *Statistical learning theory*, Wiley, New York, 1998.

77. J.-P. Vert, Méthodes statistiques pour la modélisation du langage naturel, *thèse de l'Université Paris 6,* (mars 2001) `http://www.dma.ens.fr/~vert/`
78. J.-P. Vert, Double mixture and universal inference, *preprint*, 2000.
79. J.-P. Vert, Adaptive context trees and text clustering, *IEEE Trans. Inform. Theory* **47** (2001), no. 5, 1884–1901
80. J.-P. Vert, Text categorization using adaptive context trees, *Proceedings of the CICLing-2001 conference,* A. Gelbukh (Ed.), LNCS 2004, Springer-Verlag Berlin Heidelberg, pp. 423-436, 2001.
81. E. T. Whittaker and G. N. Watson *A course of mordern analysis*, Cambridge University Press, 1927.
82. F. M. J. Willems, Y. M. Shtarkov and T. J. Tjalkens The Context-Tree Weighting Method: Basic Properties, *IEEE Trans. Inform. Theory,* **41** (1995), no 3.
83. F. M. J. Willems, Y. M. Shtarkov and T. J. Tjalkens Context Weighting for General Finite-Context Sources, *IEEE Trans. Inform. Theory,* **42** (1996), no 5.
84. Q. Xie and A. Barron, Asymptotic minimax regret for data compression, gambling and prediction, *IEEE Trans. Inform. Theory.* **46** (2000), no. 2, 431–445.
85. Y. Yang, On adaptive function estimation, *preprint* (1997) `http://www.public.iastate.edu/~stat/update/reports.html`
86. Y. Yang and A. Barron, Information-theoretic determination of minimax rates of convergence. *Ann. Statist.* **27** (1999), no. 5, 1564–1599.
87. Y. Yang, Combining Different Procedures for Adaptive Regression, *Journal of Multivariate Analysis.* **74** (2000), pp. 135-161.
88. V. V. Yurinskiĭ, Exponential estimates for large deviations (Russian), *Teor. Verojatnost. i Primenen.* **19** (1974), 152–154.
89. S. C. Zhu, Y. Wu and D. Mumford, Minimax Entropy Principle and Its Applications to Texture Modeling *Neural Computation* **9** (1997), no 8, Nov. 1997. `http://www.cis.ohio-state.edu/~szhu/publication.html`
90. S. C. Zhu, Y. N. Wu and D. Mumford, FRAME: Filters, Random field And Maximum Entropy: — Towards a Unified Theory for Texture Modeling *Int'l Journal of Computer Vision.* **27**(2) 1-20, March/April. 1998. `http://www.cis.ohio-state.edu/~szhu/publication.html`

# Index

# List of participants

| | |
|---|---|
| AMIDI Ali | Beheshti University, Tehran, Iran |
| ARNAUDON Marc | Univ. de Poitiers, F |
| ASCI Claudio | Universita degli Studi di L'Aquila, Italy |
| BAHADORAN Christophe | Univ. Blaise Pascal, Clermont-Ferrand, F |
| BALDI Paolo | Universita Roma Tor Vergata, Italy |
| BARDET Jean-Baptiste | Ecole Polytechnique Fédér. de Lausanne, CH |
| BEN AROUS Gérard | Ecole Polytechnique Fédér. de Lausanne, CH |
| BERARD Jean | Univ. Claude Bernard, Lyon, F |
| BERNARD Pierre | Univ. Blaise Pascal, Clermont-Ferrand, F |
| BOLTHAUSEN Erwin | Univ. de Zurich, CH |
| BOUGEROL Philippe | Univ. Pierre et Marie Curie, Paris, F |
| BOURRACHOT Ludovic | Univ. Blaise Pascal, Clermont-Ferrand, F |
| BRETON Jean-Christophe | Univ. Lille 1, F |
| CAMPILLO Fabien | INRIA, Marseille, F |
| CERNY Jiri | Ecole Polytechnique Fédér. de Lausanne, CH |
| CHAMPAGNAT Nicolas | Ecole Normale Supérieure de Paris, F |
| CLIMESCU-HAULICA Adriana | Comm. Research Center, Ottawa, Canada |
| DA SILVA Soares Ana | Univ. Libre de Bruxelles, Belgique |
| DARWICH Abdul | Univ. d'Angers, F |
| DEMBO Amir | Stanford University, USA |
| DJELLOUT Hacene | Univ. Blaise Pascal, Clermont-Ferrand, F |
| DUDOIGNON Lorie | INRIA, Marseille, F |
| FEDRIGO Mattia | Scuola Normale Superiore di Pisa, Italy |
| FERRIERE Régis | Ecole Normale Supérieure de Paris, F |
| GIACOMIN Giambattista | Univ. Denis Diderot, Paris, F |
| GILLET Florent | Univ. Henri Poincaré, Nancy, F |
| GROSS Thierry | Univ. Denis Diderot, Paris, F |
| GUILLIN Arnaud | Univ. Blaise Pascal, Clermont-Ferrand, F |
| GUIONNET Alice | Ecole Normale Supérieure de Lyon, F |
| HOFFMANN Marc | Univ. Denis Diderot, Paris, F |
| KERKYACHARIAN Gérard | Univ. Denis Diderot, Paris, F |
| KISTLER Nicolas | Univ. de Zurich, CH |
| LAREDO Catherine | INRA, Jouy-en-Josas, F |

| | |
|---|---|
| LOECHERBACH Eva | Univ. de Paris Val de Marne, F |
| LOPEZ-MIMBELA Jose Alfredo | CIMAT, Guanajuato, Mexico |
| LORANG Gerard | Centre Universitaire de Luxembourg |
| MILLET Annie | Univ. de Paris 10, F |
| MOUSSET Sylvain | Univ. Pierre et Marie Curie, Paris, F |
| NICAISE Florent | Univ. Blaise Pascal, Clermont-Ferrand, F |
| NUALART Eulalia | Ecole Polytechnique Fédér. de Lausanne, CH |
| OCONE Daniel | Rutgers University, Piscataway, NJ, USA |
| PARDOUX Etienne | Univ. de Provence, Marseille, F |
| PAROISSIN Christian | Univ. René Descartes, Paris, F |
| PIAU Didier | Univ. Claude Bernard, Lyon, F |
| PICARD Dominique | Univ. Denis Diderot, Paris, F |
| PICARD Jean | Univ. Blaise Pascal, Clermont-Ferrand, F |
| PLAGNOL Vincent | Ecole Normale Supérieure de Paris, F |
| RASSOUL-AGHA Firas | Courant Institute, New York, USA |
| ROUAULT Alain | Univ. de Versailles-St Quentin, F |
| ROUX Daniel | Univ. Blaise Pascal, Clermont-Ferrand, F |
| ROZENHOLC Yves | INRA, Paris, F |
| SKORA Dariusz | University of Wroclaw, Poland |
| SOOS Anna | Babes-Bolyai Univ., Cluj-Napoca, Romania |
| SORTAIS Michel | Ecole Polytechnique Fédér. de Lausanne, CH |
| WU Liming | Univ. Blaise Pascal, Clermont-Ferrand, F |

## List of short lectures

Claudio ASCI, Generating uniform random vectors.
Christophe BAHADORAN, Boundary conditions for driven conservative particle systems.
Jean-Baptiste BARDET, Limit theorems for coupled analytic maps.
Jean BÉRARD, Genetic algorithms in random environments.
Erwin BOLTHAUSEN, A fixed point approach to weakly self-avoiding random walks.
Philippe BOUGEROL, A path representation of the eigenvalues of the GUE random matrices.
Jean-Christophe BRETON, Multiple stable stochastic integrals: representation, absolute continuity of the law.
Jiři ČERNÝ, Critical path analysis for continuum percolation.
Adriana CLIMESCU-HAULICA, Cramér decomposition and noise modelling: applications from/to communications theory.
Ana DA SILVA SOARES, Files d'attente fluides.
Amir DEMBO, Random polynomials having few or no real zeros.
Mattia FEDRIGO, A multifractal model for network data traffic.
Florent GILLET, Algorithmes de tri: analyse du coût, stabilité face aux erreurs.
Alice GUIONNET, Enumerating graphs, matrix models and spherical integrals.
Eva LÖCHERBACH, On the invariant density of branching diffusions.
José Alfredo LÓPEZ-MIMBELA, A proof of non-explosion of a semilinear PDE system.
Florent NICAISE, Infinite volume spin systems: an application to Girsanov results on the Poisson space.
Eulalia NUALART, Potential theory for hyperbolic SPDE's.
Dan OCONE, Finite-fuel singular control with discretionary stopping.
Didier PIAU, Mutation-replication statistics of polymerase chain reactions.
Yves ROZENHOLC, Classification trees and colza diversity.
Michel SORTAIS, Large deviations in the Langevin dynamics of short range disordered systems.

Printing and Binding: Strauss GmbH, Mörlenbach

# Lecture Notes in Mathematics

For information about Vols. 1–1675
please contact your bookseller or Springer

Vol. 1676: P. Cembranos, J. Mendoza, Banach Spaces of Vector-Valued Functions. VIII, 118 pages. 1997.

Vol. 1677: N. Proskurin, Cubic Metaplectic Forms and Theta Functions. VIII, 196 pages. 1998.

Vol. 1678: O. Krupková, The Geometry of Ordinary Variational Equations. X, 251 pages. 1997.

Vol. 1679: K.-G. Grosse-Erdmann, The Blocking Technique. Weighted Mean Operators and Hardy's Inequality. IX, 114 pages. 1998.

Vol. 1680: K.-Z. Li, F. Oort, Moduli of Supersingular Abelian Varieties. V, 116 pages. 1998.

Vol. 1681: G. J. Wirsching, The Dynamical System Generated by the 3n+1 Function. VII, 158 pages. 1998.

Vol. 1682: H.-D. Alber, Materials with Memory. X, 166 pages. 1998.

Vol. 1683: A. Pomp, The Boundary-Domain Integral Method for Elliptic Systems. XVI, 163 pages. 1998.

Vol. 1684: C. A. Berenstein, P. F. Ebenfelt, S. G. Gindikin, S. Helgason, A. E. Tumanov, Integral Geometry, Radon Transforms and Complex Analysis. Firenze, 1996. Editors: E. Casadio Tarabusi, M. A. Picardello, G. Zampieri. VII, 160 pages. 1998.

Vol. 1685: S. König, A. Zimmermann, Derived Equivalences for Group Rings. X, 146 pages. 1998.

Vol. 1686: J. Azéma, M. Émery, M. Ledoux, M. Yor (Eds.), Séminaire de Probabilités XXXII. VI, 440 pages. 1998.

Vol. 1687: F. Bornemann, Homogenization in Time of Singularly Perturbed Mechanical Systems. XII, 156 pages. 1998.

Vol. 1688: S. Assing, W. Schmidt, Continuous Strong Markov Processes in Dimension One. XII, 137 page. 1998.

Vol. 1689: W. Fulton, P. Pragacz, Schubert Varieties and Degeneracy Loci. XI, 148 pages. 1998.

Vol. 1690: M. T. Barlow, D. Nualart, Lectures on Probability Theory and Statistics. Editor: P. Bernard. VIII, 237 pages. 1998.

Vol. 1691: R. Bezrukavnikov, M. Finkelberg, V. Schechtman, Factorizable Sheaves and Quantum Groups. X, 282 pages. 1998.

Vol. 1692: T. M. W. Eyre, Quantum Stochastic Calculus and Representations of Lie Superalgebras. IX, 138 pages. 1998.

Vol. 1694: A. Braides, Approximation of Free-Discontinuity Problems. XI, 149 pages. 1998.

Vol. 1695: D. J. Hartfiel, Markov Set-Chains. VIII, 131 pages. 1998.

Vol. 1696: E. Bouscaren (Ed.): Model Theory and Algebraic Geometry. XV, 211 pages. 1998.

Vol. 1697: B. Cockburn, C. Johnson, C.-W. Shu, E. Tadmor, Advanced Numerical Approximation of Nonlinear Hyperbolic Equations. Cetraro, Italy, 1997. Editor: A. Quarteroni. VII, 390 pages. 1998.

Vol. 1698: M. Bhattacharjee, D. Macpherson, R. G. Möller, P. Neumann, Notes on Infinite Permutation Groups. XI, 202 pages. 1998.

Vol. 1699: A. Inoue,Tomita-Takesaki Theory in Algebras of Unbounded Operators. VIII, 241 pages. 1998.

Vol. 1700: W. A. Woyczyński, Burgers-KPZ Turbulence, XI, 318 pages. 1998.

Vol. 1701: Ti-Jun Xiao, J. Liang, The Cauchy Problem of Higher Order Abstract Differential Equations, XII, 302 pages. 1998.

Vol. 1702: J. Ma, J. Yong, Forward-Backward Stochastic Differential Equations and Their Applications. XIII, 270 pages. 1999.

Vol. 1703: R. M. Dudley, R. Norvaiša, Differentiability of Six Operators on Nonsmooth Functions and p-Variation. VIII, 272 pages. 1999.

Vol. 1704: H. Tamanoi, Elliptic Genera and Vertex Operator Super-Algebras. VI, 390 pages. 1999.

Vol. 1705: I. Nikolaev, E. Zhuzhoma, Flows in 2-dimensional Manifolds. XIX, 294 pages. 1999.

Vol. 1706: S. Yu. Pilyugin, Shadowing in Dynamical Systems. XVII, 271 pages. 1999.

Vol. 1707: R. Pytlak, Numerical Methods for Optimal Control Problems with State Constraints. XV, 215 pages. 1999.

Vol. 1708: K. Zuo, Representations of Fundamental Groups of Algebraic Varieties. VII, 139 pages. 1999.

Vol. 1709: J. Azéma, M. Émery, M. Ledoux, M. Yor (Eds.), Séminaire de Probabilités XXXIII. VIII, 418 pages. 1999.

Vol. 1710: M. Koecher, The Minnesota Notes on Jordan Algebras and Their Applications. IX, 173 pages. 1999.

Vol. 1711: W. Ricker, Operator Algebras Generated by Commuting Projećtions: A Vector Measure Approach. XVII, 159 pages. 1999.

Vol. 1712: N. Schwartz, J. J. Madden, Semi-algebraic Function Rings and Reflectors of Partially Ordered Rings. XI, 279 pages. 1999.

Vol. 1713: F. Bethuel, G. Huisken, S. Müller, K. Steffen, Calculus of Variations and Geometric Evolution Problems. Cetraro, 1996. Editors: S. Hildebrandt, M. Struwe. VII, 293 pages. 1999.

Vol. 1714: O. Diekmann, R. Durrett, K. P. Hadeler, P. K. Maini, H. L. Smith, Mathematics Inspired by Biology. Martina Franca, 1997. Editors: V. Capasso, O. Diekmann. VII, 268 pages. 1999.

Vol. 1715: N. V. Krylov, M. Röckner, J. Zabczyk, Stochastic PDE's and Kolmogorov Equations in Infinite Dimensions. Cetraro, 1998. Editor: G. Da Prato. VIII, 239 pages. 1999.

Vol. 1716: J. Coates, R. Greenberg, K. A. Ribet, K. Rubin, Arithmetic Theory of Elliptic Curves. Cetraro, 1997. Editor: C. Viola. VIII, 260 pages. 1999.

Vol. 1717: J. Bertoin, F. Martinelli, Y. Peres, Lectures on Probability Theory and Statistics. Saint-Flour, 1997. Editor: P. Bernard. IX, 291 pages. 1999.

Vol. 1718: A. Eberle, Uniqueness and Non-Uniqueness of Semigroups Generated by Singular Diffusion Operators. VIII, 262 pages. 1999.

Vol. 1719: K. R. Meyer, Periodic Solutions of the N-Body Problem. IX, 144 pages. 1999.

Vol. 1720: D. Elworthy, Y. Le Jan, X-M. Li, On the Geometry of Diffusion Operators and Stochastic Flows. IV, 118 pages. 1999.

Vol. 1721: A. Iarrobino, V. Kanev, Power Sums, Gorenstein Algebras, and Determinantal Loci. XXVII, 345 pages. 1999.

Vol. 1722: R. McCutcheon, Elemental Methods in Ergodic Ramsey Theory. VI, 160 pages. 1999.

Vol. 1723: J. P. Croisille, C. Lebeau, Diffraction by an Immersed Elastic Wedge. VI, 134 pages. 1999.

Vol. 1724: V. N. Kolokoltsov, Semiclassical Analysis for Diffusions and Stochastic Processes. VIII, 347 pages. 2000.

Vol. 1725: D. A. Wolf-Gladrow, Lattice-Gas Cellular Automata and Lattice Boltzmann Models. IX, 308 pages. 2000.

Vol. 1726: V. Marić, Regular Variation and Differential Equations. X, 127 pages. 2000.

Vol. 1727: P. Kravanja M. Van Barel, Computing the Zeros of Analytic Functions. VII, 111 pages. 2000.

Vol. 1728: K. Gatermann Computer Algebra Methods for Equivariant Dynamical Systems. XV, 153 pages. 2000.

Vol. 1729: J. Azéma, M. Émery, M. Ledoux, M. Yor (Eds.) Séminaire de Probabilités XXXIV. VI, 431 pages. 2000.

Vol. 1730: S. Graf, H. Luschgy, Foundations of Quantization for Probability Distributions. X, 230 pages. 2000.

Vol. 1731: T. Hsu, Quilts: Central Extensions, Braid Actions, and Finite Groups. XII, 185 pages. 2000.

Vol. 1732: K. Keller, Invariant Factors, Julia Equivalences and the (Abstract) Mandelbrot Set. X, 206 pages. 2000.

Vol. 1733: K. Ritter, Average-Case Analysis of Numerical Problems. IX, 254 pages. 2000.

Vol. 1734: M. Espedal, A. Fasano, A. Mikelić, Filtration in Porous Media and Industrial Applications. Cetraro 1998. Editor: A. Fasano. 2000.

Vol. 1735: D. Yafaev, Scattering Theory: Some Old and New Problems. XVI, 169 pages. 2000.

Vol. 1736: B. O. Turesson, Nonlinear Potential Theory and Weighted Sobolev Spaces. XIV, 173 pages. 2000.

Vol. 1737: S. Wakabayashi, Classical Microlocal Analysis in the Space of Hyperfunctions. VIII, 367 pages. 2000.

Vol. 1738: M. Émery, A. Nemirovski, D. Voiculescu, Lectures on Probability Theory and Statistics. XI, 356 pages. 2000.

Vol. 1739: R. Burkard, P. Deuflhard, A. Jameson, J.-L. Lions, G. Strang, Computational Mathematics Driven by Industrial Problems. Martina Franca, 1999. Editors: V. Capasso, H. Engl, J. Periaux. VII, 418 pages. 2000.

Vol. 1740: B. Kawohl, O. Pironneau, L. Tartar, J.-P. Zolesio, Optimal Shape Design. Tróia, Portugal 1999. Editors: A. Cellina, A. Ornelas. IX, 388 pages. 2000.

Vol. 1741: E. Lombardi, Oscillatory Integrals and Phenomena Beyond all Algebraic Orders. XV, 413 pages. 2000.

Vol. 1742: A. Unterberger, Quantization and Non-holomorphic Modular Forms.VIII, 253 pages. 2000.

Vol. 1743: L. Habermann, Riemannian Metrics of Constant Mass and Moduli Spaces of Conformal Structures. XII, 116 pages. 2000.

Vol. 1744: M. Kunze, Non-Smooth Dynamical Systems. X, 228 pages. 2000.

Vol. 1745: V. D. Milman, G. Schechtman (Eds.), Geometric Aspects of Functional Analysis. Israel Seminar 1999-2000. VIII, 289 pages. 2000.

Vol. 1746: A. Degtyarev, I. Itenberg, V. Kharlamov, Real Enriques Surfaces. XVI, 259 pages. 2000.

Vol. 1747: L. W. Christensen, Gorenstein Dimensions. VIII, 204 pages. 2000.

Vol. 1748: M. Ruzicka, Electrorheological Fluids: Modeling and Mathematical Theory. XV, 176 pages. 2001.

Vol. 1749: M. Fuchs, G. Seregin, Variational Methods for Problems from Plasticity Theory and for Generalized Newtonian Fluids. VI, 269 pages. 2001.

Vol. 1750: B. Conrad, Grothendieck Duality and Base Change. X, 296 pages. 2001.

Vol. 1751: N. J. Cutland, Loeb Measures in Practice: Recent Advances. XI, 111 pages. 2001.

Vol. 1752: Y. V. Nesterenko, P. Philippon, Introduction to Algebraic Independence Theory. XIII, 256 pages. 2001.

Vol. 1753: A. I. Bobenko, U. Eitner, Painlevé Equations in the Differential Geometry of Surfaces. VI, 120 pages. 2001.

Vol. 1754: W. Bertram, The Geometry of Jordan and Lie Structures. XVI, 269 pages. 2001.

Vol. 1755: J. Azéma, M. Émery, M. Ledoux, M. Yor (Eds.), Séminaire de Probabilités XXXV. VI, 427 pages. 2001.

Vol. 1756: P. E. Zhidkov, Korteweg de Vries and Nonlinear Schrödinger Equations: Qualitative Theory. VII, 147 pages. 2001.

Vol. 1757: R. R. Phelps, Lectures on Choquet's Theorem. VII, 124 pages. 2001.

Vol. 1758: N. Monod, Continuous Bounded Cohomology of Locally Compact Groups. X, 214 pages. 2001.

Vol. 1759: Y. Abe, K. Kopfermann, Toroidal Groups. VIII, 133 pages. 2001.

Vol. 1760: D. Filipović, Consistency Problems for Heath-Jarrow-Morton Interest Rate Models. VIII, 134 pages. 2001.

Vol. 1761: C. Adelmann, The Decomposition of Primes in Torsion Point Fields. VI, 142 pages. 2001.

Vol. 1762: S. Cerrai, Second Order PDE's in Finite and Infinite Dimension. IX, 330 pages. 2001.

Vol. 1763: J.-L. Loday, A. Frabetti, F. Chapoton, F. Goichot, Dialgebras and Related Operads. IV, 132 pages. 2001.

Vol. 1764: A. Cannas da Silva, Lectures on Symplectic Geometry. XII, 217 pages. 2001.

Vol. 1765: T. Kerler, V. V. Lyubashenko, Non-Semisimple Topological Quantum Field Theories for 3-Manifolds with Corners. VI, 379 pages. 2001.

Vol. 1766: H. Hennion, L. Hervé, Limit Theorems for Markov Chains and Stochastic Properties of Dynamical Systems by Quasi-Compactness. VIII, 145 pages. 2001.

Vol. 1767: J. Xiao, Holomorphic Q Classes. VIII, 112 pages. 2001.

Vol. 1768: M.J. Pflaum, Analytic and Geometric Study of Stratified Spaces. VIII, 230 pages. 2001.

Vol. 1769: M. Alberich-Carramiñana, Geometry of the Plane Cremona Maps. XVI, 257 pages. 2002.

Vol. 1770: H. Gluesing-Luerssen, Linear Delay-Differential Systems with Commensurate Delays: An Algebraic Approach. VIII, 176 pages. 2002.

Vol. 1771: M. Émery, M. Yor (Eds.), Séminaire de Probabilités 1967-1980. A Selection in Martingale Theory. IX, 553 pages. 2002.

Vol. 1772: F. Burstall, D. Ferus, K. Leschke, F. Pedit, U. Pinkall, Conformal Geometry of Surfaces in $S^4$. VII, 89 pages. 2002.

Vol. 1773: Z. Arad, M. Muzychuk, Standard Integral Table Algebras Generated by a Non-real Element of Small Degree. X, 126 pages. 2002.

Vol. 1774: V. Runde, Lectures on Amenability. XIV, 296 pages. 2002.

Vol. 1775: W. H. Meeks, A. Ros, H. Rosenberg, The Global Theory of Minimal Surfaces in Flat Spaces. Martina Franca 1999. Editor: G. P. Pirola. X, 117 pages. 2002.

Vol. 1776: K. Behrend, C. Gomez, V. Tarasov, G. Tian, Quantum Comohology. Cetraro 1997. Editors: P. de Bartolomeis, B. Dubrovin, C. Reina. VIII, 319 pages. 2002.

Vol. 1777: E. García-Río, D. N. Kupeli, R. Vázquez-Lorenzo, Osserman Manifolds in Semi-Riemannian Geometry. XII, 166 pages. 2002.

Vol. 1778: H. Kiechle, Theory of K-Loops. X, 186 pages. 2002.

Vol. 1779: I. Chueshov, Monotone Random Systems. VIII, 234 pages. 2002.

Vol. 1780: J. H. Bruinier, Borcherds Products on O(2,1) and Chern Classes of Heegner Divisors. VIII, 152 pages. 2002.

Vol. 1781: E. Bolthausen, E. Perkins, A. van der Vaart, Lectures on Probability Theory and Statistics. Ecole d' Eté de Probabilités de Saint-Flour XXIX-1999. Editor: P. Bernard. VIII, 466 pages. 2002.

Vol. 1782: C.-H. Chu, A. T.-M. Lau, Harmonic Functions on Groups and Fourier Algebras. VII, 100 pages. 2002.

Vol. 1783: L. Grüne, Asymptotic Behavior of Dynamical and Control Systems under Perturbation and Discretization. IX, 231 pages. 2002.

Vol. 1784: L.H. Eliasson, S. B. Kuksin, S. Marmi, J.-C. Yoccoz, Dynamical Systems and Small Divisors. Cetraro, Italy 1998. Editors: S. Marmi, J.-C. Yoccoz. VIII, 199 pages. 2002.

Vol. 1785: J. Arias de Reyna, Pointwise Convergence of Fourier Series. XVIII, 175 pages. 2002.

Vol. 1786: S. D. Cutkosky, Monomialization of Morphisms from 3-Folds to Surfaces. V, 235 pages. 2002.

Vol. 1787: S. Caenepeel, G. Militaru, S. Zhu, Frobenius and Separable Functors for Generalized Module Categories and Nonlinear Equations. XIV, 354 pages. 2002.

Vol. 1788: A. Vasil'ev, Moduli of Families of Curves for Conformal and Quasiconformal Mappings.IX, 211 pages. 2002.

Vol. 1789: Y. Sommerhäuser, Yetter-Drinfel'd Hopf algebras over groups of prime order. V, 157 pages. 2002.

Vol. 1790: X. Zhan, Matrix Inequalities. VII, 116 pages. 2002.

Vol. 1791: M. Knebusch, D. Zhang, Manis Valuations and Prüfer Extensions I: A new Chapter in Commutative Algebra. VI, 267 pages. 2002.

Vol. 1792: D. D. Ang, R. Gorenflo, V. K. Le, D. D. Trong, Moment Theory and Some Inverse Problems in Potential Theory and Heat Conduction. VIII, 183 pages. 2002.

Vol. 1793: J. Cortés Monforte, Geometric, Control and Numerical Aspects of Nonholonomic Systems. XV, 219 pages. 2002.

Vol. 1794: N. Pytheas Fogg, Substitution in Dynamics, Arithmetics and Combinatorics. Editors: V. Berthé, S. Ferenczi, C. Mauduit, A. Siegel. XVII, 402 pages. 2002.

Vol. 1795: H. Li, Filtered-Graded Transfer in Using Noncommutative Gröbner Bases. IX, 197 pages. 2002.

Vol. 1796: J.M. Melenk, hp-Finite Element Methods for Singular Perturbations. XIV, 318 pages. 2002.

Vol. 1797: B. Schmidt, Characters and Cyclotomic Fields in Finite Geometry. VIII, 100 pages. 2002.

Vol. 1798: W.M. Oliva, Geometric Mechanics. XI, 270 pages. 2002.

Vol. 1799: H. Pajot, Analytic Capacity, Rectifiability, Menger Curvature and the Cauchy Integral. XII,119 pages. 2002.

Vol. 1800: O. Gabber, L. Ramero, Almost Ring Theory. VI, 307 pages. 2003.

Vol. 1801: J. Azéma, M. Émery, M. Ledoux, M. Yor (Eds.), Séminaire de Probabilités XXXVI. VIII, 499 pages. 2003.

Vol. 1802: V. Capasso, E. Merzbach, B.G. Ivanoff, M. Dozzi, R. Dalang, T. Mountford, Topics in Spatial Stochastic Processes. Martina Franca, Italy 2001. Editor: E. Merzbach. VIII, 253 pages. 2003.

Vol. 1803: G. Dolzmann, Variational Methods for Crystalline Microstructure - Analysis and Computation. VIII, 212 pages. 2003.

Vol. 1804: I. Cherednik, Ya. Markov, R. Howe, G. Lusztig, Iwahori-Hecke Algebras and their Representation Theory. Martina Franca, Italy 1999. Editors: V. Baldoni, D. Barbasch. X, 103 pages. 2003.

Vol. 1805: F. Cao, Geometric Curve Evolution and Image Processing. X, 187 pages. 2003.

Vol. 1806: H. Broer, I. Hoveijn. G. Lunther, G. Vegter, Bifurcations in Hamiltonian Systems. Computing Singularities by Gröbner Bases. XIV, 169 pages. 2003.

Vol. 1807: V. D. Milman, G. Schechtman (Eds.), Geometric Aspects of Functional Analysis. Israel Seminar 2000-2002. VIII, 429 pages. 2003.

Vol. 1808: W. Schindler, Measures with Symmetry Properties.IX, 167 pages. 2003.

Vol. 1809: O. Steinbach, Stability Estimates for Hybrid Coupled Domain Decomposition Methods. VI, 120 pages. 2003.

Vol. 1810: J. Wengenroth, Derived Functors in Functional Analysis. VIII, 134 pages. 2003.

Vol. 1811: J. Stevens, Deformations of Singularities. VII, 157 pages. 2003.

Vol. 1812: L. Ambrosio, K. Deckelnick, G. Dziuk, M. Mimura, V. A. Solonnikov, H. M. Soner, Mathematical Aspects of Evolving Interfaces. Madeira, Funchal, Portugal 2000. Editors: P. Colli, J. F. Rodrigues. X, 237 pages. 2003.

Vol. 1813: L. Ambrosio, L. A. Caffarelli, Y. Brenier, G. Buttazzo, C. Villani, Optimal Transportation and its Applications. Martina Franca, Italy 2001. Editors: L. A. Caffarelli, S. Salsa. X, 164 pages. 2003.

Vol. 1814: P. Bank, F. Baudoin, H. Föllmer, L.C.G. Rogers, M. Soner, N. Touzi, Paris-Princeton Lectures on Mathematical Finance. X,172 pages. 2003.

Vol. 1815: A. M. Vershik (Ed.), Asymptotic Combinatorics with Applications to Mathematical Physics. St. Petersburg, Russia 2001. IX, 246 pages. 2003.

Vol. 1816: S. Albeverio, W. Schachermayer, M. Talagrand, Lectures on Probability Theory and Statistics. Ecole d'Eté de Probabilités de Saint-Flour XXX-2000. Editor: P. Bernard. VIII, 296 pages. 2003.

Vol. 1817: E. Koelink (Ed.), Orthogonal Polynomials and Special Functions. Leuven 2002. X, 249 pages. 2003.

Vol. 1818: M. Bildhauer, Convex Variational Problems with Linear, nearly Linear and/or Anisotropic Growth Conditions. X, 217 pages. 2003.

Vol. 1819: D. Masser, Yu. V. Nesterenko, H. P. Schlickewei, W. M. Schmidt, M. Waldschmidt, Diophantine Approximation. Cetraro, Italy 2000. Editors: F. Amoroso, U. Zannier. XI,353 pages. 2003.

Vol. 1820: F. Hiai, H. Kosaki, Means of Hilbert Space Operators. VIII, 148 pages. 2003.

Vol. 1821: S. Teufel, Adiabatic Perturbation Theory in Quantum Dynamics. VI, 236 pages. 2003.

Vol. 1822: S.-N. Chow, R. Conti, R. Johnson, J. Mallet-Paret, R. Nussbaum, Dynamical Systems. Cetraro, Italy 2000. Editors: J. W. Macki, P. Zecca. XIII, 353 pages. 2003.

Vol. 1823: A. M. Anile, W. Allegretto, C. Ringhofer, Mathematical Problems in Semiconductor Physics. Cetraro, Italy 1998. Editor: A. M. Anile. X, 143 pages. 2003.

Vol. 1824: J. A. Navarro González, J. B. Sancho de Salas, $C^{\infty}$ - Differentiable Spaces. XIII, 188 pages. 2003.

Vol. 1825: J. H. Bramble, A. Cohen, W. Dahmen, Multiscale Problems and Methods in Numerical Simulations, Martina Franca, Italy 2001. Editor: C. Canuto. XIII, 163 pages. 2003.

Vol. 1826: K. Dohmen, Improved Bonferroni Inequalities via Abstract Tubes. Inequalities and Identities of Inclusion-Exclusion Type. VIII, 113 pages, 2003.

Vol. 1827: K. M. Pilgrim, Combinations of Complex Dynamical Systems. IX, 118 pages, 2003.

Vol. 1828: D. J. Green, Gröbner Bases and the Computation of Group Cohomology. XII, 138 pages, 2003.

Vol. 1829: E. Altman, B. Gaujal, A. Hordijk, Discrete-Event Control of Stochastic Networks: Multimodularity and Regularity. XIV, 313 pages, 2003.

Vol. 1830: M. I. Gil', Operator Functions and Localization of Spectra. XIV, 256 pages, 2003.

Vol. 1831: A. Connes, J. Cuntz, E. Guentner, N. Higson, J. E. Kaminker, Noncommutative Geometry, Martina Franca, Italy 2002. Editors: S. Doplicher, L. Longo. XIV, 349 pages. 2004.

Vol. 1832: J. Azéma, M. Émery, M. Ledoux, M. Yor (Eds.), Séminaire de Probabilités XXXVII. XIV, 448 pages. 2003.

Vol. 1833: D.-Q. Jiang, M. Qian, M.-P. Qian, Mathematical Theory of Nonequilibrium Steady States. On the Frontier of Probability and Dynamical Systems. IX, 280 pages, 2004.

Vol. 1834: Yo. Yomdin, G. Comte, Tame Geometry with Application in Smooth Analysis. VIII, 186 pages, 2004.

Vol. 1835: O.T. Izhboldin, B. Kahn, N.A. Karpenko, A. Vishik, Geometric Methods in the Algebraic Theory of Quadratic Forms. Summer School, Lens, 2000. Editor: J.-P. Tignol. XIV, 190 pages. 2004.

Vol. 1836: C. Năstăsescu, F. Van Oystaeyen, Methods of Graded Rings. XIII, 304 pages, 2004.

Vol. 1837: S. Tavaré, O. Zeitouni, Lectures on Probability Theory and Statistics. Ecole d'Eté de Probabilités de Saint-Flour XXXI-2001. Editor: J. Picard. VII, 315 pages. 2004.

Vol. 1838: A.J. Ganesh, N.W. O'Connell, D.J. Wischik, Big Queues. XII, 254 pages, 2004.

Vol. 1839: R. Gohm, Noncommutative Stationary Processes. VIII, 170 pages, 2004.

Vol. 1840: B. Tsirelson, W. Werner, Lectures on Probability Theory and Statistics. Ecole d'Eté de Probabilités de Saint-Flour XXXII-2002. Editor: J. Picard. VII, 200 pages. 2004.

Vol. 1841: W. Reichel, Uniqueness Theorems for Variational Problems by the Method of Transformation Groups, XIII, 152 pages. 2004.

Vol. 1842: T. Johnsen, A.L. Knutsen, K3 Projective Models in Scrolls. VIII, 164 pages. 2004

Vol. 1843: B. Jefferies, Spectral Properties of Noncommuting Operators, VIII, 184 pages. 2004

Vol. 1844: K.F. Siburg, The Principle of Least Action in Geometry and Dynamics, XII, 128 pages. 2004.

Vol. 1845: Min Ho Lee, Mixed Automorphic Forms, Torus Bundles, and Jacobi Forms, X, 239 pages. 2004

Vol. 1846: h. Ammari, H. Kang, Reconstruction of Small Inhomogeneities from Boundary Measurements, IX, 238 pages. 2004

Vol. 1848: M. Abate, J. E. Fornaess, X. Huang, J. P. Rosay, A. Tumanov, Real Methods in Complex and CR Geometry, Martina Franca, Italy 2002. Editors: D. Zaitsev, G. Zampieri. IX, 219 pages. 2004.

Vol. 1849: Martin L. Brown, Heegner Modules and Elliptic Curves, X, 517 pages. 2004

Vol. 1850: V. D. Milman, G. Schechtman (Eds.), Geometric Aspects of Functional Analysis. Israel Seminar 2002-2003. X, 301 pages. 2004.

Vol. 1851: O. Catoni, Statistical Learning Theory and Stochastic Optimization, VIII, 273 pages. 2004.

Vol. 1852: A.S. Kechris, B.D. Miller, Topics in Orbit Equivalence, X, 134 pages. 2004

## Recent Reprints and New Editions

Vol. 1200: V. D. Milman, G. Schechtman, Asymptotic Theory of Finite Dimensional Normed Spaces. 1986. – Corrected Second Printing. X, 156 pages. 2001.

Vol. 1471: M. Courtieu, A.A. Panchishkin, Non-Archimedean L-Functions and Arithmetical Siegel Modular Forms. – Second Edition. VII. 196 pages. 2003.

Vol. 1618: G. Pisier, Similarity Problems and Completely Bounded Maps. 1995 – Second, Expanded Edition VII, 198 pages. 2001.

Vol. 1629: J.D. Moore, Lectures on Seiberg-Witten Invariants. 1997 – Second Edition. VIII, 121 pages. 2001.

Vol. 1638: P. Vanhaecke, Integrable Systems in the realm of Algebraic Geometry. 1996 – Second Edition. X, 256 pages. 2001.

Vol. 1702: J. Ma, J. Yong, Forward-Backward Stochastic Differential Equations and Their Applications. 1999. – Corrected Second Printing. XIII, 270 pages. 2000.